Interdisciplinary Contributions to Archaeology

For further volumes:
http://www.springer.com/series/6090

Vera Tiesler

The Bioarchaeology of Artificial Cranial Modifications

New Approaches to Head Shaping and its Meanings in Pre-Columbian Mesoamerica and Beyond

with a chapter contribution by Pilar Zabala

Springer

Vera Tiesler
Facultad de Ciencias Anthropologicas
Universidad Autonoma de Yucatan Facultad de Ciencias Anthropologicas
Mérida, Yucatán
Mexico

with a chapter contribution by Pilar Zabala

ISSN 1568-2722
ISBN 978-1-4614-8759-3 ISBN 978-1-4614-8760-9 (eBook)
DOI 10.1007/978-1-4614-8760-9
Springer New York Heidelberg Dordrecht London

Library of Congress Control Number: 2013948226

Printed on acid-free paper

Springer is part of Springer Science+Business Media (www.springer.com)

Foreword

Cranial modification practices are among the themes most scrutinized by physical anthropologists and bioarchaeologists throughout the world. Earlier studies in the nineteenth century focused their attention on the development of typologies and classification systems whereas recent multidisciplinary approaches have attempted to understand how and why modification practices were enacted. These studies suggest that these labor-intensive practices are complex and have multiple social, ideological, and aesthetic meanings. Used as an individual or collective identifier, cranial modification was imposed by society at the birth of an infant and represents an indelible marker of an individual's identity throughout her/his life. Unlike other forms of imposed body modifications, the shape of the head created a unique visual cue that could be enhanced through the use of headdresses and creatively disguised by hats, but never fully modified or eliminated. During one's life, this permanent modification would have withstood the malleability of other changeable identity markers. After death, head shapes, along with contextual archaeological data, provide researchers with a powerful tool to elucidate factors involved in the complex construction of such practices and to address indigenous worldviews that are otherwise inaccessible.

In *The Bioarchaeology of Artificial Cranial Modifications: New Approaches to Head Shaping and its Meanings in Pre-Columbian Mesoamerica and Beyond*, the prolific scholar Vera Tiesler constructs an exemplary treaty of the study of cranial modification practices among the Maya and Mesoamerican peoples in general. Although her approach focuses on the pre-Hispanic and colonial indigenous heritage, this rich and synthetic contribution transcends any geographical and disciplinary boundaries, as her visionary treaty on the subject incorporates archaeological, osteological, ethnohistorical, historical, and medical research from a wide range of contexts. Furthermore, as most of her sources are not in English, she opens a plethora of new references and studies for the English-speaking community concerned with the study of head-shaping practices and physical embodiment, extending our understanding of the body alterations of the past and the present.

This book will surely appeal to a wide audience of those interested in Mesoamerican ideology and culture, especially to (bio) archaeologists, iconographers, and physical anthropologists, as well as historians. Those who study cranial anatomy and craniofacial growth may encounter valuable insights for the application of medical models. Finally, this volume is surely suitable to upper-division courses on bioarchaeology and anthropological theory and embodiment and identity.

The book starts with a discussion about the myriad of terms that have been used to describe such practices. Within this context, "cranial deformation" is eliminated due to its negative implication, and a useful standardized nomenclature is provided in order to unify the terms for description and to produce comparable systems of analysis across different datasets. In the following chapters, Tiesler embeds her research within contemporary theoretical approaches to the study of the biological and social body, arguing that the concept of "the body" represents different realities and perceptions. In Mesoamerica, such body practices were inextricably linked to deeply embedded, head-centric worldviews. Here, the head with its outer insignia was used as a metaphor for designating the individual, the person, and the "self", which, in a broader sense, served as a model for the indigenous universe. These and other emblematic connotations of head treatments are carefully explored by the contextualized study of modified skulls and other sources of material and intellectual indigenous culture. Tiesler also contends that such visual cues were used as metaphors for rites of passage associated with social and biological life cycles throughout the life of an individual. For instance, the integration of a newborn into the native world was crystallized by the careful reshaping of the head, a societal investment that lasted many years, while dental modification marked the transition from childhood to adulthood. It is within this lens that she moves beyond traditional typologies and examines individual and collective pathways that structured past societies.

Modified head shapes perplexed Europeans at the time of the conquest of the Americas. Chroniclers considered these practices barbaric, and in the Andes such accounts included gruesome images, such as the extrusion of the brain from ears of infants due to extensive pressure. While the direct effects of the pressure applied to the pliable head have not been previously treated, Tiesler provides a detailed analysis of the biological and health effects that cranial shaping may have had in the past. Within this context, she acknowledges some health risks but argues that such risks were minimized, as female practitioners were specialists and were extremely careful with these modeling practices, as they were done on a daily basis, and in many cases executed in order to protect newborns from injury. Equally important in this book is the systematic anthology of ethnohistorical references provided by Tiesler's historian collaborator Pilar Zabala. This extensive section covers an impressive selection of eyewitness testimonies of the performance of native head traditions from the time of the European contact up to the twentieth century. This extensive section includes the Mexican end of North America, the Caribbean region, and South America. All of these accounts attest to the deeply rooted importance and prevalence of head shaping in precolonial America and offer the reader exquisite sources that will serve as the basis of anthropological interpretation and future studies.

Within this rich theoretical, medical, and historical backdrop, Tiesler moves to the analytical part of her study, where she offers the reader a diachronic perspective of head-shaping practices in Mesoamerica, providing an interpretation of meanings against changes within the sociopolitical landscape during Olmec-dominated Preclassic period, the regional developments of the Classic period, pan-Mesoamerican Post-Classic culture, and the adjustments after the Iberian Contact. As Tiesler notes, the archaeological context, along with iconographic depictions, is critical to the interpretation of such practices, and after systematically analyzing about 2,000 skulls and reviewing additional scholarship on the practice, she is able to carefully dissect the meanings of this long-standing indigenous custom, one that survived many years despite the prohibition by Europeans. Body modification has been inextricably linked to the human condition, and with treaties like this one, we are bound to gain a better understanding not only of past treatments of the head and their subtle and profound meanings, but also of our own collective experiences that shape and reshape our bodies and lives.

The University of Chicago María Cecilia Lozada

Preface

With this work, I wish to share with a broad readership an encompassing empirical and interpretational synopsis of infant head modeling in Mesoamerica any beyond. The research is the outcome of decades of active, but slowly evolving quest for ancient Mesoamerican cranial modifications. The nature of this topic, and my personal research interests and diversified academic background, made it seem suitable to align different frames of reference from anthropological theory, as well as regionally based ideological and archaeological concepts. These are cornerstones in a survey of the Mesoamerican cranial record, iconography and historical testimonies. The latter have been patiently accrued and rigorously commented by my historian colleague, Pilar Zabala, who has participated actively in the endeavor of this work. The combination of different sources of information has been enormously facilitated in the data-rich Maya study environment, where integrated research provides the foundation for tracing head shapes to deeply embedded ideological schemes and their changes over time. During the years that I have dedicated to the study of Maya cranial modifications, I have not lost my sense of curiosity to learn more about this body practice. I keep being impressed by the coherence of the cultural elements involved in its performance and their adherence to deeply rooted social and ideological undercurrents. These have accompanied the unfolding of the broad historical processes that span the centuries and millennia of cultural evolvement in this sphere.

Milestones during these years of (re)search have been a Master and a Ph.D. thesis, two single-authored books and a sabbatical year of research dedicated to this topic. In 2012, our alma mater, the autonomous University of Yucatán, hosted an international colloquium on Mesoamerican head-shaping practices in our hometown, Mérida, Mexico, with the explicit goal to bring to the table, selected local and international scholars who are conducting research on American cranial modification. Conference sessions, rounds of discussion, and a practical workshop identified new venues of studying this and other native body practices. A need for standardized taxonomic criteria and active research was equally endorsed by the participants of the conference, which examined the social meanings and visible expressions of head modeling in different areas and at different moments within and around Mesoamerica. The contributions of the encounter are now in the process of publication as an edited volume.

I also hope that this work may lay groundwork towards a balanced, empirical, analytical, and conceptual perspective for future anthropological investigations on ancestral head-modeling practices in the Mesoamerican sphere and beyond. On an analytical level, this volume seeks to be a resource book. In this spirit, each chapter was written as an independent, get articulated unit. Part I of this book delineates a set of interdisciplinary, basic analytical elements anchored in concepts derived from craniometrics, concepts in neural growth and adapted classification criteria, to be applied and interpreted in Part II of this volume. Although the research is anchored in physical anthropology and archaeology (specifically bioarchaeology), this book also integrates knowledge derived from anatomy and human physiology, historical and iconographic sources, linguistics (polisemy), and ethnography. Given its interdisciplinary focus, this volume fits well within Springer's *Interdisciplinary Contributions to Archaeology* series as it seeks integrated interpretations on the social and cultural roles of ancient head modifications (and body practices in general).

My thankful recognition go to all institutions, projects, and colleagues who have engaged actively in the academic endeavor of this work, and to those who have facilitated information or access to valuable resources for the research that supports it. I am grateful to Nene Lozada and William Duncan for their useful advice and their reviews. My thanks go to Pilar Zabala, my colleague, historian, and co-author of this book, with whom I have the pleasure of sharing this ongoing quest and research. I am also greatly indebted to my all-time academic mentor, Professor Arturo Romano of the National Institute of Anthropology, for his continued support for this research and for sharing his broad experience and knowledge of cranial examination in the fine "Old School" manner. I am also grateful to Andrea Cucina, for proofreading and commenting on preliminary drafts of this manuscript but, above all, for his encouragement and support at all stages of writing.

In general, I wish to credit and at the same time thank all the many projects and institutions for kindly providing access to the skeletal series illustrated or mentioned in this review. These include the Dirección de Antropología Física (José Antonio Pompa, INAH), the National School of Anthropology and History (INAH), the INAH Centers of Yucatán, Quintana Roo, Oaxaca, and Campeche. Similarly, the University of Yucatán, has been very supportive in this research, as has been the Na Bolom Museum in Chiapas, and the Claustro de Sor Juana in Mexico City. In Guatemala, the Museo Nacional de Arqueología y Etnología (IDAEH, Guatemala) and the Atlas Arqueológico de Guatemala (Juan Pedro Laporte, IDAEH) and the French Mission (CEMCA) kindly assisted this study in the form of collaboration and study permits. The state authorities of Honduras and their collaboration with an ongoing Japanese initiative at the site of Copán, facilitated access to the skeletal series from the Copán Valley of Honduras. My thanks also go to Jorge Gómez-Valdéz and Teresina Jaén Esquivel for sharing their work and illustrations. The Popol Vuh Museum of the Universidad Francisco Marroquin kindly granted permission to use images taken of its exquisite figurine collection.

The *Peabody Museum of Harvard University* and the Middle American Research Institute of Tulane University granted access to bibliographic sources and skeletal material. The University of Chicago has funded research at the Sierra Lacandona in

Chiapas, which includes research on historical head practices. My thanks also go to the community of Mensabak who kindly shared knowledge and gave logistical help. The University of La Sapienza, which curates the Sergii skeletal collection, and the Musée de l'Homme in Paris, France, have been helpful in providing illustrations and advice. in the early stages of this research, which have been financed over the years by the *Deutscher Akademischer Austauschdienst*, PIFI, PROMEP, CONACYT, UADY, and Wenner Gren funds. A full list of project credits is published in my book (2012) *Transformarse en maya. El modelado cefálico entre los mayas prehispánicos y coloniales*. Universidad Nacional Autónoma de México, Universidad Autónoma de Yucatán, México, D.F.

Other colleagues, friends and family, who kindly helped with this research are Linda Manzanilla, Carlos Serrano, Jane E. Buikstra, Grégory Pereira, Cármen Valverde, and Laura Sotelo, Mercedes de la Garza, Rosaura Yepez, José Gamboa and Ana Leslie Escalante, Thelma Sierra, Allan Ortega, Nora López, Luis Felipe Bate, and Nydia Lara. Thanks also go to Attilio Cucina (†), Ekkehard Tiesler (†),Luz Evelia Campaña and Karina Romero, to Fernando Robles, Rafael Cobos, Adri-anaVelázquez, Enrique Nalda (†) and Sandra Balanzario, Ramón Carrasco, William Sanders and Rebecca Story, Juan Pedro Laporte (†) and William Folan. More recently, this work has benefited enormously from the generous intellectual input and encouragement by Nene Lozada, John Verano, Virginia Miller, Takeshi Inomata, Oswaldo Chinchilla, Grégory Pereira, T. Douglas Price, Lilia Escorcia, Bill Duncan, Andrew Scherer, Gabriel Wrobel, Rosaura Yépez, and Erik Velázquez. On an ideological plane, the advice by Joel Palka, Gabrielle Vail, Victoria Bricker, Ana García, andAlfonso Lacadena has been instrumental in assigning meanings to Maya head terminology and their input is gratefully appreciated here. The comments and suggestions of the editorial board of Springer Press have been instrumental during the proposal stage abd writing process. Titanic logistical help was provided by Khali Ashton-Lomax, Belem Ceballos, Kadwin Pérez, Mirna Sánchez, and Shintaro Suzuki. Thanks for your patience!

About the Author

Vera Tiesler is currently a research professor at the University of Yucatan, Mexico. She holds a Ph.D. in anthropology (National University of Mexico, Mexico), a B.A. degree in art history (Tulane University, New Orleans, USA), B.A. and M.A. degrees in archaeology (Escuela Nacional de Antropología e Historia, Mexico), with five years of additional training in human medicine (Medizinische Hochschule Hannover, Germany, and Politécnico Nacional, Mexico). Tiesler is member of the Mexican National Research Foundation and of the Mexican National Academy of Sciences. Her research is grounded in the integrated study of human remains within their archaeological, cultural, and social undercurrents and focus on general health and dietary conditions, along with physical embodiment and mortuary traditions among pre-Hispanic and colonial Maya. She has participated in several field projects and has studied hundreds of skeletal collections from Mesoamerica, the Caribbean, Andes, and Europe, including forensic research. Recent publications include the co-edited books: *Janaab' Pakal of Palenque. Life and Death of a Maya Ruler* (Arizona University Press), *Natives, Europeans, and Africans in Colonial Campeche. History and Archaeology* (University Press of Florida), and *New Perspectives on Human Sacrifice and Ritual Body Treatments in Ancient Maya Society* (Springer Press)

Facultad de Ciencias Anthropologicas,
Universidad Autonoma de Yucatan,
PO Box 76, 97000 Mérida, YUC, Mexico
e-mail: vtiesler@uady.mx

Contents

List of Figures

List of Tables

Chapter 1
Introduction

People have, for centuries, modified their bodies for a plethora of motives and motifs that range from mundane aesthetics, to political and ideological ideals, to gender identity, to ethnicity, and to other cultural distinctions. Actually, head shaping constitutes one of the most ubiquitous biocultural practices of the past, which has been documented in all continents. Across the globe, head splinting and binding practices and cradling traditions have evolved over long periods of time within defined cultural territories. Cranial-vault modification also appears to be a very ancient practice that is not even limited to modern humans, but goes back to hominid forms predating *Homo sapiens* (Trinkaus 1982, 1983; Weidenreich 1938).[1]

Cultural cranial modification has grown into an established subject of interest to anthropological and health related sciences alike and has received plenty of attention both in popular and scientific writing. While cultural cranial modifications are recognized archaeologically in most parts of the world, its application probably attained its greatest incidence in the Americas, specifically in the Andes and pre-Columbian Mesoamerica. Among pre-Columbian Mesoamericans, to whom this book is dedicated, artificial head shaping counts as one of the most long-standing and at the same time diversified traditions, a daily activity that positioned itself at the "hard core" of its sophisticated cosmology and cultural expression (López-Austin 2001). This custom accompanied its millennial development from its emergence before and during the Early Preclassic, and, in the centuries that followed, turned into a widespread and at the same time diversified tradition. In most of its territories, both the elites and the mass of society, in settlements large and small, exhibited artificially enhanced cranial appearances (Romano 1973; Tiesler 1999, pp. 327–330, 2012a, b). The diversity in artificial head shapes was already noted by the American anthropologist Earnest Hooton who described the artificial formations in the skulls

[1] Two Neanderthal skulls from Shanidar, Iraq, evidence an artificial flattening of the forehead and slight transversal grooving of the vault behind the coronal sulcus (Trinkaus 1982, 1983). These go back to 45,000 years B.P. Another early finding of a similarly grooved cranium from the caves of Chou-Kou-Tien, in East China, dates to 30,000 years B.P. (Romano 1974, p. 197). In all these cases, the intentionality of the cranial modifications cannot be ascertained, since slight frontal flattening and postcoronal grooving could be equally the result of carrying heavy loads during early childhood and therefore be purely occupational.

V. Tiesler, *The Bioarchaeology of Artificial Cranial Modifications*,
Interdisciplinary Contributions to Archaeology 7, DOI 10.1007/978-1-4614-8760-9_1,

Fig. 1.1 Map of Mesoamerica showing broad cultural divisions and sites mentioned in the text. (Drawing by B. Ceballos and V. Tiesler)

retrieved from the Sacred Cenote of Chichén Itzá, noting that "the varieties of cranial modification are so numerous that they are bewildering" (Hooton 1940, p. 273). This appraisal has been echoed by more recent Mesoamerican scholars (Stewart 1974, p. 222; Tiesler and Romano 2008).

1.1 Cultural Frames

Mesoamerica denotes a geocultural space that was, and still is, home to a group of highly complex native cultures and peoples who maintain close cultural and historical ties among each other (Fig. 1.1). Today, this cultural tapestry encompasses modern Mexico, Guatemala, Belize, Honduras, El Salvador, Nicaragua, and northern Costa Rica with their rich heritage of native cultures. Among them are the Maya, Zapotec and Mixtec, Purepecha, Totonac, Huastec, and Nahua (Manzanilla and López-Luján 1994; Sharer and Traxler 2006) (Fig. 1.1). A set of subregions within Mesoamerica is defined by an aggregate of geographic and cultural attributes. These areas include Central Mexico, West Mexico, the Gulf Coast Lowlands, Oaxaca, the Southern Pacific Lowlands, and the Southeast Mesoamerican borderlands of Honduras, Nicaragua, and northern Costa Rica. The Maya area is generally divided into Lowland and Highland regions; the latter is home to inland Petén and Yucatecan groups.

Mesoamerica was defined as such during the first half of the last century by Paul Kirchhoff (1943). Following the academic tradition of cultural history then in vogue, Kirchhoff defines this cultural sphere still rather statically by its suite of shared material and ideological elements, which include pantheistic religious concepts, a cosmological notion of balance with the sacred world, and that of a cyclical evolution of the native universe. Most Mesoamerican peoples produced sophisticated art and music. A vigesimal numeric system, calendrical counts, and glyphic writing were shared by most, as was a tradition of ball playing and a distinct style of religious construction which is seen throughout the area. The combined cultivation of subsistence products such as maize, squash, beans, amaranth and chile, once defined the subsistence basis that defines Mesoamerica. In pre-Hispanic times, a number of its cultural territories harbored state level political structures, grounded on social stratification, urbanization, and economic surplus.

Mesoamerican civilization looks back to a continuous millenary cultural past. Here, incipient agriculture began almost 10,000 years ago, defining the Archaic period (8000–2000 B.C.), which led to the establishment of sedentary agricultural villages. The 2,000 years that encompass the Formative period (2000–0 B.C.), saw the rise of first hierarchically organized social organizations, like the Gulf Coast Olmecs that extended ties with other peoples north along the coast and towards the Highlands, and the Maya area eastwards across the Isthmus of Tehuantepec. These networks, together with the beliefs and rituals that accompanied them, laid the basis for subsequent Mesoamerican cultural evolvement during the Classic period, which roughly spans the first millennium A.D. The Early Classic was dominated by Teotihuacán and Monte Albán in Central Mexico and Lowland Maya society to the east. The latter was organized along extended family ties and shifting territorial units. The political landscape was always dynamic, while the distribution of power fluctuated, as the dynastic histories of all major centers among the city-states show. Maya leaders were related through family ties, which were constantly reinforced through strategic marriage bonds (Martin and Grube 2008).

After the tumult of Teotihuacán's collapse around 600 A.D., rivalry ensued between several important Highland centers. Epi-Classic period Nahua groups started to migrate from the north, while merchant folk from the Gulf Coast became prominent around the shores of Mesoamerica. The close of the first millennium was marked by the irrevocable decline of many important Classic period nations, such as Zapotec Monte Albán or the Lowland Maya inland hegemonies.

Regarding the capital topic of this volume, we will examine Preclassic and Classic period cephalic practices in Chaps. 7, 8, and 9. Chapter 10 covers Postclassic head-shaping trends, which are distinct from earlier cultural expressions and reflect broader pan-Mesoamerican ideological shifts and a Nahua dominated streamlining of ideological systems. Oligarchic councils now operated in much of Central Mexico and in Yucatán and dual rulership was to replace the more theocratic governments of previous times. Evermore powerful, wide-ranging networks of *pochteca* merchants dominated the economic life toward and during the Postclassic era, while military orders gained prominence and, toward the close of the Postclassic period, culminated in the Aztec empire. The distinct Mesoamerican cultural tradition started to be abandoned after the Spanish conquest in the early sixteenth century, as explored in

Chap. 10. Over the next centuries, Mesoamerican indigenous cultures were gradually subjected to Spanish colonial rule and native traditions, such as cranial flattening practices, transmuted. Some aspects of ancestral head practices still survive among the indigenous peoples who inhabit Mesoamerica today.

1.2 Basic concepts

Generally speaking, cranial modification was conducted by female adults (kin or women called upon for their experience) on newborns, and sometimes continued in older babies and toddlers (Dingwall 1931). This biological time frame denotes a stage in child development when the skull is still malleable. When the child reaches the age of 2–3 years, the cranium hardens and the artificial head form becomes permanent. The external redirection of the neurocranial vectors does not change the brain volume or functioning, but changes its growth vectors towards those surfaces that are not restrained. The practitioners accomplished cephalic constriction and compression by conducting daily massages or by applying hard compression devices or soft constricting wraps, bandages, and hats.

Upon examination, the anthropological qualities of cranial shaping acquire many shades, dimensions, and cultural undertones, as the chapters of this volume will highlight again and again. Some of their inherent attributes are purely behavioral; some are daily, private, and ephemeral; others relate to the produced head morphology, which is indelible, permanent, and displayed in public. The individual and collective enactment of this practice involves different age groups, namely, second and third adult generation practitioners and their infants who would display the resulting artifical morphology for the remainder of their lives. This quality, by itself, renders head shaping important as potential gender expression and child socialization and at the same time confers a conservative, generation-bridging role to it. The protracted quality of cranial modification raises its cultural importance above that of more transitory body fashions and turns it into an enormously useful point of departure for examining gradual collective cultural dynamics of social integration and embodied group identity, up to long-term expressions of ethnicity and group ancestry.

A few words on terminology and semantics: in this book, the terms "compression," "shaping," "molding," "flattening," "modeling," "modification," "alteration," "manipulation," and "transformation" of the "head," the "cranium," the "skull cap," or the "cranial vault" are used as synonyms to denote the body practice per se and specifically its quotidian performance by putting pressure on the infants' heads day after day. Specific techniques are identified here by their proper names, such as the practices of "cradleboard use," "head splinting," "head binding," or "swaddling." The visible outcome of the compression process, i.e., the artificial head morphology, will be described as "headwear," "artificial look," "cranial form," "head sculpture," or "head mold". Adjectives like "cultural," "shaped," "molded," or "artificial" denote the nonnatural and noncircumstantial nature of the practice. Note that I will avoid the terms "deformed" or "deformation" for sake of impartiality and academic correctness.

Although currently still in usage, at least among my fellow Latin American colleagues, these and other expressions (such as "disfiguration," "aberration," "unnatural," etc.) may convey preconceptions of abnormality and project a denigrating attitude towards indigenous body modifications, which is in no way intended in this work.

A semantic problem of a different kind emerges when attributes such as "intentional," "conscious," "deliberate," "willful," or "voluntary" are assigned to the interpretation of the cranial record and herewith conjure up the presumed thought processes and purposes of the head modifiers that cannot longer be ascertained from the skeletal record. This cautionary remark of mine obviously does not imply that there was no intentionality involved. But, given the diversity of culturally sanctioned behaviors that were involved in the shaping of the infant head—many of them triggered by doxic (i.e., commonsensical, spontaneous, or nondiscursive) referents—I consider it problematic to distinguish a priori between intentional and unintentional modifications, as has been advocated by some authors (Neumann 1942; Saul 1972).

Another caveat lies in the materiality of the data sets. Therefore, the analytical approaches that impose notions of intention to the cranial record are potentially at risk of oversimplifying and thereby biasing the interpretation of the dynamic and complex often multilayered meanings and roles the practice once held, as I will argue for the ideological framework of Mesoamerica (Chap. 6). Here, infant head manipulation expresses varied goals and meanings, some of them protective, others preventive, and phenomenological, but most of them unrelated to the visible end result (Tiesler 2011, 2012a). In fact, it might even be misleading to conceive the artificial skull morphologies as the expression of one single practice, especially as we know that several techniques, implements, and practitioners could interfere with daily Mesoamerican head procedures.

Keeping with the terminology, the enactment and daily performance of compressing the infant cranium, is designated here as a "practice" or "practices," or more generally "custom" or "tradition." The semantic connotations of "custom" and more so that of "tradition" underscore its collective performance within a cogent ideological frame. I use "tradition" here as a created, shared, and transferred intellectual property, which is prone to be modified socially. In the words of Alfredo López-Austin, "its representations and forms of enactment contain ideas and forms of conduct with which the members of a society confront different situations in life; these expressions might be individual or collectively shared, mental or externalized" (see López-Austin and López-Luján 1996, p. 62). As a long-standing custom whose roots run deep within Mesoamerican culture, López-Austin and López-Luján (1996, p. 62) highlight the notion of cranial modeling as a tradition, and concede that the behavioral and mental notions underlying this practice very likely expressed a complex set of values and beliefs (which will be discussed later). These were shared and passed down from generation to generation, as part of a collective concept, which formed a long-lasting group identity. In this quality, the custom was capable of being adjusted to social changes by reconstructing and transforming it throughout the centuries. Considering the idea that cranial modifications persisted during at least seven millennia, it should be considered one of the territory's oldest and most enduring traditions, strongly embedded within the intimate structure of the "nuclear core" of Mesoamerica's ideological framework and belief system (López-Austin 2001).

Examined from the archaeological record, the in vivo tradition of head shaping is prone to leave tangible expressions in skeletal populations, namely, their skulls, the contextualized examination of which is suitable to disentangle frequencies, social dimensions, and geographic and temporal trends. The irreversible, permanent nature of this body practice, its transgenerational quality, and its manifold morphological results in the skull concede specifically bioarchaeological approaches like the one that anchors this work. Bioarchaeology is a suitable point of departure, as it is inherently interdisciplinary when contextualizing skeletal data sets with archaeological information (Sofaer 2006). The material record allows skeletal studies to combine different scopes of time and geography. The regional scale, advocated here, relates to overarching society or cultures; from here we might zoom in on cities or communities with documented, representative, and mortuary records. And from here, excavated house units grant glimpses of head shaping as a family practice, specifically in its expression of gender and its role in upbringing and social integration.

Scrutinizing head-shaping practices from the lense of bioarchaeology is especially rich when examined jointly with additional sources of nonarchaeological information, gleaned from health sciences and cultural studies. While health sciences contribute with knowledge on cranial plasticity, neurology, growth, and morphology, the humanities warrant closer looks at underlying ideological aspects and the body, gender, ritual performance and social structure, and even at ancient aesthetics and constructed beauty. The feasibility of including nonarchaeological cultural data sets in this type of research naturally stands or falls with the availability of materialized cultural expressions, written sources, and the permanence of the past society under examination. In the case of the data-rich Mesoamerican study environment, interdisciplinary cultural examination is facilitated by a host of additional sources to draw from, namely, ethnography, iconography, glyphic, ethnohistorical, and semiotic studies. Here, the combination of the bioarchaeology of head shaping with other sources of information is a suitable analytical point of departure to understanding the culturally embedded meanings of skull modifications. Beyond that, they are capable of tracing the underlying, long-lasting social dynamics and cultural and ethnic undercurrents that head shapes may express in the form of popularity vs. scarcity, variety vs. homogeneity, abandonment vs. continuity, equality vs. distinction. It is this double goal—of comprehending the roles of head-modification practices per se and exploring the underlying social and ideological trends they express in Mesoamerica (and Latin America in general)—which sets the stage and tempo of this book.

1.3 Organization of This Volume

The flow of the contents follows the central goals and the interdisciplinary approach of the book to benefit comparisons with other areas of the world and an integrated understanding of the practice itself. Chapter 2 provides different conceptual platforms, anchored in social and cultural frames of reference, for examining and interpreting ancient head-shaping practices and other permanent body modifications in Mesoamerica and beyond. Chapter 3 provides relevant groundwork on embryology and infant maturation in humans, applicable to the dynamics operating in artificial head flattening. Specifically, unaltered and altered neural and cranial growth dynamics are relevant to explain the physiopathological processes involved in infant cranial modifications, to conjecture about the natural limitations and possibilities of modifying the shape of infants' heads during peak expansion, and lastly to explore the morphological effects as contingent on the duration of compression. The treatment of the morphological end results introduces the secondary side effects of extrinsic compression. I will refer specifically to the potential systemic and neurological health risks and local reactions of the organic tissues during the compression process. In all probability, these health implications triggered a host of measures taken by the pre-Columbian women practitioners in order to prevent harm to their little ones.

Chapter 4 introduces the criteria employed in the identification and classification of head-shaping practices from the archaeological cranial record. After a brief review of the anthropological literature and their undercurrents, I will focus on the seminal work by José Imbelloni (Dembo and Imbelloni 1938), which counts as a global benchmark for anthropological studies of artificial cranial modifications. Imbelloni's classification system is still widely employed in current research on Mesoamerican head forms. This system allows inferences on the type of techniques and head instruments that once produced the visible head shape. Craniometric studies complement and enrich observational criteria and benefit the standardizing and objectifying of Mesoamerican—and in general American—typologies. Again, complementary sources of information are examined, such as the Mesoamerican body of figurines and so-called suprainiac lesions on the occiput. These are helpful in the recreation of the lived head practice and its technicalities.

Head shaping as a lived practice is the central theme of the Chap. 5, compiled and commented by my historian colleague, Pilar Zabala. She assembles some hundred ethnohistorical testimonies of indigenous head compression in historical Latin America under the umbrella of colonialism and more recent independence and nationalism. This is an anthology of citations that covers five centuries (fifteenth to twentieth) of lived head-shaping practices in Hispanic America. The historical extracts lend a voice both to native body practitioners and Iberian eye witnesses. They let government analysts, adventurers, and naturalist bystanders, and lastly, modern anthropologists, who describe the last bastions of head traditions still being practiced in the Amazon Basin and parts of Mesoamerica during the twentieth century, speak. Their surveys introduce the second, more specific, part of the book, which is about pre-Hispanic Mesoamerica.

Part II combines different data sets in order to trace head shapes and their cultural meanings within the Mesoamerican sphere, specifically the data-rich Maya realm, the only cultural area that has been systematically studied on a regional scale (Tiesler 1998, 2012b). In Chap. 6, I return to body and gender theory and the broad set of ideological schemes of native cosmology and thought for understanding the nuances of different ideological dimensions of head-modification practices. The examination of the regional record with the general conceptual and analytical parameters, traces a trajectory for cephalic modeling on different scales of approximation that go from family spheres to neighborhoods and communities and zoom out to focus on regional expression and others that define pan-Mesoamerican cultural heritage. The elements discussed here for Mesoamerica transcend ideological aspects and highlight broader notions of cultural, political, and economic evolvement of the areas and eras that encompass native society during different times of pre-Hispanic, colonial, and modern life. The volume closes with a synthetic chart of Mesoamerican head-shaping traditions, of their ideological undercurrents and their meanings, along with present and future venues for their study.

References

Dembo, Adolfo, and José Imbelloni. 1938. Deformaciones intencionales del cuerpo humano de carácter étnico. Buenos Aires: Humanior.

Dingwall, Eric John. 1931. Artificial cranial deformation; a contribution to the study of ethnic mutilations. London: Bale & Sons & Danielsson.

Hooton, Earnest Albert. 1940. Skeletons from the Cenote of Sacrifice at Chichen Itza. In The Maya and their neighbors: Essays on Middle American anthropology and archaeology, ed. Clarence L. Hay, Ralph L. Linton, Samuel K. Lothrop, Harry L. Shapiro, and George C. Vaillant, 272–280. New York: Appleton-Century.

Kirchhoff, Paul. 1943. Mesoamérica. Acta Americana 1:92–103.

López-Austin, Alfredo. 2001. El núcleo duro, la cosmovisión y la tradición mesoamericana. In Cosmovisión, ritual e identidad de los pueblos indígenas de México, ed. Johanna Broda and Félix Báez-Jorge, 47–65. Mexico: Consejo Nacional para la Cultura y las Artes, Fondo de Cultura Económica.

López-Austin, Alfredo, and Leonardo López-Luján. 1996. El pasado indígena. Mexico: Colegio de México, Fideicomiso Historia de la Américas, Fondo de Cultura Económica.

Manzanilla, Linda, and Leonardo López-Luján (ed.). 1994. Historia antigua de México, 3 vols. Mexico: Instituto Nacional de Antropología e Historia, Universidad Nacional Autónoma de México.

Martin, Simon, and Nikolai Grube. 2008. Chronicle of the Maya kings and queens. New York: Thames and Hudson.

Neumann, Georg K. 1942. Types of artificial cranial deformation in the Eastern United States. American Antiquity 3:307–318.

Romano, Arturo. 1973. Los cráneos deformados de El Zapotal I, Veracruz. Antropología Física 1:57–64.

Romano, Arturo. 1974. Deformación cefálica intencional. In Antropología física, época prehispánica, ed. Juan Comas, 197–227. Mexico: Instituto Nacional de Antropología e Historia.

Saul, Frank P. 1972. The human skeletal remains of Altar de Sacrificios. Cambridge: Papers of the Peabody Museum of Archaeology and Ethnology 63(2), Harvard University.

Sharer, Robert, and Loa Traxler. 2006. The ancient Maya. Sixth edition. Stanford: Stanford University.

Sofaer, Joanna R. 2006 The body as material culture. A theoretical osteoarchaeology. Cambridge: Cambridge University.

Stewart, T. Dale. 1974. Human skeletal remains from Dzibilchaltun, Yucatan, Mexico, with a review of cranial deformity types in the Maya region. In Middle American Research Institute, no. 31, 199–225. New Orleans: Tulane University.

Tiesler, Vera. 1998. La costumbre de la deformación cefálica entre los antiguos mayas: Aspectos morfológicos y culturales. Mexico: Instituto Nacional de Antropología e Historia.

Tiesler, Vera. 1999. Rasgos bioculturales entre los antiguos mayas: Aspectos arqueológicos y sociales. Ph.D. diss., Universidad Nacional Autónoma de México.

Tiesler, Vera. 2011. Becoming Maya: Infancy and upbringing through the lens of pre-Hispanic head shaping. Childhood in the Past 4:117–32.

Tiesler, Vera. 2012a. Studying cranial vault modifications in ancient Mesoamerica. Journal of Anthropological Sciences 90:1–26.

Tiesler, Vera. 2012b. Transformarse en maya. El modelado cefálico entre los mayas prehispánicos y coloniales. Mexico: Universidad Autónoma de Yucatán, Universidad Nacional Autónoma de México.

Tiesler, Vera, and Arturo Romano. 2008. El modelado del cráneo en Mesoamérica. Emblemática costumbre milenaria. Arqueología Mexicana 94:18–25.

Trinkaus, Erik. 1982. Artificial cranial deformation in the Shanidar 1 and 5 Neandertals. Current Anthropology 23(2):198–199.

Trinkaus, Erik. 1983. The Shanidar Neanderthals. New York: Academic Press.

Weidenreich, Franz. 1938. On the earliest representatives of modern mankind recovered on the soil of East Asia. Peking Natural History Bulletin 13:161–174.

Part I
Interdisciplinary Approaches to Artificial Cranial Modification

Part I sets the tempo for exploring ancient head-shaping practices in Mesoamerica. Here, the manifold relevant aspects of culturally induced cranial modifications are addressed from different angles, concretely, by providing groundwork in sociocultural theory and cranial physiology, the latter delineating the real and possible morphological effects on cranial growth. For this purpose, anatomical and physiopathological information on normal and the altered head growth dynamics is discussed, to ponder on potential health risks involved and counteractive measures taken by the pre-Columbian women practitioners. An extensive treatment of the classification criteria, to be used in distinguishing molded skull shapes and infer implements and techniques, employed in cranial modification, lays groundwork for standardizing head shape typologies specifically in Mesoamerican bioarchaeology and for delineating future general methodological approximations to the study of artificial head shapes. These approaches are complemented by transcriptions of historical eyewitness accounts on the head shapes, head shaped, and head shapers, compiled and interpreted by Pilar Zabala.

Chapter 2
Cultural Frameworks for Studying Artificial Cranial Modifications: Physical Embodiment, Identity, Age, and Gender

Ex visu cognoscitur vir, et ab occursu faciei cognoscitur sensatus: amictus corpois et risus dentium et ingressus hominis enunciant de illo
[Bartolomé de las Casas 1967, Chap. XIX]

The human body, with its physical and psychological properties, figures both as a basis and mediator in all cultural interactions and, as such, is also affected by the social life it supports. Thus, its anthropological study, and that of its cultural modifications, does not only inform on morphological adaptation and plasticity but equally grants glimpses on society itself. Regarding archaeologically retrieved cultures of the past, permanent body enhancement is still evident in buried human bones and teeth, two body tissues that resist decomposition much longer than soft tissues. The bioarchaeological study of artificial modifications of hard tissues therefore provides invaluable insight into ancient customs and may hint at underlying cultural and social dynamics involved in their execution. This chapter explores broad concepts of cultural and social meaning that facilitate the linkage of past head-shaping practices and body modifications in general with social processes; namely, their role in the physical embodiment of ancient society, culture, identity, gender, and age. The concepts, detailed here, anticipate the more specific interpretations of meanings in Chap. 6.

2.1 Individuals, Corporeality, and Ancient Head Practices

> El cuerpo humano es núcleo y vínculo general de nuestro cosmos, centro de nuestras concepciones, generador de nuestro pensamiento, principio de nuestra acción y rector, beneficiario y víctima de nuestras pasiones. (Alfredo López-Austin 1989, p. 7)

Anthropologists' and sociologists' inquiries on the relationship between the body and its reflective and sociocultural roles have spanned the notions of its physicality, conceptualization, self-reflection, and model for the surrounding world (Le Breton 1994; Mauss 2007; Shilling 1993; Turner 1984; see also Lock and Farquhar 2007; Sofaer 2006). Recent inquiries tend to focus on the social "construction" of the body and its implications for understanding aspects of agency on agential behaviors and sociocultural negotiations. In general, scholars have been receiving feedback in the last 30 years on a host of phenomenological, structural, and semiotic proposals, mostly gleaned from explanatory readings on cultural phenomena. These studies

V. Tiesler, *The Bioarchaeology of Artificial Cranial Modifications*,
Interdisciplinary Contributions to Archaeology 7, DOI 10.1007/978-1-4614-8760-9_2,

attempt to reproduce and recreate broader meanings at the interstice between the mind, the body, and culture, and their mutual interaction (Csordas 2003; Douglas 1973; Featherstone et al. 1991; Sandoval 1985; Sauvain-Dugerdil 1991; Shilling 1993; Sofaer 2006; for specific work on cultural vault modeling in the Americas, see also Blom 2005; Geller 2006; Yépez 2006, 2009; Yépez and Arzápalo 2009).

Although I consider that these cultural readings are innovative, some invaluable for assigning new meanings and for providing new venues for future inquiries and general cultural understanding, my own approach neither attempts any a priori "agential" narratives on infant head modeling nor do I wish to engage in semiotic readings of the patterns these left in the archaeological record. Instead, I have given priority to comprehending Native American—namely Mesoamerican—head practices from an "emic" point of view, by carefully categorizing, analyzing, and contrasting different sets of data and by conceptualizing the expressions of indigenous body notions, specifically those of the head and its vital components. A point of departure for this approach is the abstract notion of the "individual" with its indivisibly bio-psycho-social properties, inseparably entwined with its materiality and integration into the human collective. The latter embraces populations and societies, including all its cultural and specifically its ideological expressions.

2.1.1 Recognizing the Individual in the Material Record

Any attempt to assign a set of sociocultural motifs, or even just any role to the artificial head modeling among ancient Mesoamericans, would be in vain without a framework of bio-socio-cultural reference and criteria that would allow its interpretation above the interdisciplinary divides. Specifically, the skeletal approach of this work requires a framework that authorizes sociocultural evaluation from the material record, first as part of material culture and then as data directly relevant to sociocultural reconstruction and interpretation. In the case of archaeological interpretations of past Mesoamerican society and cosmovision, permanent body modifications (such as cranial-vault modifications) that leave a mark on archaeologically retrieved human remains can be directly studied from the perspectives of physical anthropology and bioarchaeology, lending to fertile interdisciplinary dialogues between human biology and archaeology.

In recent years, the analysis of skeletal materials has increasingly responded to the parameters of bioarchaeological research agendas, a term coined by Jane E. Buikstra during the 1970s (Buikstra 1991). "Bioarchaeology" broadly designates a thematic specialization in archaeology or physical anthropology that studies human remains in their context and as part of the archaeological body of information employing explicit biocultural approximations. This line of research is noteworthy for integrating skeletal research and cultural sets of data. South of the USA, Mexican "biosocial archaeology" similarly devises a series of theoretic and methodological concepts that anchors the study of human remains as an integral part of the archaeological context jointly with other cultural data (Terrazas 2000; Tiesler 2007, p. 31–40).

These concepts facilitate the interpretation of head modeling from the material record.

The following paragraphs strive to set theoretical foundations and more specific tenets on the human being and his sociocultural references, in particular, the "individual" as an abstract concept and as a basic analytical unit to translate the material record into dynamic behavior and human interaction. These are followed by speculations concerning the body and matter, the human parts in their biocultural dimension, and their meaning among ancient as well as modern Mesoamericans. Some of these ideas have already been established and widely discussed in previous works (Tiesler 1996, 1997a, b, c, 1999, 2007). For a more general review of this subject and recent literature on biosocial and biocultural inferences and their acceptance, I recommend the works of Dogan and Pahre (1991), Fox and King (2002), Goldschmidt (1993), Goodman and Leatherman (1998), Joyce (2005), Sauvain-Dugerdil (1991), Sofaer (2006), Vera (1998), and taken to a more general level, Skibo and Schiffer (2009). Today, the analysis of the "individual" or the "person" in its historical, vital, and social context has started to constitute a major concern also in mainstream archaeology.

A conceptual point of departure for this study is the abstract notion of the "individual" with its indivisibly bio-psycho-social properties and as inseparably entwined to the human collective in the form of both (biological) populations and society, which include all of its cultural and specifically its ideological expressions. For the purposes of this work, the "individual" is conceived of as a dynamic entity, as a living human that interacts actively and is formed and transformed by society. The individual in its physicality is converted after death into an object of mortuary treatments, and, still later in the timeline, becomes a study object for archaeologists.

Generally speaking, we can conceive the individual human as an organic unit, as a thinking being who reflects, socializes, produces, and reproduces. As a biological system, the individual is intertwined with the biosocial medium around him, with whom he or she forms part of different chains of relationships. In its physical body, the individual is subject to physiological and pathological changes during the life cycle and as an organic system in exchange with the environs and the mass of society that he or she is intrinsically tied to. The individual engages in constant and dynamic transformation both as a singular organism (embryogenesis, life cycle), as well as collectively (human evolution, adaptation), with the speed of those changes tending to operate at different levels and in different cycles. The quantitative physical–biological scope of individuals identifies *populations*, defined as the set of individuals who engage in biological and social relationships among them.[1] Biologically, the population is formed by individuals of both sexes in different phases of their life cycle, who interact and reproduce.

In the psychosocial sphere, the capacity of being conscious of—and of reflecting on—reality, epistemologically converts the individual into a subject, capable of

[1] The concept of population is delimited by other definitions than those given here. Many emphasize the biological aspect of the concept. Note that population, present and past, according to many authors forms the purpose and unit of analysis of physical anthropology, as a study of man and his origins, evolution, and diversity (Buettner-Janusch 1980).

acting and interacting consciously, of reflecting on him- or herself, and of knowing the physical and social environment around him or her. The subject studied here emphasizes, more than the structural relationships established by the individual with society (a social being), their articulation by system of shared ideas, prone to be affected by shared value systems, by standards or even actions instrumented by organizations or institutions (Bate 1998). We will explore the different ideological dimensions of Mesoamerican head practices extensively in Chap. 6.

It is the social dimensions that directly link the individual with the society of which he or she is a part. This relationship is dynamic, complex, and mutual, but never symmetrical. Although there can be no social history without individuals, individuals by themselves are not self-sufficient. They require society for their biological and social reproduction and to satisfy their material and psychological needs (Bate 1996, p. 60–61; Meillassoux 1987). Of course, social interaction occurs in different spheres. These range from the domestic realms (intrafamily, gender, between families), and others, established between social sectors, subcultures, cultures, and groups (Service 1971). These are frequently, although not necessarily, tied to social positions or "status." In general, the members of a society interact and integrate depending on their culturally conferred age and phase of procreative and productive life, which in turn has a biological component (age groups formed according to growth, maturation, and degeneration). This dimension is conditioned by a succession of life stages that must also have guided the course of pre-Hispanic life among ancient Mesoamericans. As in other societies, the transitions are manifested here in a fluid, yet scaled progression of stages, many acknowledged, some sanctioned collectively with ceremonies. These separated the persons from their previous (age) groups and integrated them with a more advanced age group, culturally enacting group cohesion and conferring identity to those involved. Transition rituals typically reinforced the cohesive power of the collective *Weltanschauung* (world view) by celebrating the inclusion of the individual into society, its progressive transformation within the social structure of social relationships.

On a more operative scale, the dynamic physical and social characteristics of the "individual," as an analytical unit, provide both a starting point and link for the archaeological patterning and sociocultural generalizations on head-shaping practices. They relate directly to both the performance and the outcome of the practice in the head, and from here, allow to decode and understand more collective behaviors directed to treatment of infant heads.

2.1.2 *Conceptualizing the Body*

The essential locus of the individual is the body. The host of research on body theory, materiality, and embodiment, usually "read" or reconstruct the dynamic living body with its changing intrinsic or given properties. These may vary according to physical and cultural age, sex and gender, or specific uses of the body (Csordas 2003; Lock and Farquhar 2007, p. 50–68; Mauss 1971). In this scheme, recent bioarchaeological approaches are especially well suited to examining those body attributes that leave permanent traces in the skeleton, as they are capable of granting insights on wide-ranging aspects of aesthetics and crafted beauty, on identity and culture, gender, ritual performance, and social structure. This is achieved by translating attributes from the joint contextualized evaluation of the material record. It comes as a surprise in this regard that studies on head modifications still remain only marginally treated in the bioarchaeological literature and likewise, in fact, in most resource compendia on archaeological and anthropological body theory, embodiment, and gender (see, for example, Joyce 2000; Joyce 2005; Klein and Quilter 2001; Lewis 2007; Lock and Farquhar 2007; Moore and Scott 1997; Sofaer 2006).

Most of the recent bioarchaeological scholarship on head-shaping practices that does incorporate broader concepts of the biological, social, and cultural body, advocates life history approaches and individual life narratives, or more general thoughts on agency, body theory, and embodiment (see for example, Blom 2005; Geller 2006; Lorentz 2008; Lozada 2011; Reischer and Koo 2004; Yépez and Arzápalo 2009). The underlying idea of the body in much of this work is that of Michel Foucault's docile body, a manipulated, socially constructed or "inscribed" entity that conveys social information, linked to gender, age, personhood, lived experience, identity, and embodied group affinity, thereby constituting a forum of power relationships and negotiations (Joyce 2005; Meskell 1998). Tentative hermeneutical, cognitive "readings" of the material record ascribe agential properties and intentionality to long-vanished cultural dynamics and embodied experiences of personhood. Lorentz (2008) interprets cultural cranial modeling in this vein as a form to generate physical capital, emphasizing the dynamic and mindful properties of the body. Beyond general embodiment, some of the regionally oriented (bio)cultural studies on ancient head-shaping practices on Mesoamerica and on Highland Andean head practices (see for example, Duncan 2009; Duncan and Hofling 2011; Lozada 2011; Yépez 2006) have made laudable efforts to reach a culturally sensitive, *emic* understanding of the body and its body parts, by engaging overlapping lines of arguments derived directly from native ideological frames and worldviews. This is also the line of thought advocated here and will be the focus of Chap. 6.

This work, following the approach of López-Austin's seminal work on the human body in the ideology of ancient Mesoamerica, conceives the body as both the core and link to the human cosmos, perceptions, and thoughts. It is both the originating component and the recipient of human action and interaction (López-Austin 1989, p. 7). At the same time, this entity is the immediate study object of bioarchaeologists, albeit only in its incomplete material nature and departing from its static quality, because its organic metabolism has ceased long ago with death, while its human carrier has vanished together with his or her quality as active sociocultural participant, now only hinted at faintly from the mortuary record. The object of study now contains the "frozen" corporeal properties that the body held at the time of death, such as the age-related skeletal morphology or the stage of a given disease, to name just a couple. Specifically for bioarchaeological approximations, the convenient immediate analytical unit is the single dead skeleton, which harbors the information that remits to the living individual through its biological and cultural materiality (see also Sect. 4.7).

This conceptual frame is also heuristically suited to overcome the disciplinary divide between the physicality and static nature of the skeletal record and the dynamic cultural qualities reconstructed from its material expressions (Schiffer 1987; Skibo and Schiffer 2009; Tiesler 1993).

The material record may also convey past experiences of the human life course, specifically those that are still materialized in bones and teeth. These lend themselves to the reconstruction of the behavioral components of body performance (Schiffer 1999, p. 116–120; Skibo and Schiffer 2009). Marcel Mauss (2007, 50–66) conveys the idea of "body techniques." These are sorts of acquired abilities or faculties of habits which vary according to the age, sex (or more correctly, gender), as well as to the efficiency and training within a given social fabric and, more so, between societies or education. Mauss stresses the learned behavioral component and highlights the body as a type of tool or vehicle for learned behavior. In his specific terms, the practice of cranial-vault modeling, as performed daily by female caretakers on their infant kin, projects a kind of *chaine d'operatoir* of tasks, typically transmitted by elder women and learned by younger mothers who will gradually improve their skills in modeling their baby's head.

2.2 Body Modifications of the Past: An Overview

Permanent alterations of the human anatomy are not isolated cultural phenomena of the past but identify now, more than ever, omnipresent incorporated epitomes of modern lifestyle and individual aesthetic expression and assumed identity. Apart from long-standing cultural traditions, such as religiously motivated circumcision, body sculpturing now often follows medical indications. Also, body plastic surgery for nonreligious and nonmedical purposes has reached our mainstream (post)modern society. Body piercings and tattoos are customary now alongside other, more drastic surgical body alterations. They may be carried out for the sake of aesthetics and beauty, body art and sexual enhancement, individual self-expression and, almost always, group affiliation. It is noteworthy that the state of the art of body makeovers also includes surgical transformations of adult head shape by transdermal and subdermal implants (Gump 2010).

2.2.1 Studying Artificial Body Modifications

The humanities established cultural body modifications as a sort of formal study object some time in the nineteenth century, which has since been the focus mainly of ethnology and physical anthropology. Approaches by iconographers or art historians are complementary to the different forms into which the bodies are transformed and their cultural and aesthetic connotations. Cultural body modifications are commonly defined as procedures that are carried out in order to modify the external aspect of the person (Alt et al. 1999; Brain 1979; Feest and Janata 1989; Flower 1881). This definition excludes mobile body ornaments (i.e., external adornments, like jewelry). Body modifications are distinguished as either temporary (such as body painting), or permanent alterations, such as changes in the skin, the mucous membranes, teeth, and bones. Some permanent modifications are then differentiated as artificial modifications of the body such as, for example, placing objects in the physical orifices, or changes that modify the external aspect of the body for the remainder of life. In traditional societies, these permanent modifications of the body tend to be associated with initiation rituals and rites of passage, as a requirement on the part of each individual as part of a shared cultural manifestation (Van Gennep 1960).

Changes in the appearance of the body constitute a common element in all societies, both today as in the past (Dembo and Imbelloni 1938; Feest and Janata 1989). The body procedures performed on the hair, skin, mucous membranes, teeth, or the skeleton have always been an integral part in the rich network of traditions that define personal choice and, in general, the cultural heritage of a group, sometimes marking ethnic affiliation, gender, or status statements. Some practices, such as the application of body paint or the wearing of head ornaments, are temporary, reserved for special occasions or applied on a daily basis. Others produce longer-lasting impacts on human anatomy, remaining throughout a person's life span. Many of these are strictly personal choices and adhere to individual preferences; others are more collective, as they follow fashions imposed by social dictates of the time. Still others, typically more conservative and restricted in application, may be regulated or even institutionalized.

Beneath or beyond personal and collective choices of looks, there is always visible display and ostentation involved in the performance of body modifications and ornamentation, naturally. Adornments are prone to reify social identities. These may be real or perceived, borrowed or even fictitious (Reischer and Koo 2004). The cultural adjustment of the body may be *doxic* or intentional, its communication of ascribed social identities may either be hidden and subtle or blatantly obvious and even gaudy. Miller (1982) makes this point for the surging rich from the popular sectors, the *nouveaux riches*, who attempt to establish themselves by gaudy displays of possessions and outer looks. This notion has recently been introduced in the study of pre-Hispanic Nicaraguan folk (McCafferty and McCafferty 2011). The authors explore body ornaments from burial contexts that once might have been perceived as flashy, possibly decoding aspects of ancient social revindication and claim to recognition. Also within the Mesoamerican heartlands, the ostentatious display of beauty, riches, or power was on the order of the day and could be enhanced institutionally. Here, I recall the gaudy personal cult that the Classic Maya courtiers indulged in, garnished with self-assigned attributes of perfection and the marking of the divine.

2.2.2 *Modifying the Body in Ancient Mesoamerica*

Beyond doubt, temporary and permanent artificial body modifications materialize a wide range of Mesoamerican ideological expressions. Here, many "looks" of the

Fig. 2.1 Late Classic period clay figurine from the Maya area. The head figurine displays a series of flamboyant adornments and adjustments of the skin and cartilage. (The Museo Popol Vuh, Universidad Francisco Marroquín, Guatemala, adapted and redrawn by B. Ceballos)

body and its anatomical constituents were to evolve into deeply embedded traditions that were shared by families, communities, women and men, among different age groups and social sectors (Fig. 2.1). As a result, a host of distinct forms of temporal body decoration and also some permanent body castings (such as head shaping and dental fillings) came to be popular in Mesoamerica's diversified cultural repertoires, decoding a myriad of roles and meanings in different locations and human contexts.

Regarding temporal body changes, most hairstyles and body paintings were part of everyday engagements in Mesoamerica, while others, usually more elaborate displays, were reserved for festive occasions. Body painting was widespread and could communicate kingly or warrior status, mourning or sacrifice, among other more mundane motives (Vela 2010). Other interventions, such as scars and tattoos—or cartilage piercings for holding ear plugs, labrets, and nose rings—permanently changed the aspect of those who wore them, translating into enduring expressions of cultural identity or social membership (Vela 2010). As with head shaping, permanent perforations of soft tissues, facial scarification, and dental modifications (in the form of inlaying, filling and incision), came to acquire special importance in many parts of Mesoamerica over the centuries. Unfortunately for Mesoamerican scholarship, the scars that once covered ancient living bodies have decomposed posthumously along with the skin it once marked, except for very rare examples of soft tissue preservation by natural mummification that are limited to the dry northern and western highlands. Therefore, only figurative presentations of body scarification lend to their study in the Mesoamerican sphere. Equally vanished are skin piercings and larger cartilage perforations. Their presence can only be inferred indirectly by the form and size of those personal body ornaments found together with the skeletal remains of graves.

Among Mesoamerica's permanent body modifications and apart from head shaping, only dental decoration has left "hard" evidence in the material record of Mesoamerica. This is because enamel and dentine are the body components that resist decomposition more than any other human tissues, including bone. The archaeological record of Mesoamerica shows artificially pointed teeth already by the second millennium B.C. and probably even before. These are the dates given to dental fillings, such as the ones documented from El Arbolillo in Mexico (Romero 1958; Romero 1974). By the onset of the Classic period, tooth filling had turned widespread across most of Mesoamerica's cultural landscape. This modification implied the selective attrition of the dental tissues by help of abrasive sand and stones, whereas incisions with sharp lithic instruments led to grooves on the dental labial surfaces. A third style of dental work consisted of securing hard and soft tooth inlays in drilled teeth. This procedure was much more demanding on the artisanal skills than filing, as it necessitated meticulous single or multiple perforations of enamel and dentine with subsequent precision adjustment and permanent fixation of the material to be inlayed (Ramírez et al. 2003). At the end of this operation, people would put on view in their anterior teeth semiprecious stones such as pyrite, turquoise, and jadeite. Also, fibrous filling materials could be employed to seal the dental cavity; these materials appear to have been alternative choices of fillings or were used once the inlayed stones had fallen out of their dental sockets (Romero 1958; Tiesler 2000).

Different from artificial cranial-vault modifications, dental work appears—in Mesoamerica—not to have been performed before having reached late adolescence or early adulthood. The overwhelming majority of dental modifications appear in permanent dentitions and show different degrees of postmodification physiological abrasion (Romero 1952, 1958, 1970, 1984, 1986), whereas alteration of deciduous teeth is beyond doubt infrequent (Peña 1992). This means that dental work and its observable results would have visibly exhibited acquired and achieved qualities among grown-ups. In fact, status distinctions must have played a role in the display of dental work in certain social contexts, at least among Classic period lowland Maya (Tiesler 2000). When considered jointly with other indicators of the Maya cultural record, "Ik" styles and inlays appear more frequently in privileged burials than in plain mortuary contexts (Tiesler and Benítez 2001). However, beyond this general tendency, there are no radical, socially affiliated distinctions in dental wear. Dental styles manifest themselves in the material record only in terms of preference, never denote exclusivity (exercised by some people but never by others). This lack of practitioner codes appears to indicate that dental decoration was not regulated by any explicit prohibition or strict norms among the Classic Maya, and no relationship could be established between the presence and style of the dental work and artificial head form among lowland Maya. This lack of association confirms, as do the age differences, that both body practices once responded to different cultural needs, an observation that should be analogous in other parts of Mesoamerica. (Tiesler 2000) Considered jointly with the archaeological record, the evolution of dental practices and their multifaceted visible expressions in the dentitions of their carriers manifest daily behaviors and individual, circumstantial choices besides also long-standing patterns of cultural change across the evolving Mesoamerican landscape.

Among the distinct Mesoamerican techniques employed for the dental work, incisions and incrustation were much less widespread than tooth filling in terms of cultural distribution and time depth. The latter is circumscribed to the Late and Terminal stages of the Preclassic period, a time after which it gained prominence in Veracruz, Oaxaca, and the Maya territories during the Classic period. Although no dental procedure or pattern appears as exclusive to either sex, more Maya men than women appear to have had their frontal dentition inlayed (Tiesler 2000). Similar to head shaping, in most parts of Mesoamerica, the canons of dental decorations became homogenized at the onset of the Postclassic period. During the first half of the second millennium A.D., incrustation disappeared from almost all parts of the Mesoamerican material record (Romero 1958, 1984, 1986).

2.3 Head-Shaping Practices and Identity

The following paragraphs seek points of departure in the conceptualization of ancient cranial-vault modifications from different angles of past social life. Each of these elaborates on a different component in these practices: the practitioners and the "wearer" of the custom are addressed; then the procedure's role in gender and age expression; and more collectively, as a tradition and visible emblem of beauty, identity, and ethnicity (see also Chap. 6).

2.3.1 Head Practices as Traditions

More conservative and generation bridging than most other body modifications was the artificial head molding of newborns. These were not conducted (or even influenced) by the subject him- or herself, but effected by a grown-up person, usually the parents or kin. In this procedure, the prospective "wearer" of the artificial shape had no possibility to change or avert the processes and their lifelong visible outcome on the head. Performed by second or third generation women on newborns that later bore the visual result for the rest of their lives, artificial vault modification is a practice that has transcended generations (Blom 2005; Torres-Rouff 2002; Yépez 2006). This protracted, conservative quality of head shaping identifies long-lasting cultural dynamics and raises the cultural importance of this practice above that of more ephemeral, transitory body shaping, which erroneously still permeates the literature on head modeling (Christensen 1989).

This practice, performed within the domestic confines, rested in the experienced hands of women who daily applied this technique to their infants. Beyond the mechanical quotidian routing the maneuvers on the infant heads surely established links with the cosmos for their Mesoamerican practitioners, who were in regular converse with the divine. (Chap. 6). The fact that this ultimately became a regular practice among most Mesoamerican peoples allows us to establish head modeling, by definition, as a custom (or customs), which enjoyed general approval by society, at least before the Spanish Conquest. As a long-standing and deeply rooted custom, cranial modeling actually constitutes a Mesoamerican tradition following the interpretative approach by López-Austin and López-Luján (1996), understood as "an intellectual heritage that is socially created, shared, transmitted and modified, comprised of representations and forms of action, in which ideas and rules of conduct are developed with which the members of a society either individually or collectively confront, mentally or physically, the different situations that they face in life" (López-Austin and López-Luján 1996, p. 62). Their enactment and mental background expressed a complex range of values and beliefs (as we will see later) that were shared and passed from one generation to another, thereby participating in the collective and long-lasting construction of the group identity. This custom in turn was capable of assimilating underlying social changes, and of transforming and renovating over the centuries. Considering that head modification endured for thousands of years it must be included among the most ancient and long-lasting of Mesoamerican traditions, found within the innermost spheres, the "hard core" of Mesoamerican ideology and beliefs (López-Austin 1998, 2001).

In these terms, and aware of the widespread practice in the bosom of pre-Hispanic Mesoamerican society, it follows that the practice of head modeling must have been a principal expression in daily life, family identity, and cultural belonging. It must have been a form of ideological credential—both on an emblematic as well as ritual plane—for the infants (both boys and girls who experienced this), for their preproductive and preprocreative integration with the group. In all probability, the women who practiced this custom on these infants, their own sons and daughters as well as those of others, must have engaged in its daily performance, perhaps without much reflection but following a self-evident notions of things desirable, in fulfillment of the proper way to rear a child passed on by the elderly. Apart from individual social fulfillment, these actions, formalized in practices, must have expressed broadly shared, yet changing, sometimes conflicting values.

2.3.2 *Age and Head Practices*

Cranial modification links biological and cultural aspects of age and ageing as have few other practices of the past. It is biologically conditioned, as it can only take place while the baby's skull is still malleable during the first few years of an infant's life. Most of all, the first months after birth show peak cranial growth; after that age, head expansion slows and ceases at the age of 2–3 years. Once the skull hardens, the resulting shape becomes permanent. This physiological sequence, recorded roughly here in terms of months and a couple of years (see also Chap. 3), puts natural constraints on the maximum duration of infant head molding. It allocates compression practices between the developmental stages of baby and toddler age, which by themselves designate progressive stages of gaining independence and growing up (Lewis 2007). But what about the cultural correlates of biological growth? Native Mesoamerican cultures specifically consecrated the milestones of infant maturation, such as sitting on the hip, walking, or eating maize for the first time. These were in tandem with education, evolving personhood, and progressive social and economic integration, sometimes consecrated by initiation ceremonies (see Chap. 6).

Although the infant transition rites and their performance were varied within Mesoamerica and should have differed from analogous age ceremonies held in other areas, the phenomena of rites of passage that deal with birth, adolescence, and death, by themselves constitute universal manifestations (Van Gennep 1960). It is noteworthy in this regard that the ethnographic literature identifies many permanent body transformations in these types of festivities, specifically initiation rites during puberty (Feest and Janata 1989, p. 211; Dembo and Vivante 1945; see also Dembo and Imbelloni 1938). Regarding the role of head-shaping practices in transition ceremonies, it appears that the procedures span the time between postpartum rituals and later infancy rites among a surprising number of ethnic groups (Dingwall 1931). Some areas mark the beginning of head-compression devices with induction rituals, others consecrate its finalization. For example, among Inca Peruvians, the first placement of the newborn into the cradle device was an occasion of joyful gathering among family and kin. This was the time the crib was presented to the family Huaca or totem, that was believed to protect the little one from harm (Latcham 1929, p. 542; Latcham 1937; Purizaga 1991, p. 43–45).

As in all other societies, the increasing locomotive abilities, mental maturation, and independence of the little ones are vocalized by the sequence of successive infant age categories that are identified in Mesoamerica. Here, one important maturation category is the spiritual heat or energy (*calor*) and the prospect of becoming a person (Furst 1995; see Chap. 6). Younger infants especially, regardless of sex, were deemed frail and spiritually vulnerable among many Mesoamerican native groups, as they were believed to be at risk of losing their vital energy because of extrinsic or intrinsic harm. They were in need of constant protection against malignant influences and of positive spiritual and natural reinforcement. Therefore, mothers were induced to apply carefully a set of measures and prohibitions during the first weeks, months and years of their little ones' lives (Bonavides 1992; Tiesler 2011). This care would feel for the mothers like a direct progression of the care taken during pregnancy (López-Austin 1989; Nájera 2000).

Alfredo López-Austin (1989, p. 322–328) has delineated age progression through semantic attributes in the native Aztec (*Nahua*) languages, which align roughly with past and modern concepts of growing up in the Mesoamerican world. Within this scheme, native terminology labels distinctly those children who are nursed and those that still do not talk. Older child age categories distinguish ages below and above 6 years of age, a transition marked by the children's gradual gendered incorporation into the household duties in native society, although age references vary (see, for example, Ardren and Hutson 2006, p. 8–9; Boremanse 1997; 1998, p. 80–81; Farriss 1984, p. 135–136; Kramer 2005). The successive stages of infancy and later childhood define their role and needs within the family, their duties and rights, and ultimately, their social and economic integration.

2.3.3 Gender and Head Practices

The archaeology of gender seeks to understand female (or perhaps also male) roles and forms of involvement in performing child-rearing practices, such as head modeling. Unfortunately, the mortuary record itself, at least in the Mesoamerican cultural sphere, does not hold in situ information on the gender of those who managed the techniques and implements used for compression, which have vanished like most other organic vestiges. More eloquent is the material record of ceramic figurines in the pre-Hispanic Mesoamerican legacy, which includes probably hundreds of cradleboard and splinting scenes from different epochs and cultural areas. In all Mesoamerican figurines of this sort under study ($N = 88$), the baby or toddler is depicted either by itself, with a mystical creature or animal, or together with a female caretaker. Conversely, no scene shows any male practitioner positioned together with the minor (see Sect. 4.4). In the adult–child pairs, babies most often rest in a cradleboard or crib device; some of the body kits include either a separate head wrap or head splint. The age of the female adult varies, suggesting that not only the presumed mother but also other older female kin in postreproductive age, possibly midwifes or respected elders were actively involved in the daily procedures on the baby (see Chap. 6).

The above glimpses of the ancient practitioners enable us to cautiously relate the enactment of infant head shaping to womanhood and female gender expression within a broader context of social theory and regional interpretation. As in most ancient societies of the past, the role of Mesoamerica's women was more circumscribed to the house and its immediate environs, while men worked outside and at a distance. In this ambit, women were in charge of domestic chores, such as food processing, weaving, house maintenance, and child rearing (Claasen and Joyce 1997; Klein 2001). In Mesoamerican thought, for example, the gendered identity and place in the community, society, and in the Maya cosmos itself, was considered as essentially complementary. Male contributions were predetermined as the production of crude material, while the role of women in society was prescribed to the transformation of crude mass into objects of use.

Also, the genders of the infants who experienced the head procedures, who grew up and later in life still displayed the insignia imprinted by their mothers, provide a starting point to explore the role of head shaping and its resulting head form in signaling "girlhood" vs. "boyhood" in the early stages of life. For the Mesoamerican sphere, it is noteworthy that the *Nahua* terms for preadolescent individuals are rather vague with no clear linguistic distinction even between prepartum and postpartum periods or between boys and girls (López-Austin 1989, p. 321). Only when deemed necessary in conversation, a term was added to the word "baby" or "child" to designate its gender. Also among Mayan speakers, the designations for male and female babies tends to be applied indistinctly and changed only after entering the toddler age and beyond, a time span marked by transition ceremonies, such as naming rites or *hetzmek* festivities (Boremanse 1998, p. 80; see also Sect. 6.4). These ceremonies could well have marked the end of infant protection and molding. Analytically, these may set stones of departure in exploring early age progression and evolvement of gendered personhood, "womanhood" vs. "manhood", in the Mesoamerican value system. If we believe Landa, Yucatecan Maya *hetzmek* (which means "sitting on the hip for the first time") ceremonies appear to have sanctioned the onset of distinct, gendered life trajectories among colonial Yucatecan Maya, and in fact still do in traditional Maya communities (Cervera 2007; Marion 1994; Redfield and Villa Rojas 1967; Villa Rojas 1978).

The gender of head "wearers" also matters when the produced artificial head shapes are compared among grown-up men and women, as documented from the (bio)archaeological record. Biological sex identification from the skeleton, as the probable biological expression of gender, is the necessary point of departure to establish headwear worn by females vs. males (Sofaer 2006, p. 89–101). Surprisingly, most ancient cultures (Dingwall 1931) did not distinguish men and women or, more correctly, boys and girls, by their head shape. Mesoamerica is not the exception. Here, maybe subtle distinctions in terms of the degree of morphological change or in terms of asymmetry may have stood for distinctive amounts of work initiated, or the care and experience inculcated by the mothers in the heads of their boys and girls; however, there is no exclusive shape or preference noted among the sexes for those areas systematically covered (Tiesler 2012). Instead, the patterns of head shape point much more to equality in the head treatment of boys and girls than distinction. It appears that a baby girl's head was to be protected and modeled the same way as that of a baby boy. I deduce from this that the head practice would have constituted a nongendered tradition, which is consonant with the nongendered quality of babies, projected in most Mesoamerican languages (personal communication, Alfonso Lacadena 2010). It was only later that the children were to take different paths and follow gendered destinies. Probably, also the fact that the practitioners were female, played a role in the similarities between female and male head silhouette.

2.3.4 Head Practices, Beauty, and Identities

Apart from the active role of head modeling as an infant body practice, there are also more emblematic meanings to this modification, which in most of the societies who practiced it, epitomized notions of culturally defined beauty, ideological emulation, social distinction, or simply group identity and integration. These generally relate to the outcome of head compression, i.e., the visible transformation of the back, the crown, and the front of the head, including the face. This externally visible result is not superficial by any means but holds deeper significance, especially in Mesoamerican thought. This is communicated also linguistically by many Mesoamerican languages, which use the head with its outer insignia as a metaphor for designating the individual, the person, and the "self," as argued by Stephen Houston and his colleagues for the Classic Period Maya (Houston et al. 2006, p. 28; Houston and Stuart 1998, p. 83–85).

On a more general note, philosophical and aesthetic concepts of human beauty convey categorically notions of visually pleasing attributes. There are some very broad undercurrents of beauty ideals that express harmony, symmetry, and certain proportions (Forth 2010). Some of these are considered universal elements of beauty.

Yet, it is also true that the social perceptions of physical beauty have evolved over time and change according to the culture in which they are embedded, consonant with culture-specific values and conventions. Most of these standards go beyond superficial, outer features and identify epitomized inner beauty, such as grace, integrity, elegance or serenity, and others deemed desirable by the community.

Physical beauty may still be enhanced by body adornments in the form of cosmetics and accessories or directly by body modifications. Head shaping is a plastic form of transformation to align to beauty ideals. In fact, the notions of beauty or prettiness resonate heavily in the historical accounts on head-shaping practices and their motives (Dingwall 1931; see also Chap. 5). Sometimes, the hand-crafted head shapes were further emphasized by specific hair arrangements or eye-catching headwear that drew attention to the permanently transformed organic substrate (Dingwall 1931; Stresser-Péan 2011, p. 136). Also in the Mesoamerican sphere, such head-form-adapted headdresses were common among the Preclassic Olmec, Classic Mixtequilla people and Classic period Maya, as figurines and vase paintings testify (Acosta et al. 1992; Cheetham 2008; Taube and Taube 2008; Tiesler 2010).

Already among the Gulf Coast Olmecs, artificially contrived, pear-shaped head looks appear to be highlighted by shaving (Tiesler 2010). During the next millennium, Classic Maya small-scale portraiture still adheres to this morphological exaltation when representing strongly reclined heads partly or completely foliated. Other conventions draw the reclined head profiles with pulled-back hair, in a seeming effort to emulate the maize god (García and Tiesler 2011; Houston et al. 2006, p. 45; Taube 1996). Among Classic period Totonac folk, top-flattened heads were framed by spherical head rims, and still later in time, the artificially contrived wedged foreheads of Huastec women were emphasized by their hair parting (Stresser-Péan 2011, p. 136–137). Naturally, for the scholar who is interested in the forms of physical embodiment, these and other visible combo arrangements make worthwhile starting points for exploring the venues of crafted beauty and social identities among ancient Mesoamericans.

Beyond portraiture, we may assume that the forms of facial representations would have reflected or even exaggerated the preferences of the portrayed subjects, which beyond individual choices would have materialized the culturally desirable attributes, as will be explored in Chaps. 7, 8, and 9 for different Mesoamerican cultural environs and time frames. Specifically during the Classic period, artificial head transformations appear prominently represented in those areas with figurative imagery. Different from Andean head forms, which acquired notions of exclusiveness and prerogative (Chap. 5), it appears that in Mesoamerica, more unifying ideas operated in the form of individual integration, group identity, and potentially ethnicity, as we argue further on in this book. Ethnicity is understood here as the pertinence to a population in which its members identify with each other, usually on the foundations of a common genealogy and ancestry (presumed or real), in addition to other historical ties (Hicks 2001; see also Díaz-Andreu et al. 2005 and Jones 1997 for a broader discussion). Those groups that hold an affinity of this type, tend toward cultural cohesion and to express themselves through common cultural practices, in language and shared ideological beliefs, sometimes in confrontation with others.

References

Acosta, Agustín, Michael Coe, Felipe Solís, and Beatriz de la Fuente. 1992. Museo de Antropología de Xalapa. Mexico: Gobierno del Estado de Veracruz, Beatriz Trueblood.

Alt, Kurt W., F. Parsche, W.M. Pahl, and G. Ziegelmayer. 1999. La deformación de la dentadura como "decoración corporal"; distribución, motivos y aspectos culturales. Antropológicas 15:51–58.

Ardren, Tracy, and Scott R. Hutson (ed.). 2006. The social experience of childhood in ancient Mesoamerica. Boulder: University Press of Colorado.

Bate, Luis Felipe. 1996. Una posición teórica en arqueología. Ph.D. diss., Universidad de Sevilla.

Bate, Luis Felipe. 1998. El proceso de investigación en arqueología. Barcelona: Editorial Crítica, Grijalvo-Mondadori.

Blom, Deborah E. 2005. Embodying borders: Human body modification and diversity in Tiwanaku society. Journal of Anthropological Archaeology 24:1–24.

Bonavides, Enrique. 1992. Ritos de pasaje entre los mayas antiguos. Estudios de Cultura Maya 19:397–425.

Boremanse, Didier. 1997. The faith of the real people: The Lacandon of the Chiapas rain forest. In South and Meso-American native spirituality: From the cult of the feathered serpent to the theology of liberation, ed. G.H. Gossen, 324–351. New York: The Crossroad Publishing Company.

Boremanse, Didier. 1998. Hach Winik. The Lacandon Maya of Chiapas, southern Mexico. Albany: University of Albany.

Brain, Robert. 1979. The decorated Body. New York: Harper & Row.

Buettner-Janusch, John. 1980. Antropología Física. México: Limusa.

Buikstra, Jane E. 1991. Out of the appendix and into the dirt: Comments on thirteen years of bioarchaeological research. In What mean these bones? Studies in southeastern bioarchaeology, ed. Mary L. Powell, Patricia S. Bridges, and Ann Marie Mires, 172–188. Tuscaloosa: University of Alabama.

Casas, Bartolomé de las. 1967. Apologética historia sumaria, ed. Edmundo O'Gorman, Vols. 1 and 2. Mexico: Instituto de Investigaciones Históricas, Universidad Nacional Autónoma de México.

Cervera, María Dolores. 2007. El hetsmek' como experiencia simbólica de la construcción de los niños mayas yucatecos como personas. Revista Digital Pueblos y Fronteras, 4:1–31.

Cheetham, David. 2008. Early Olmec figurines from two regions. In Mesoamerican figurines ed. Christina T. Halperin, Katherine A. Faust, Rhonda Taube, and Giguet Aurore, 149–179. Gainesville: University of Florida.

Christensen, Wes. 1989. A fashion for ecstasy: Ancient Maya body modifications. Modern primitives. Research Publication 12:79–91.

Claasen, Cheryl, and Rosemary Joyce. 1997. Women in prehistory. North America and Mesoamerica. Philadelphia: University of Pennsylvania.

Csordas, Thomas J. (ed.). 2003. Embodiment and experience: The existential ground of culture and self. New York: Cambridge Studies in Medical Anthropology.

Dembo, Adolfo, and José Imbelloni. 1938. Deformaciones intencionales del cuerpo humano de carácter étnico. Buenos Aires: Biblioteca Humanior.

Dembo, Adolfo, and Armando Vivante. 1945. La moda de las deformaciones corporales. Buenos Aires: Las Dos Estrellas.

Díaz-Andreu, Margarita, Sam Lucy, Staša Babić, and David N. Edwards. 2005. The archaeology of identity: Approaches to gender, age, status, ethnicity and religion. New York: Routledge.

Dingwall, Eric J. 1931. Artificial cranial deformation. A contribution to the study of ethnic mutilations. London: Bale & Sons & Danielsson.

Dogan, Matei, and Robert Pahre. 1991. Las nuevas ciencias sociales, la marginalidad creadora. Mexico: Grijalba.

Douglas, Mary. 1973. Natural symbols: Explorations in cosmology. New York: Vintage.

Duncan, William N. 2009. Cranial modification among the Maya: Absence of evidence or evidence of absence?. In Bioarchaeology and identity in the Americas, ed. Kelly Knudson and Christopher Stojanowski, 177–93. Gainesville: University Press of Florida.
Duncan, William N., and Charles A. Hofling. 2011. Why the head? Cranial modification as protection and ensoulment among the Maya. Ancient Mesoamerica 22:199–210.
Farriss, Nancy M. 1984. Maya society under colonial rule. The collective enterprise of survival. Princeton: Princeton University.
Featherstone, Mike, Mike Hepworth, and Bryan S. Turner (ed.). 1991. The body, social processes and cultural theory, Vol 7. London: Sage.
Feest, Christian F., and Alfred Janata. 1989. Technologie und Ergologie in der Völkerkunde, Vol. 2. Berlin: Dietrich Reimer Verlag.
Flower, William H. 1881. Fashion in deformity. London: Nature Series.
Forth, Christopher E. 2010. Beauty and concepts of the ideal. In A cultural history of the human body in the modern age, ed. Ivan Crozier, 227–281. Oxford: Berg.
Fox, Richard, and Barbara King (ed.). 2002. Anthropology beyond culture. Oxford: Berg.
Furst, Jill L. 1995. The natural history of the soul in ancient Mexico. New Haven: Yale University.
García, Ana, and Vera Tiesler. 2011. El aspecto físico de los dioses mayas. Arqueología Mexicana 112:59–63.
Geller, Pamela L. 2006. Altering identities: Body modifications and the pre-Columbian Maya. In Social archaeology of funerary remains, ed. Rebecca Gowland and Christopher Knüsel, 279–291. Oxford: Oxbow.
Goldschmidt, Walter. 1993. On the relationship between biology and anthropology. Man 28:341–59.
Goodman, Alan H., and Thomas L. Leatherman (ed.). 1998. Building a new biocultural synthesis: Political-economic perspectives on human biology. Michigan: University of Michigan.
Gump, William. 2010. Modern induced skull deformity in adults. Neurosurgical Focus 29(6):1–6.
Hicks, Frederic. 2001. Ethnicity. In The Oxford encyclopedia of Mesoamerican cultures. The civilizations of Mexico and Central America, Vol. 1, ed. David Carrasco, 388–392. Oxford: Oxford University Press.
Houston, Stephen, and David Stuart. 1998. The ancient Maya self: Personhood and portraiture in the Classic period. RES: Anthropology and Aesthetics 33:73–101.
Houston, Stephen, David Stuart, and Karl A. Taube. 2006. The memory of bones. Body, being and experience among the Classic Maya. Austin: University of Texas.
Jones, Siân. 1997. The archaeology of ethnicity. Constructing identities in the past and present. New York: Routledge.
Joyce, Rosemary. 2000. Girling the girl and boying the boy: The production of adulthood in ancient Mesoamerica. World Archaeology 31(3):473–83.
Joyce, Rosemary. 2005. Archaeology of the body. Annual Review of Anthropology 34:139–158.
Klein, Cecilia F. 2001. Conclusions: Envisioning pre-Columbian gender studies. In Gender in pre-Hispanic America. A Symposium at Dumbarton Oaks, ed. Cecilia F. Klein, 263–386. Washington: Dumbarton Oaks Research Library and Collection.
Klein, Cecilia F., and Jeffrey Quilter. 2001. Gender in pre-Hispanic America. Washington: Dumbarton Oaks Research Library and Collection.
Kramer, Karen L. 2005. Maya children. Helpers at the farm. Cambridge: Harvard.
Latcham, Ricardo. 1929. Creencias religiosas de los antiguos peruanos. Santiago de Chile: Bacells.
Latcham, Ricardo. 1937. Deformación del cráneo en la región de los Atacameos y Diaguitas. Anales del Museo Argentino de Ciencias Naturales 39:105–123.
Le Breton, David. 1994. Corps et societés. Paris: Meridiens Klincksieck.
Lewis, Mary E. 2007. The bioarchaeology of children. Perspectives from biological and forensic anthropology. New York: Cambridge University.
Lock, Margaret, and Judith Farquhar (ed.). 2007. Beyond the body proper: Reading the anthropology of material life. Durham: Duke University Press.
López-Austin, Alfredo. 1989. Cuerpo humano e ideología. Las concepciones de los antiguos nahuas. Mexico: Universidad Nacional Autónoma de México.

López-Austin, Alfredo. 1998. Breve historia de la tradición religiosa mesoamericana. Mexico: Universidad Nacional Autónoma de México.
López-Austin, Alfredo. 2001. El núcleo duro, la cosmovisión y la tradición mesoamericana. In Cosmovisión, ritual e identidad de los pueblos indígenas de México ed. Johanna Broda and Félix Báez-Jorge, 47–65. Mexico: Consejo Nacional para la Cultura y las Artes, Fondo de Cultura Económica.
López-Austin, Alfredo, and Leonardo López-Luján. 1996. El pasado indígena. Mexico: Colegio de México, Fideicomiso Historia de la Américas, Fondo de Cultura Económica.
Lorentz, Kirsi. 2008. From bodies to bones and back: Theory and human bioarchaeology. In Between biology and culture ed. Holger Schutkowski, 273–303. Cambridge: Cambridge University Press.
Lozada, María Cecilia. 2011. Marking ethnicity through premortem cranial modification among the Pre-Inca Chiribaya, Perú. In The bioarchaeology of the human head. Decapitation, decoration, and deformation, ed. Michelle Bonogofsky, 228–240. Gainesville: University Press of Florida.
Marion, Marie-Odile. 1994. Identidad y ritualidad entre los mayas. Mexico: Instituto Nacional Indigenista.
Mauss, Marcel. 1971. Sociología y antropología. Madrid: Editorial Tecnos.
Mauss, Marcel. 2007. Techniques of the body. In Beyond the body proper: Reading the anthropology of material life, ed. Margaret Lock and Judith Farquhar, 50–68. Durham: Duke University Press.
McCafferty, Geoffrey, and Sharisse McCafferty. 2011. Bling things: Ornamentation and identity in Pacific Nicaragua. In Identity crisis: Archaeological perspectives on social identity, ed. Lindsey Amond Sen-Meyer, Nicole Angel, and Sean Pickotag, 243–252. Calgary: Chacmool Archaeological Association, University of Calgary.
Meillassoux, Claude. 1987. Mujeres, graneros y capitales. Mexico: Siglo XXI.
Meskell, Lynn. 1998. The irresistible body and the seduction of archaeology. In Changing bodies, changing meanings. Studies on the human body in antiquity, Dominique Montserrat, 139–161. London: Routledge.
Meskell, Lynn M., and Rosemary Joyce. 2003. Embodied lives: Figuring ancient Maya and Egyptian experience. London: Routledge.
Miller, Daniel. 1982. Structures and strategies: An aspect of the relationship between social hierarchy and cultural change. In Symbolic action and structural archaeology, ed. Ian Hodder, 89–98. Cambridge: Cambridge University Press.
Moore, Jenny and Eleanor Scott (ed.). 1997. Invisible people and processes. Leicester: Leicester University.
Nájera, Martha Ilia. 2000. El umbral hacia la vida. El nacimiento entre los mayas contemporáneos. Mexico: Universidad Nacional Autónoma de México.
Peña, Rosa María. 1992. Un especimen singular que muestra mutilación dentaria infantil. Revista Mexicana de Estudios Antropológicos 38:107–118.
Purizaga, Medardo. 1991. El rito del nacimiento entre los Inca. Lima: Universidad de San Martín de Porres.
Ramírez, Marco, Vera Tiesler, Iván Oliva, and Guillermo Mata. 2003. Posibles técnicas empleadas en la decoración dental en la Mesoamérica prehispánica. Un estudio experimental de instrumentos y superficies. Estudios de Antropología Biológica 11:983–998.
Redfield, Robert, and Alfonso Villa Rojas. 1967. Chan Kom A Maya Village. A classic study of the basic for culture in a village in eastern Yucatan. Chicago and London: University of Chicago Press.
Reischer, Erica, and Kathryn S. Koo. 2004. The body beautiful: Symbolism and agency in the social world. Annual Review of Anthropology 33:27–317.
Romero, Javier. 1952. Los patrones de la mutilación dentaria prehispánica. Anales del Instituto Nacional de Antropología e Historia 4(32):177–221.
Romero, Javier. 1958. Mutilaciones dentarias prehispánicas de México y América en general. Mexico: Instituto Nacional de Antropología e Historia.

Romero, Javier. 1970. Dental mutilation, trephination, and cranial deformation. In Handbook of Middle American Indians, Vol. 9, 5–67. Austin: University of Texas.
Romero, Javier. 1974. La mutilación dentaria. In Antropología física, época prehispánica ed. Javier Romero, 229–250. Mexico: Colección México: Panorama Histórico y Cultural, Secretaría de Educación Pública, Instituto Nacional de Antropología e Historia.
Romero, Javier. 1984. Incrustaciones y mutilaciones dentarias. In Historia general de la medicina en México, Vol.1, 323–327. Mexico: Universidad Nacional Autónoma de México.
Romero, Javier. 1986. Catálogo de la colección de dientes mutilados prehispánicos IV parte. Mexico: Colección Fuentes, Instituto Nacional de Antropología e Historia.
Sandoval, Alfonso. 1985. Estructura corporal y diferenciación social. Mexico: Universidad Nacional Autónoma de México.
Sauvain-Dugerdil, Claudine. 1991. El hombre irreductible. Mexico: Universidad Nacional Autónoma de México.
Schiffer, Michael B. 1987. Formation processes of the archaeological record. Albuquerque: University of New Mexico.
Schiffer, Michael B., and Andrea R. Miller. 1999. The material life of human beings. Artifacts, behavior, and communication. London, New York.
Service, Elman. 1971. Primitive social organization. New York: Random House.
Shilling, Chris. 1993. The body and social theory. London: Sage.
Skibo, James M., and Michael B. Schiffer. 2009. People and things. A behavioral approach to material culture. New York: Springer.
Sofaer, Joanna R. 2006. The body as material culture. A theoretical osteoarchaeology. Cambridge: Cambridge University Press.
Stresser-Péan, Claude. 2011. De la vestimenta y los hombres. Una perspectiva histórica de la indumentaria indígena en México. Mexico: Fondo de Cultura Económica, Centro de Estudios Mexicanos y Centroamericanos, Fundación Alfredo Harp Helú, Museo Textil de Oaxaca.
Taube, Karl A. 1996. The Olmec maize god: The face of corn in Formative Mesoamerica. In Res 29/30, 39–81. Philadelphia: University of Pennsylvania Museum Publications.
Taube, Rhonda, and Karl A. Taube. 2008. The beautiful, the bad and the ugly: Aesthetics and morality in Maya figurines. In Mesoamerican Figurines, ed. Christina T. Halperin, Katherine A. Faust, Rhonda Taube, and Giguet Aurore, 236–258. University of Florida, Gainesville.
Terrazas, Alejandro. 2000. Teoría de coevolución humana: Una posición teórica en antropología física. Master diss., Universidad Nacional Autónoma de México.
Tiesler, Vera. 1993. Algunos conceptos y correlatos para la consideración del individuo en arqueología. Boletín de Antropología Americana 28:5–16.
Tiesler, Vera. 1996. Restos humanos como fuente de información arqueológica. Aplicaciones en la investigación mayista. Cuicuilco 6(2):213–222.
Tiesler, Vera. 1997a. El esqueleto muerto y vivo: algunas consideraciones para la evaluación de restos humanos como parte del contexto arqueológico. In El cuerpo humano y su tratamiento mortuorio, ed. Elsa Malvido, Gregory Pereira, and Vera Tiesler, 77–89. Mexico: Colección Científica, Instituto Nacional de Antropología e Historia.
Tiesler, Vera. 1997b. Rasgos bioculturales en la población prehispánica del sureste del Petén, Guatemala. In Memorias del X Simposio de investigaciones Arqueológicas en Guatemala. Museo Nacional de Arqueología y Etnología, ed. Juan Pedro Laporte and Héctor Escobedo, 573–585. Guatemala: Ministerio de Cultura y Deporte, Instituto de Antropología e Historia, Asociación Tikal.
Tiesler, Vera. 1997c. La arqueología biosocial: Bases conceptuales para la evaluación de restos humanos en arqueología. B.A. diss., Escuela Nacional de Antropología e Historia/Instituto Nacional de Antropología e Historia.
Tiesler, Vera. 1999. Rasgos bioculturales entre los antiguos mayas: Aspectos arqueológicos y sociales. Ph.D. diss., Universidad Nacional Autónoma de México.
Tiesler, Vera. 2000. Decoraciones dentales entre los antiguos mayas. Mexico: Ediciones Euroamericanas, Instituto Nacional de Antropología e Historia.

Tiesler, Vera. 2007. Bases conceptuales para la evaluación de restos humanos en arqueología. Mérida: Universidad Autónoma de Yucatán.
Tiesler, Vera. 2010. "Olmec" head shapes among the Preclassic period Maya and cultural meanings. Latin American Antiquity 21(3):290–311.
Tiesler, Vera. 2011. Becoming Maya: Infancy and upbringing through the lens of pre-Hispanic head shaping. Childhood in the Past 4:117–132.
Tiesler, Vera. 2012. Studying cranial vault modifications in ancient Mesoamerica. Journal of Anthropological Sciences 90:1–26.
Tiesler, Vera, and Ruth Benítez. 2001. Head shaping and dental decoration: Two biocultural attributes of cultural integration and social distinction among the ancient Maya. American Journal of Physical Anthropology, Annual Meeting Supplement 32:149.
Torres-Rouff, Christina. 2002. Cranial vault modification and ethnicity in middle horizon San Pedro de Atacama, North Chile. Current Anthropology 43:163–171.
Turner, Victor. 1984. The body and society: Explorations in social theory. Oxford: Basil Blackwell.
Van Gennep, Arnold. 1960. The rites of passage. Chicago: University of Chicago.
Vela, Enrique. 2010. Tatuajes. Decoración corporal prehispánica, Catálogo visual. Arqueología Mexicana (Edición Especial) 37:56–61.
Vera, José Luis. 1998. El hombre escorzado. Un estudio sobre el concepto de eslabón perdido en evolución humana. Mexico: Instituto de Investigaciones Antropológicas, Universidad Nacional Autónoma de México.
Villa Rojas, Alfonso. 1978. Los elegidos de dios: Etnografía de los mayas de Quintana Roo. Mexico: Instituto Nacional Indigenista.
Yépez, Zoila R. 2006. La práctica cultural de modelar la cabeza en dos culturas andinas del Antiguo Perú: Paracas y Chancay. Un estudio de los procesos de significación de la cabeza modelada intencionalmente. Ph.D. diss., Universidad Nacional Autónoma de México.
Yépez, Rosaura. 2009. El simbolismo de la modificación cultural de la cabeza en la cultura andina de Paracas del antiguo Perú. Estudios de Antropología Biológica 14:526–543.
Yépez, Rosaura, and Ramón Arzápalo. 2009. La práctica cultural de modificar el cuerpo como un texto de información e interpretación social para la antropología física. Una perspectiva semiótica. Papeles de trabajo 15:75–108.

Chapter 3
Cranial Expansion and Artificial Vault Modifications

In those societies that modeled the heads of their infants, modification was achieved by massage or by adjusting compression or constriction devices right after birth (Dingwall 1931). Depending on the practitioners (the mothers or midwives), these procedures lasted days, weeks, or months and could even be prolonged over years. Upon conclusion of the head molding, the skull vault became narrow or wide, long or elevated. The formational change in the skull itself was accomplished not by passive tissue retreat but by compensatory neurocranial expansion. The final volume of this equals that of unmodified individuals. Therefore, head shaping cannot be understood as a unilateral or static mechanical phenomenon but needs to be viewed jointly with the intrinsic dynamics of morphogenesis and postnatal craniofacial growth, which is the central focus of this chapter. Linked with external head-shaping procedures, cranial embryology and postnatal development provide useful points of departure to delve into natural limitations and morphological impacts of ancient shaping procedures, along with health risks and possibly harmful side effects.

Many of the sources, consulted for this chapter, involve applied anatomy, orthodontics, and medicine. Pediatric, orthopedic, and neurological research has been using artificially produced cranial formations for examining the relationships of calvaric, basicranial, and facial growth. Other work focuses specifically on the relevance of cultural shaping for examining paleoneurosurgery and evolutionary morphogenesis (Enchev et al. 2010; Lieberman et al. 2000). Some of these inquiries apply to experimental growth restrictions in rats (Pucciarelli 1974, 1978). Additional studies that are also relevant to this chapter are found in biological anthropology. They deal with developmental juvenile osteology (Scheuer and Black 2000), evolutionary anthropology, and craniometry.

In keeping with current anatomical terminology and for the sake of consistency, the following paragraphs will refer to the *skull* as the *cranium* and the *mandible*. The skull supports the structures of the face and encloses the brain (Larsen 2003; Testut and Latarjet 1985; White and Folkens 2000). The facial skeleton is referred to as the *splanchnocranium* or the *viscerocranium*, whereas the *neurocranium* is identified as the *braincase*. The *calvaria*, *skullcap* or *skull vault* distinguish the upper part of the neurocranium from the *basicranium* or *cranial base*, which encapsulates the brain from beneath.

V. Tiesler, *The Bioarchaeology of Artificial Cranial Modifications*, Interdisciplinary Contributions to Archaeology 7, DOI 10.1007/978-1-4614-8760-9_3,

3.1 Basics in Cranial Maturation and Growth

3.1.1 Principles of Physiological Head Morphogenesis

The manipulated infant head is not a passive substrate, but is a complex living functional unit, which is capable of dynamically adjusting its growth process to intrinsic and extrinsic factors. Therefore, the shaping process acquires the form of a dynamic interaction between culturally induced extrinsic mechanical stress and intrinsic compensatory forces, which redirect head expansion. The active properties of growing tissues imply that the pressure exerted during head modeling does not cause any reduction in neurocranial size, of course. Instead, the external restriction induces the internal, neurological, and connective tissues to grow toward the uncompressed areas of the skull. This organic response is possible given the plasticity of the cephalic tissues during the postnatal stage. In humans, artificial modification is possible only during the first 3 years of life (first infancy) and, to a much lesser extent, during second infancy (up to 6 years in age). Active cerebral expansion, which stimulates the formational adjustment in cranial growth, is almost complete in humans at the end of their third year of life.

To better understand the components involved in artificially restrained cranial growth, it is helpful to trace some basic ontological, physiological, and pathological notions of the cranium and the growing head. These address a number of basic questions:

- What are the factors and mechanisms that determine brain and bone growth in humans?
- Which are the parts of the cranial vault that enable modifications in shape? Which parts resist extrinsic forces?
- What are the physiological compensation processes and mechanisms induced by cranial compression? How do these phenomena present themselves in the cranial substrate?
- What are the possible osteological and neurological alterations associated with cranial modification? What are the growth dynamics that favor or limit its outcome?
- How important are systemic and local health risks?

3.1.2 Cranial Anatomy and Embryological Development

Not unlike other tissues, the formation of the human skull is conditioned by genetic factors, and, to a lesser degree, by environmental ones (Kohn 1991, pp. 261–278). Its development results in 28 cranial bones in the adult (Testut and Latarjet 1985). These are divided into 14 facial bones (or splanchnocranium): the 8 segments that make up the neurocranium (occipital and frontal bones, sphenoid and ethmoid bones, paired parietal and temporal bones) and 3 pairs of minute ear ossicles. The adult skull hinges with the lower jaw. Cranial bones are held together by sutures (contours of cartilage fibers). Generally, these sutures are more visible in young adults and gradually fade as the individual ages.

The development of the human cranium begins with the condensation of the mesenchyme, or mesenchymal connective tissue, which surrounds the notocordium (Carlson 2004; Larsen 2003, pp. 356–357; Patten 1960). From here, intrauterine growth can be divided into three stages. Two of these are always present, whereas the third one only occurs in endochondral ossification, to be explained below. During the fifth embryonic stage, the mesenchyme compacts itself around the neural tube to form the primitive meninges. From its internal strata, the *pia madre* is formed, whereas the external layer constitutes the *dura madre* and the majority of the cranium's bones. By the end of the sixth embryonic stage, a layer of mesenchymal tissue *has formed* and now surrounds both the occipital and frontal areas. It is united anteriorly with the first two primitive brachial segments.

Cartilage may be formed during the second stage of embryonic ossification (Carlson 2004; Scheuer and Black 2000). This occurs only during the process of endochondral ossification, one of the two essential processes during fetal development. In the human neurocranium, this form of ossification only affects the segments of its base, namely, the bone squama around the occipital orifice. Also, the sphenoid and ethmoid bones go through this second embryological stage (although in some areas also desmal ossification intervenes). These segments gradually turn to cartilage while the chondral centers expand and fuse.

During the seventh intrauterine week, the first ossified centers appear at the basicranium and generally ossify before the calvarian bones. It is worth remembering that the oval foramen, which constitutes the exit point of the fifth or trigeminal nerve, has not fully ossified in newborns. The trigeminal nerve is the largest of the cranial nerves and is responsible for facial sensation and other locomotive functions, such as biting and chewing. The mastoid processes do not develop in utero, which is why the seventh cranial nerve of the newborn (a facial nerve with a sensory and a motor portion) remains rather exposed around the internal acoustic canal. Conversely, the bones that make up the cranial vault or calvaria do not undergo any chondrification process, but instead create bone tissue directly by way of intramembranous (or desmal) ossification. This type of ossification is found in the mesenchyme that covers the brain and later forms the flat bones of the skull cap, such as the parietals and the frontal bone, which are purely membranous. Also the upper portion of the temporal bones and the upper, interparietal regions of the occipital bone are desmal.

3.1.3 Postnatal Growth

During intrauterine life, and to a lesser extent also during first infancy, the membranes that act as connecting tissues still separate the bones that form the cranium (Scheuer and Black 2000). At the moment of birth, the cranium remains in the middle of

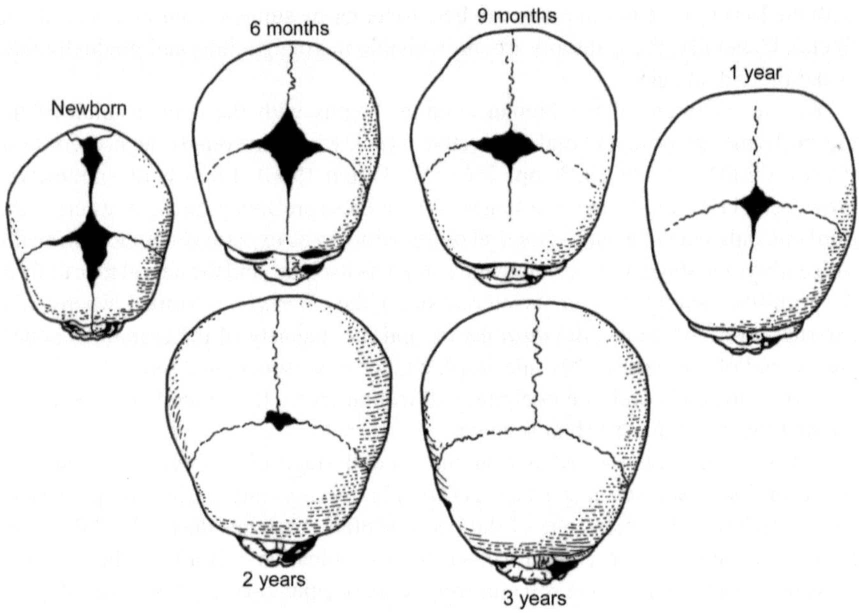

Fig. 3.1 Developmental stages of fontanelle closure between birth and 3 years of age (adapted from Vyslozil and Slavicek 2001, p. 255; drawing by B. Ceballos)

a developmental process, which now enters a phase of extrauterine expansion and maturation.

3.1.3.1 Membranes, Fontanelles, and Cartilage

Those ossification centers of the skull vault that are created by a desmal process are still very thin and soft at birth. These so-called squama are united to one another through membranes. Apart from the linear membranes, extensive connective tissues, called fontanelles, connect the osseous squama of the perinatal cranial vault (Fig. 3.1). At birth, these fibrous areas are elastic, necessary for passing the birth canal. After birth, both the bones and the fibrous membranes are still malleable and allow the head to alter its shape, an ability that now guarantees a harmonious ossification process during cerebral expansion (Campillo 1997; Palacios and Games 1988).

The sutures and fontanelles function during the first years of life as growth and contact areas between the vault's bones and therefore play a key role in cranial expansion. At birth, there are still six fontanelles: one anterior (or frontal), one posterior (or occipital), two sphenoid (or anterolateral), and two mastoid (or posterolateral) ones (Herring 1972, p. 245; Scheuer and Black 2000). An additional small linear sagittal fontanelle opens along the posterior portion of the sagittal suture. The ossification of the cranial bones causes the posterior fontanelle to close by 2–3 months. The anterior fontanelle remains open until the infant has reached toddler age, to vanish by 18–24 months of an infant's life (Scheuer and Black 2000). When open, these "soft spots" of the infant cranium may pulsate because of the vascularization of adjacent meningeal and brain tissues. During the first year of life, the fibrous tissue of the fontanelles is gradually invaded and replaced by newly formed bone (Fig. 3.1). During that time also the frontal or metopic suture, which in the newborn still divides the two halves of the bony forehead, disappears in most children with the two halves of the frontal bone being fused together. The interparietal suture usually closes during the first year as well. Sutural bones may persist and vary in frequency in major human population groups (Hanihara and Ishida 2001).

The rapidly spreading bone squama of the calvarium must adjust their shape to allow harmonious cranial expansion along the sutures and fontanelles. Formation of the bone plates usually leads to the straightening of the bone plate and/or unbending of its central eminence. This is induced by tensions in the tissues and implies the coordinated reabsorption and apposition of bone on the endosteal, diploic, and exosteal surfaces of each bone plate. Taken together, marginal ossification and central morphological adjustment form important mechanisms of bone expansion that are also reconciled through tensions when force is applied externally (Cheverud et al. 1992; Moss 1958).

Unlike the skull vault, the cranial base (or chondrocranium) is created by an endochondral ossification process (Scheuer and Black 2000, p. 43). The cartilage that connects the three main bones of the skull base (occipital, mastoid, and sphenoid) forms the epiphyseal discs. Here, bone expansion is made possible through interstitial growth in the chondral regions and along the lines of cartilage. The skull base hardens earlier than the remainder of the cranium and is therefore more rigid when the baby is born. Evolutionarily, this aspect may have to do with the need to provide early protection and rigid support for vessels and cephalic nerves crossing the skull base.

3.1.3.2 Growth Induction

The bones that encapsulate the head's soft tissues expand in a relatively passive manner following neural expansion. This dependence was discovered a century ago, when Nobel Prize winner Hans Spemann brought about the formation of a neural tube in parts of an amphibic body with chordamesoderm (Carlson 2004; Starck 1965). During the experiments, by Spemann, the manipulated neural tube induced the growth of its own osteological capsule. This relationship between bone growth and cephalic expansion is also patent in the case of microcephaly, for instance. Here, the retardation of cephalic growth conditions a reduction in calvarian expansion. In the particular case of anencephaly, the cranial vault does not develop at all (acrania). Conversely, hydrocephalic skull vaults show thin but voluminous cranial capsules (Farmer 1983; Richards and Anton 1991).

The stimuli of neurally conditioned cranial growth are transmitted mechanically through the *dura madre* from the base of the cranium to the interior layer of the

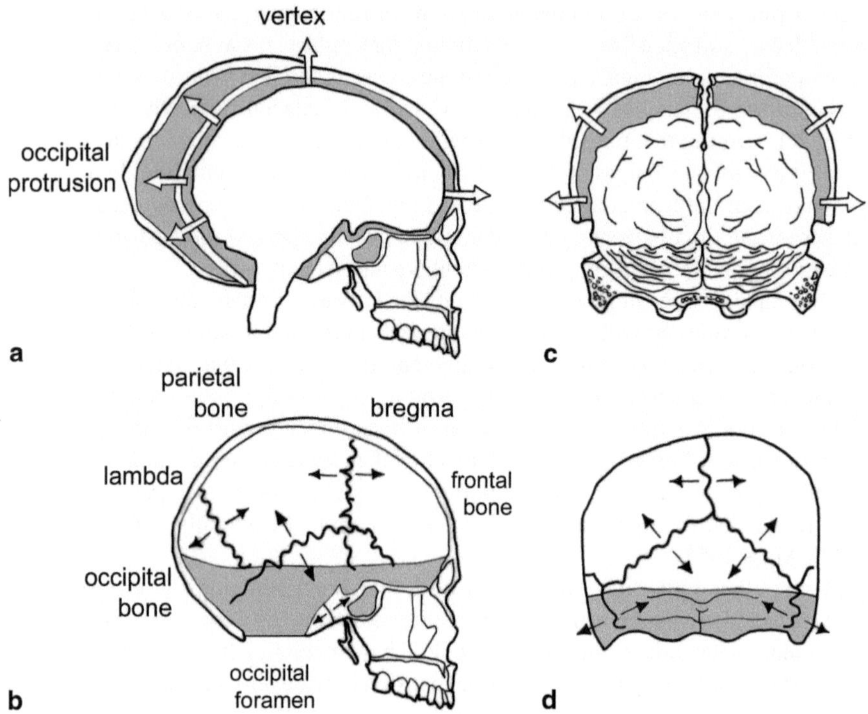

Fig. 3.2 Growth model of neurobasicranial complex (NBC), adapted from Donald H. Enlow. Here, brain expansion (*open arrows*) and sutural and synchondral growth vectors (*closed arrows*) are displayed in lateral views (**a/b**) and posterior views (**c/d**). In this scheme, neurocranial intrasutural and synchondral growth responds to tension in the basicranium and below the circumcranial reverse line by drift (areas shaded gray) (adapted from Liebermann et al. 2000, p. 294; and Enlow and Roger 1990; drawing by B. Ceballos)

calvarian bone. The *dura madre*, the skull vault and the periosteum form the embryonal neurocranial capsule. Later in development, all three form a functional system (McCarthy et al. 1990; Moss 1958). It appears that the basicranium takes a decisive role in craniofacial expansion and largely determines the cranial vault's formation during physiological growth. It is also conditioned by the effects of neurocranial growth, although to a lesser extent (Cheverud and Midkiff 1992; Cheverud et al. 1992; Enlow and Roger 1990; Kohn et al. 1993; Lieberman et al. 2000; Moss 1958). In this process, the endochondral cranial base grows in tandem with the mesodermal neurocranium, forming a highly coordinated morphological growth unit, called "neurobasicranial complex" (NBC) in D. H. Enlow's model of ontogenetic interaction (Fig. 3.2; Enlow and Roger 1990; Lieberman et al. 2000).

Also, muscular stimulus impacts cranial development, although this feedback appears to be irrelevant prior to second infancy or at least before deciduous teeth have appeared (Planas 2008). In this process, mechanical stress induces the differentiation of the exterior bone layer and thereby models its formal characteristics. More than the development of the neurocranium, it is facial growth (of the splanchnocranium) that is dependent on muscular stimulation. Here, the masticatory muscles (buccal and masseter) generate tension on the upper and lower jaws. To a lesser degree, this induces the corresponding segments of the temporal, frontal, and parietal bones of the neurocranium to generate osteological counter strengths (Cheverud et al. 1992; Lippert 1990; Ogura et al. 2006; Planas 2008).

3.1.3.3 Head Expansion and Neurological Maturation

The growth curve of the cranial vault follows normal infantile cerebral expansion (*vida supra*), which already comes close to adult proportions by the age of 2 or 3 years, even though its volume increases slightly still beyond third infancy (Campillo 1997; Kräußel 1979). Unsurprisingly, the period of major cranial growth during the first years of life coincides with the timing of infant head modeling. According to the American National Center for Health Statistics (2013), the average cranial circumference averages 35.81 cm in newborn boys and 34.7 cm in girls. During the first six months, circumference increases to some 43 cm and by the end of the first year some 46.5 cm; by the end of the second year it has reached 48.5 cm in boys and 47.5 cm in girls. A relatively insignificant amount of additional growth occurs during the subsequent years. During the first year of life, most of the skull's expansion takes place in the frontal region, whereas between the ages of 2 and 6 years, the cranium tends to expand backward, although growth is much reduced at that point (Kräußel 1979, pp. 44–45). This aspect could be important to the head-modeling process and holds implications for the change in vectors of bone mass expansion during head shaping.

Note that during the first four years of life, the cranial base expands proportionally less than the remainder of the neurocranium. At this time, the metopic suture, which still divides the frontal bone at birth, is obliterated (although in some cases it may persist). Also, the fontanelles and the interparietal suture close; the latter is referred to as *Inca* bone if it persists past infancy. World populations show varied frequencies of natural occurrence of Inca bones (Hanihara and Ishida 2001, p. 145). Infantile facial expansion follows a distinct pattern and sets in relatively late when compared with cerebral development (Enlow and Roger 1990). The face presents continual growth until adulthood, assimilating postcranial bone growth to some extent (Kräußel 1979, pp. 1–2).

The adequate neurological maturation and cranial expansion are also direct indications of the integrity of the central nervous system. In the living infant, these go in tandem with its progress in locomotive and language skills, in visual precision and hearing. Babies learn to control their gross motor abilities, such as neck control and grasping objects, during the first months of life. Other developmental milestones are the abilities to sit up, to crawl, to stand, and to walk. Babies and toddlers learn to communicate with others through the language they hear. Learning and maturation progress in a certain pattern and age range; however, each one develops on its own timeline. In healthy children, the perinatal extensor reflex on the sole of the foot

(Babinski), a telling sign of neural differentiation and therefore neurological maturation, disappears at around 12–24 months of age. The vision does not fully mature until after the third year of life (Farmer 1983; Palacios and Games 1988).

This brief outline of prenatal and postnatal head growth in humans has identified and described some key principles, which gain importance as active intrinsic factors in cultural head modeling. In synthesis, we have underscored here that the head's different anatomical elements do not develop independently but constitute a functional dynamic unit that is capable of responding to intrinsic or extrinsic modifications through the readjustment of its elements during growth. This includes the development of the bony face. After birth, the cranium expands most visibly during the first year of life. During intrauterine life, and to a lesser extent during first infancy, membranes separate bones as squama. In infants, we know of two types of membranes: those known as membranous sutures and six areas of extensive fibers known as fontanelles. Because membrane areas (sutures and fontanelles) are contact zones during growth, they play a decisive role in infant cranial expansion. During this process, the bony braincase expands as the result of a coordinated mechanism of apposition and reabsorption that changes each bone in form and size (unarching) (Starck 1965, p. 574).

3.2 Altered Postnatal Cranial Expansion

It is clear that maturing cranial bones do not constitute autonomous anatomical elements; these are interdependent instead and act together as one functional system (Cheverud and Midkiff 1992, pp. 167–171; Cheverud et al. 1992, pp. 323–345; Enlow and Roger 1990; Oetteking 1924). It is important to keep this relationship in mind when analyzing the dynamics of extrinsically and intrinsically induced restraints. Depending on the type of artificial modeling,[1] the local tensions of externally exerted pressure are biologically assimilated by broader structural compensation processes. This is why artificial neurocranial modification also results in changes of those cranial tissues that are not directly affected by the mechanical force, such as the lower jaw, the facial skull, and the basicranium (Björk and Björk 1964, pp. 353–362; Cheverud and Midkiff 1992; Cheverud et al. 1992; McNeill and Newton 1965, pp. 241–245; Moss 1958).

Moss' groundbreaking model of cranial growth (Moss 1958; Moss and Young 1960; see also Enlow and Roger 1990) supports the theory that osteological alterations in the neurocranium are intimately correlated to encephalic mass restructuring. From this point of view, it is clear that the imposed cultural modeling in fact induces changes in the neural growth vectors and these adjust bone expansion in turn. In this dynamic, tension stimuli are transmitted through the *dura madre* and the sutures,

[1] Unfortunately, most of the literature on the morphological implications of artificial head modeling does not differentiate between annular and tabular modeling, or between tabular oblique and erect shapes.

which constitute adherence zones between the *dura madre* and the vault's bones, which are connected to each other. As said, the base of the cranium holds a determining role in physiological cranial growth and also takes an active part when cranial expansion is intrinsically altered. This is the case in stenosis, for instance, or when the skull cap is extrinsically conditioned by compression (Moss 1958). We will return to this idea in a moment.

3.2.1 Intrinsic Alterations of Cranial Growth

At this point, it is pertinent to examine some pathological conditions that restrain skull growth in order to draw cautiously formulated presumptions on the growth dynamics induced by cultural compression. Naturally, there are limitations and biases implied in drawing parallels between pathological adjustments and those that are induced by applying external pressure on a healthy infant cranium. In any case, exploratory comparisons of this kind make suitable points of departure for further studies because they draw the attention to potential functional and dysfunctional compensatory responses to artificial vault compression.

Abnormal skull shape is usually the result of an unbalanced relationship between the expansion of the brain and cranial growth during infancy. This is the case in craniosynostoses or craniostenoses for example. These are conditions in which one or more of the cranial sutures obliterates prematurely by ossification, while the skull is still in its growth process (Campillo 1997; Delashaw et al. 1989, pp. 159–165). Because the cranium cannot expand perpendicular to an ossified suture, adaptive compensatory growth happens parallel to the closed suture. In extreme cases of craniosynostoses, i.e., when more than one suture is obliterated and when closure is only one sign of severe, usually congenital defects, the development of the brain and sensory organs may be at risk. This is unlikely in isolated single-suture primary synostosis, which we consider more relevant with regards to our topic. In the latter, the encephalic mass enlarges continually from redirecting its growth.

The increase of intracranial pressure, which neuropediatricians often find related to synostoses, may by itself function as a stimulant for osteological growth (Fehlow 1988a, b, 1990). Generally, the risk of intracranial hypertension increases with the age of the baby and the number of obliterated sutures. Specifically, local hypertension on the floor of the cranial cavity constitutes a considerable health hazard (as exemplified by brachycephalic pathologies), because it may easily interrupt blood irrigation and the circulation of cerebrospinal liquid in this anatomical region of transit (Fehlow 1987). Acute surges in intracranial hypertension may lead to death, whereas a more chronic progression may imply parenchymal cerebral damage or optic atrophy when unattended.

Secondary cranial readaptation depends on the region and extension of the sutural restriction (Campillo 1997; Dufier et al. 1986; Fehlow 1988a, b, c, d; Morax 1984; Rougerie and Tessier 1981; Tischler 1979). The following compensation mechanisms are reported for cranial stenosis: In the partially synostosed neurocranial expansion, vectors are altered, with concurrent changes in the positions of the skull base's orifices. Additional adjustment in the *silla turca*, basilar kyphosis (bulging of the skull base), and platybasia (abnormal flattening of the skull base) may occur depending on the place of synostosis. This is the case with an increase in local arterial pressure and changes in vascular organization.

Occasionally, increased digital impressions on the interior wall of the neurocranium appear in radiographies of synostosed patients. Also, bipolar asymmetry (plagiocephaly) is usually enhanced beyond the scale of physiological asymmetry between both hemispheres (Galaburda et al. 1978, pp. 852–856). Likewise, the facial portion of the skull is affected by the shifts in neurocranial growth. Often-cited facial adjustments relate to plagioprosopia (facial asymmetry), orbital changes (for example, exophthalmia or asymmetry), nasal protrusion, and alveolar prognathism.

We may presume that some of the local bone readjustments brought about by craniostenoses were also relevant when forces were applied extrinsically. This is apparent in bipolar asymmetry (of both the neural and the facial portion) of artificially modeled skulls. Asymmetry between both parietal lobes is generally enhanced and may be extreme when rigid tablets are used as compression devices, and even more so in full-body cradling (Tiesler 2012; see Sect. 3.3). While oblique molding may lead to an artificial flattening of the skull base, as expressed by its increased angle, severe anteroposterior flattening may result in the bulging of the dorsal part of the cranial base (Moss 1958; but see also McNeill and Newton 1965). Outjutting facial features characterize the physiognomy of extremely elongated skulls. These are confirmed by craniometrics performed on skull series, showing increased facial depth in artificially elongated heads (Tiesler 1998, pp. 161–163). Conversely, skullcaps with strong anteroposterior shortening also show a reduced facial depth (Cheverud et al. 1992; Kohn et al. 1993).

More systemic side effects from cranial synostosis can ensue when multiple sutures are closed prematurely or when the stenosis forms part of severe systematic congenital defects. Among the side effects predominate migraines, epilepsy, delayed locomotive development, hypertensive frontal syndrome, meningeal or cerebral hernias, atrophy of the optics and nerves I and VIII, as well as the limitation and alteration of ocular movement that result in strabismus, ptosis, and visual hypermetropia. Renier et al. (1982) note an inverse association between the intensity of and prolongation of intracranial pressure and intellectual development. Harmful systemic consequences of craniostenosis are rare in primary single suture closures, especially in the perinatal age with its enhanced physiological ability to react to mechanical stress, such as by the release of catecholamines, which assures greater cerebral irrigation during birth and beyond (Brailowsky et al. 1992; Creutzfeld 1988; Lagercrantz and Slotkin 1986). Most probably, this physiological protection played out as a major preventive role in the incipient stages of extrinsic head compression.

3.2.2 Extrinsic Changes to Vault Expansion

Naturally, the inherent morphologial properties and dynamic interactions among the tissues of the head are decisive also during extrinsic modifications by means of culturally induced head compression. For the reasons provided earlier, the flexible desmal skullcap is easier to manipulate in the baby than is its rigid endochondral base. In fact, flattening and subsequent brachycephalism or plagiocephaly are not even restricted to head compressors but may result simply when the infant sleeps on its back (Graham et al. 2005). This condition is called primary postural occipital flattening, which should not be confused with cranial sutural synostosis or flattening induced by artificial anteroposterior compression. Primary postural flattening usually consists of the reduction of the baby's occipital bun as the result of habitually sleeping in the identical supine position on a relatively hard surface. In recent years, this condition has received more and more attention among pediatricians, because parents are increasingly laying their babies to rest on their backs, as this position has been shown to prevent sudden infant death syndrome (Graham et al. 2005, pp. 255; Tubbs et al. 2006).

Let us turn our attention to the compression dynamic itself. What type of mechanical loading do the implements transmit? What are the assimilation and compensation mechanisms in the bones of the vault while being constrained? What are the changes in sutural growth? Regarding mechanical loads, one should bear in mind that only right after birth is moderate active head compression by bending feasible, a time when the skull and the tissues that are contained therein are still highly flexible. Past the perinatal age, the pressure on the skullcap quickly turns isometric and now restricts active cranial expansion by transmitting tension. Tension induces the intrinsic growth vectors to redirect toward unrestricted areas. In this process, tension stimuli are transmitted through the *dura madre* to connective tissues of the fontanelles, when open, and the sutures, all of which constitute adherence zones between the *dura madre* and the bone. The resulting intrinsic growth response tends to be perpendicular to the vectors of extrinsic compression.

This reaction is felt most keenly when the external pressure converges on the prominent center of a bone. In oblique front-and-back modeling, for example, the central occiput and frontal bone are pressed, which subsequently expand toward the sides and upward. The distribution of tension is more complex when the margins of the compressors come close to fontanelles and sutures, as is the case in all annular head constrictions or most tabular anteroposterior modifications. In particular, when the compression is severe and prolonged past fontanelle closure, sutural growth tends to produce compensatory bulging on the margin toward the compressed side and will generate a furrow on the other, unrestricted side (Fig. 3.3).

In severely altered skulls, a distinctive groove may surround the entire frontal compression plane behind the coronary suture (Fig. 3.3). Others may show bilateral furrows or a depression behind *bregma*, as Stewart (1948, p. 71) describes for a series of culturally modified skulls and interprets the grooves as growth distortions under artificial restraint. Far from being produced by any cultural binding, these depressions are really the end product of growth tensions and assimilation that occur when each side of a suture expands at a distinctive pace.

This idea is confirmed by a recent study that focused on the association between postcoronary sulcus expression and degree and type of artificial flattening, and infant age at death (Tiesler 1999, pp. 216–217, 2006). The results indicate an almost lineal

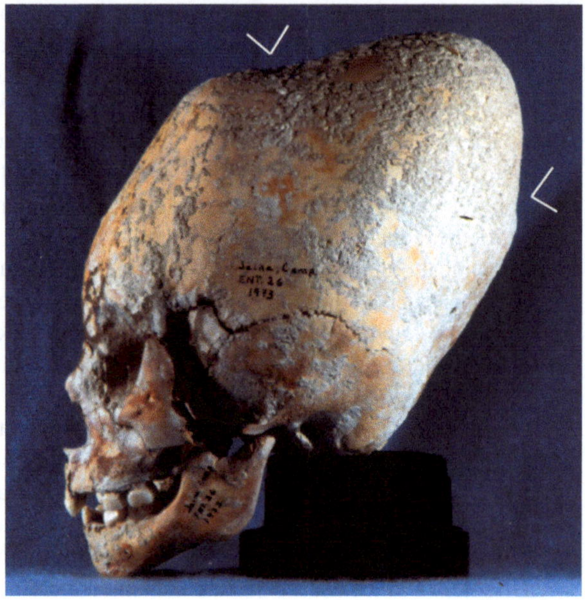

Fig. 3.3 Severely elongated infant skull displaying postcoronary groove and depression above lambda, as indicated by white arrows (Jaina, Campeche, Mexico (DAF/INAH); photo by V. Tiesler)

association between the degree of modification and the presence of a postcoronal grooving. Conversely, none of the nonmodified skulls of the sample displayed any continuous postcoronary depression. The presence and expression of the furrow also depend on the individual's age: it is absent or reduced before the age of 1 year, probably a time of increased adaptive response before the closure of the anterior fontanelle. Analogous in origin to postcoronary depressions are supralambdoid grooves that may hollow the skull above the point of convergence of the lambdoid and sagittal suture (*lambda*). These grooves usually appear on severely constricted, elongated cranial vaults in which the posterior compression plane does not reach lambda.

3.2.3 Impact on Sutural Closure and Ossicles

The effect that extrinsic mechanical loads have on sutural growth and the origin of sutural ossicles is amply documented in the scholarly literature (O'Loughlin 2004; El-Najjar and Dawson 1977; Gerszten 1993; Ossenberg 1970; Pucciarelli 1974; Sánchez-Lara et al. 2007; White 1996). Although there is controversy, most researchers would coincide on the positive impact that vault compression has on the presence and number of sutural ossicles, also called wormian bones, specifically when the area above the suture is compressed. Wormian bones are isolated excess ossicles that appear in addition to the common ossification centers of the neurocranium. Most of their research has focused on lambdic ossicles and the etiology of Inca bones in skull series with external posterior flattening.

Although not endorsed explicitly by scholarship, the pattern of infant suture closure is likely affected by external head compression, concretely by way of shifted tensile stimuli. The transmitted mechanical load is prone to increase in some sutural segments and to diminish in others. If the latter happens, sutures that do not experience further stretch may react by fusing. Conversely, the metopic suture may remain open beyond infancy when strain persists on the forehead, such as in anterior cultural flattening or in coronary synostoses (Sullivan et al. 1990). It is in that vein that both the premature obliteration of the sagittal suture (in artificially elongated calvaria) and the delay in frontal suture closure are considered as possible side effects of intrinsic compensatory efforts to counteract externally applied forces, although the findings that sustain this idea are inconsistent. Such is the argument made by White (1996), who infers that populational brachycephalism may bear an increased risk for premature sagittal synostosis in ancient Maya populations. Her findings also indicate that the artificially shortened series of crania under study showed a significantly higher proportion of premature sagittal synostosis, which she attributes to the adaptive response to the artificially changed tensile forces toward the parietal bones during peak growth.

3.2.4 Extrinsic Changes of the Basicranium

As was argued, the cranial base of the newborn is more rigid than the upper part of the neurocranium because of its advanced ossification and its endochondral quality. This aspect is important, especially during and after birth, because it provides a necessary rigid support for the vessels, marrow, and nerves in this transit area of the skull. Here, on the floor of the endocranium, the optical nerve fibers cross over and reach the areas of the visual cortex in the occipital part of the brain. But note that in newborns, the oval foramen, the exit point for the fifth cranial nerve, has still not fully ossified. As was pointed out earlier, the mastoid processes also do not develop in utero. Therefore, right after birth, the seventh cranial nerve (facial nerve) remains rather exposed at its point of departure around the internal acoustic meatus. We might wonder therefore if this exposure could have led to complications during head splinting or binding right after birth.

Given the intrinsic attributes of skull growth, it is clear that the head compressors needed to sit tight on the occipital bun in order to accomplish equivalent morphological change, especially when applied perpendicularly to the stiff basicranium, as occurs in most tabular-erect modifications. By restricting dorsal expansion of the brain, erect modifications directly oppose the cranial cephalic growth vectors. In general, it should have been the base of the skull more than other regions, which posed physical limitations to the degree of flattening.

One expression of this phenomenon is posterior basicranial kyphosis, i.e., the downward bulging of the occipital bone behind the occipital foramen. Extremely shortened skulls show this phenomenon when compressed horizontally (Fig. 3.4; Moss 1958; but see Anton 1989, for different conclusions). Already, Moss' seminal study (1958, pp. 279–280) comments on erect anteroposterior compressions, in

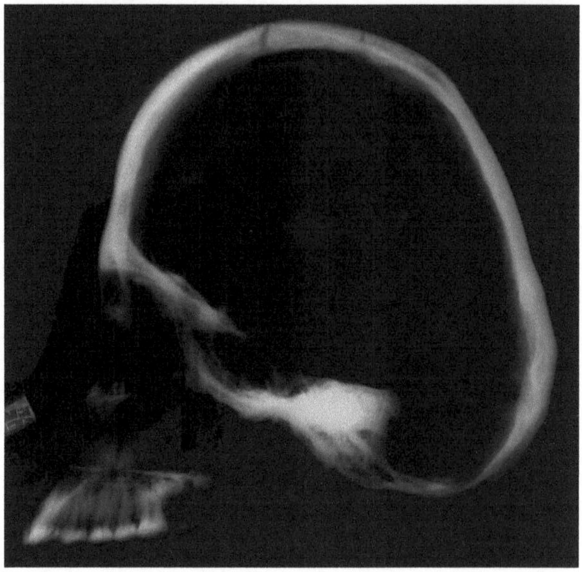

Fig. 3.4 Lateral x-ray of a severely shortened cranium of a young adult female (San Gervasio, Quintana Roo, Mexico (INAH Center Yucatán) with shifts in the *silla turca* and post foraminal bulging of the occiput (photo by V. Tiesler)

which "... the direction of the externally applied force is roughly parallel to the plane of the foramen magnum and to the eye-ear plane ..." that "... the growing neurocranium is unable to expand posteriorly and the intrinsic growth forces seek resolution in other directions. The most obvious clinical alteration, the elevation and lateral expansion of the neurocranium, does not provide sufficient room for the rapidly growing posterior cerebral fossa. With posterior expansion impossible, there is no alternative except expansion anteriorly. The newly directed force of the expanding cerebellum is directed against the anteriorly situated basioccipital portions of the skull base secondarily producing a basal kyphosis" (Moss 1958, pp. 279–280).

Other forms of external occipital restraint, as those applied in oblique constrictions, lead not to bulging but to flattening of the occipital bone or (pseudo)platybasia (Kohn et al. 1993). Moss (1958, pp. 281) suggests that platybasia is a result of the occipital backward and upward movement caused by the adjusted internal growth vectors. This effect also explains the posterior positioning of the occipital foramen in oblique crania. Consequently, slanted modifications allow a more intense formational change than the erect types because they allow a backward and upward expansion of the neural tissues encapsulated by the occipital bone.

3.2.5 Secondary Changes in Facial Growth and Proportions

Also the splanchnocranium (that is to say, the facial portion of the cranium) participates in the compensation process during altered growth. Because of this, skulls with neurocranial asymmetry show a proportionately decreased imbalance in the splanchnocranium (Björk and Björk 1964). This is owing to compensatory growth reduction in the upper and lower jaw areas of the flattened side. The compensatory remodeling of the bony face does not appear to intervene with normal maxillofacial functioning. The results of several studies did not point to any restrictions as in masticatory function or alterations in dental occlusion patterns (Björk and Björk 1964; Cheverud and Midkiff 1992; Cheverud et al. 1992; Kohn et al. 1993; Jiménez et al. 2012; Ogura et al. 2006; Vyslozil and Slavicek 2001).

Beyond compensatory expansion, scholarship has also demonstrated a host of more specific changes in the mandible and the facial cranium (Björk and Björk 1964; Cheverud and Midkiff 1992; Cheverud et al. 1992; Moss 1958; Ogura et al. 2006; Rhode and Arriaza 2006; Vyslozil and Slavicek 2001; but see also Helmuth 1970). Naturally, the amount and form of facial change depend on the degree and type of extrinsic pressure. In anteroposterior cranial-vault flattening, the documented side effects include the shortening of facial depth and the reduction of the mandibular angle. As a byproduct of redirected neurocranial growth, the face becomes either flat and wide when the skull is shortened sagittally or outjutting and prognathic when culturally elongated and bent back.

This phenomenon is described extensively for head elongations that were still practiced in southern and western France during the eighteenth and nineteenth centuries (Fabié 1968). Also, when measured in a series of 155 adult Maya skulls (Tiesler 1998, pp. 161–163), the concurrent facial changes express shifts in facial protrusion angles. The superior facial index is reduced on average (mesene, i.e., broader face) among tabular erect skulls and is increased in artificially reclined heads (leptene, i.e., relatively narrow face). In particular, buccal prognathism is the result of severe and extreme oblique modifications induced by tabular or annular growth restrictions. The prognathic index by Flower confirms this tendency for the same series of pre-Hispanic Maya crania. When compared with physically rounded skulls (with an index of 99.6 ± 3.26, on average mesognate), tabular oblique heads showed an augmented index of 104.83 ± 5.58 (on average prognate), which was still more pronounced in severe pseudoannular elongations (106.32 ± 5.4 prognates). Severe modifications may also lead to morphological shifts in the orbital area, such as exophthalmia, orbital asymmetry and increase in orbital height. These phenomena have been cited repeatedly in the medical literature (Anton 1989; Cheverud and Midkiff 1992; Cheverud et al. 1992; see also Bautista et al. 2003).

The mentioned facial adaptations naturally take influence on the external appearance of the face, although its effects are manifested differently in each person and population and should only be conceived as trends and not as absolute changes. Some of the patent tendencies have been noted in systematic facial reconstruction drawings of 25 Maya skulls, some of which were artificially flattened or elongated (Sánchez 2008). Planimetric (profile) facial reconstructions were obtained from all specimens, using forensic reference points and soft tissue thickness. The facial measurements and reconstructed profile drawings correlate neurocranial modification positively with facial prognathism, particularly for the nose and the alveolar area. In this dynamic, the nasal bridge tends to acquire an aquiline outline and the physiological depression at the root of the nose (between glabella and nasion) is reduced and, in some cases, nearly eliminated. This trend is most noticeable in severely oblique modifications, such as the ones from the Classic Maya case of a feminine dignitary from Palenque, Chiapas, who the press coined the "Red Queen," for being covered with thick layers of vermilion cinnabar (see Sect. 9.4.1 of this volume).

3.2.6 *Changes in Head Posture*

In the living person, the weight redistribution caused by the alteration of the natural shape should have shifted the balance of the head and affected natural head and neck posture, depending on the type and degree of modification. What was the comfortable head position of individuals with artificially elongated heads? What was the inclination of those faces that belonged to flattened heads? Up to today, only minor attention has been bestowed to any potential compensation processes in head posture (but see Björk 1968; Björk and Kuroda 1968; Solow and Tallgren 1976) after Imbelloni pointed to related shifts in head posture when arguing about kyphoid angles of the foramen magnum in artificially shortened heads and pithecoid angles in reclined heads (Dembo and Imbelloni 1938, p. 254, Fig. 98). Recent studies of the same phenomenon in Maya skull series show a similar average tendency when comparing the angle formed between the Frankfurt plane and the occipital foramen between shaped and unshaped skulls and between erect and oblique forms. These results are inconclusive, however, in the presence of diversity (Tiesler 1998, p. 164).

In one other study, Björk and Kuroda (1968) found a positive relationship between the flattening of the cranial base and a backward inclination of the basicranium, while the bending of the occipital foramen leads the individuals to carry their heads with the face lowered. In this line of research, systematic examinations of intracranial weight shifts and basal inclinations will surely lead to more conclusive insights regarding rebalancing and potentially shifted head postures. It is probably no coincidence that individuals with extremely elongated heads, such as the recent Mangbetu of Africa, are photographed with their faces pointed upward (Trias et al. 1927, p. 99). Also, the Classic Maya portraiture of extremely inclined head profiles tend to represent them as reclined backward, granting the impression that, apart from looking at ease, this conformed to the ideal of beauty (Fig. 3.5; see also Chaps. 2 and 6).

3.2.7 *Intrinsic Recompensation After Cultural Compression*

Finally, what happens to craniofacial growth after the compression period? Once the head-shaping devices are removed, osteological and cerebral expansion vectors will reassimilate their foreseen growth directions by way of reverted tension stimuli. Postcompression compensatory growth tends to neutralize the artificial shape, a phenomenon that benefits postoperatory growth after correcting sutural synostoses (Littlefield 2004). The degree of postcompression compensatory growth depends on

Fig. 3.5 Female attendant with reclined, elongated head, displayed in Classic Maya palace scene (Maya vase painting, Kerr archive 0511; redrawn by M. Sánchez). Note the backward inclination of the woman's face and head

the age at which the head procedure finalized in tandem with the state of progression of craniofacial growth.

Already, in the 1930s, José Imbelloni alluded to this phenomenon in the following manner:

> Among the crania that represent the same type of deformation, it is possible to distinguish infantile stages (the full expression of the conventional form) and adult stages (an imperfect realization that is somewhat erased), as the age and plethoric state of the bone tissue (heads that are heavy and robust up to the paquicephalic extreme) constitute forces that work in opposition to those of deformation, and thereby cancel its effects to some degree. (Imbelloni in Dávalos 1965, p. 36)

In synthesis, this section has made its point arguing that artificial neurocranial compression does not reduce growth but instead redirects its vectors without affecting cranial capacity. During this process, the degrees of neuronal and osteological expansion condition directly and indirectly the extent of extrinsic transformation. Because postnatal cranial expansion peaks during the first months, it is this time span in which the skullcap is most malleable. There are no significant increments in head circumference after the age of 3 years. After birth, the most malleable part of the neurocranium is its upper, desmal portion. The compression techniques condition the artificial morphology of the skullcap: the use of boards for front-and-back compression produces bilateral bulging (brachycephalic), whereas annular constriction results in the narrowing and elongation of the cranial vault either in an upward or backward direction. Unlike the calvarium, the more rigid endochondral basicranium tends to resist the unnatural compression process. Depending on the pressure vectors and their topographic location, the lower part of the occipital bone responds with flattening or, alternatively, with compensatory bulging or kyphosis.

Potential secondary effects in the splanchnocranium include flattening or prognathism. The head posture of extrinsically altered skulls may rest in their physiological posture or may acquire either a downward or upward reorientation depending on the balance. After the extrinsic compression ceases, the cranial vault's natural tendency to expand toward areas with lesser resistance is prone to lead to the gradual neutralization of the artificially produced shape; the degree of redress depends on the age in which the head procedures are finalized (with fewer physiological adjustments the more advanced the age).

3.3 Side Effects and Health Risks Involved in Head Shaping

As is perceived from the aforementioned examination of extrinsically altered craniofacial expansion, most of the reformations are compensatory and per se do not reflect negatively on the functionality of head tissue growth and neurological maturation. However, there are other intrinsic responses that are less functional or possibly less appealing in physical appearance (such as severe facial asymmetry). Also, some organic responses to the imposed growth restriction bear clear adverse effects on the health and life of the baby. Naturally, these reactions depend on the manifold circumstances that once determined infant head manipulation and their agencies. If we believe the ethnohistorical sources and the cranial record, many of the side effects appear either related to the molding technique and the head implements or the ways and care in which the daily procedures were enacted (see Chap. 4). Other risks had to do with the amount of pressure exerted on the infant skull and the timing and extent of its exposure; still others surely depended on the constitution of the child itself.

3.3.1 Morphological Consequences

Here, I wish to explore some of these potentially adverse results of artificial head modification; namely, culturally induced (pseudo)plagiocrania and facial asymmetry, exophthalmia and orbital asymmetry (Bautista et al. 2003). Alterations in the orbital shapes, for example, exophthalmia and increase in orbital height, have been documented in addition to nasal protrusion and alveolar prognathism in the case of severe circular elongation (Anton 1989). I will also touch upon exocranial and endocranial bleedings in extreme modifications. Additional problems are directly associated with the compressed surfaces, where mechanical pressure, lack of hygienic access, and lack of circulation and airflow sometimes led to local tissue gangrene and secondary infections. Unfortunately, most of these secondary effects have not been examined systematically, a shortcoming that limits generalizations on risks, risk management, and cures by the practitioners.

Fig. 3.6 Top view of extremely shortened (tabular erect) adult cranium, displaying strong bipolar asymmetry (Type III). Argelia, La Angostura, Chiapas, Mexico (DAF/INAH); photo by V. Tiesler)

The adjustment of hard shaping devices on the infant's head vault predispose to secondary cranial asymmetry, which is frequently noted in studies of cranial series once submitted to anteroposterior compression (Björk and Björk 1964; Dembo and Imbelloni 1938; Kohn et al. 1995).[2] Unlike soft and flexible head shapers (constricting bands or tight caps), rigid devices do not transmit pressure evenly on the calvarium. If not adapted with cushions or held in place by semirigid pads and rings, this unyielding quality may result in secondary cranial asymmetry, as the ensuing compensatory cranial expansion will be greatest in the less restricted areas of the vault. This is especially the case when the compressor is not adjusted in a precise bilateral or anteroposterior position. The possibility of its becoming loose and sideslip during use is even higher when the compression tablets restrict the natural movements of the baby's head and body. By definition, cradleboards fall into this category because the compression planes are tightened to a whole-body infant carrier. We remind that these multifunctional units were employed not only as head shapers but also for the infant's transportation, cleaning, and nursing, thereby increasing the chances of slackening (Fig. 3.6; Dingwall 1931; Gervais 1989; Romano 1974; see also Chap. 4).

The side-shifted compression of the skull vault and its asymmetric result is perceived either as bulging of the less compressed side or as a laterally displaced flattening of the occiput or forehead (Kohn et al. 1995; Tiesler 1998). In a Peruvian sample of 149 adult crania, Björk and Björk (1964) found that laterally displaced occipital flattening led to the shortening of the ipsilateral basicranium in artificially

[2] Also, unshaped heads tend to be slightly asymmetrical when both brain hemispheres are compared (Galaburda et al. 1978; Geschwind and Levitsky 1968; Geschwind et al. 1978). Slight shifts toward both the right and the left side occur. The specific side is associated to the individual's particular left- or right-handedness. Thus, the global cerebral shape of right-handed individuals shows a slight increase in the occipital and parietal lobes toward the left. The opposite trend was noted in left-handed individuals.

modeled skulls. The degree and side of bipolar asymmetry and its correlation to different forms and degrees of artificial cranial modeling was the focus of another more recent metric and visual examination of 106 adult Maya skulls (Tiesler 1998, 1999). The results confirm that cranial asymmetry (or pseudoplagiocrania), secondary to modeling,[3] is closely related to the extrinsic compression technique. So is the degree of resulting morphological alteration (Fig. 3.6). Some 62 % of all skullcaps showed laterally displaced occipital flattening, which was ranked from slight (Type I), to moderate (Type II), to severe (Type III).

According to other results of this study (Tiesler 1998, 1999), the ratio of sheer presence vs. absence of asymmetry and also its sidedness was similar when the group of artificially modified specimens was compared with naturally grown skull vaults. As expected, only the intensity of disproportion varies significantly. The splinted (tabular oblique) crania present a 40 % incidence of noticeable asymmetry (Type II and above) compared with the higher 70 % (Type II and above) found in tabular erect crania obtained from cradleboard use.

The increase in asymmetry that was displayed by the cradled (erect) skulls, echoes the abovementioned difficulties involved in stabilizing the baby's head with cradleboard devices. Apart from the infant's own movements, maladjustment could have also ensued from positioning the baby for periodic breastfeeding (Dembo and Imbelloni 1938; Romano 1965). Whether these factors were the cause or the result of lopsided expansion of cradled pre-Hispanic minors, we do not know for certain. What we can say is that approximately two-thirds of the cradled individuals with disproportionate cranial growth (62 vs. 38 %) must have rested with their head toward the right side.

An additional reason for increased disproportion of cradled heads could lie in the structural dynamics of cephalic growth. As described earlier, oblique tabular compression and occipital bandaging permit the backward relocation of cephalic structures, whereby in straight anteroposterior flattenings, the mechanical vectors oppose and thereby destabilize the internal growth vectors located at the cranial base. Both the frequency and intensity of asymmetry score appear elevated in the lambdoid variety, that is, when a posterior compression plane dominates. It is noteworthy that secondary asymmetry of cradled skulls increases with degree of posterior flattening, with no such correlation being recorded in the opposing frontal plane. It seems, therefore, that the frontal compression board could have come to have a stabilizing effect during the compression period.

3.3.2 *Acute and Subacute Health Hazards*

Local health nuisances, such as bleedings, infections, or necrosis of the compressed tissues, could arise during the time of the compression procedures. In absence of any modern statistics, it is difficult to estimate how important these health consequences

---

[3] Which is to be differentiated from primary plagiocephaly.

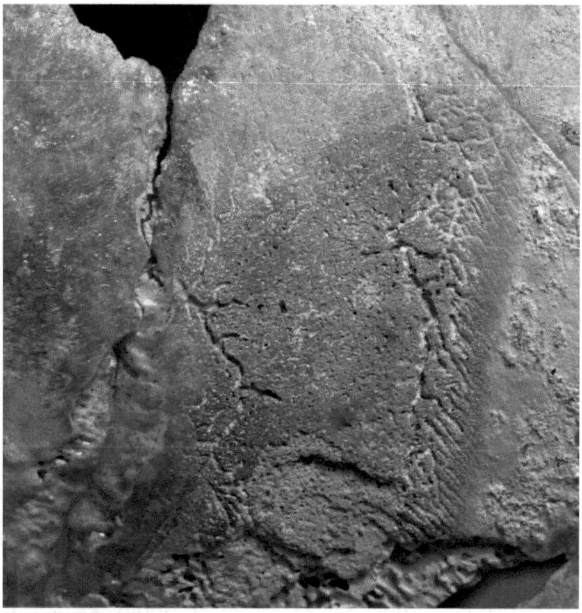

Fig. 3.7 Postcoronary area of extremely elongated skull vault of a 1-year-old infant, displaying proliferative periosteal bone reaction (Jaina, Campeche, Mexico (DAF/INAH); photo by V. Tiesler)

could have been for the average infant. Most colonial ethnohistorical sources highlight and probably exaggerate the health hazards involved and some contradict each other when confronted (see Chap. 5). While Father Bernabé Cobos asserts that the Peruvian Aymara molded the heads of their children "... with so much vigour that he knew of one child who died from the pain caused by the operation and doubtless there were others also who suffered a similar fate" (Cobo cited by Dingwall 1931, p. 214), Sala (1897, p. 80) asserts that heads were flattened without much suffering.

In the cranial record, subacute and chronic complications may show up in the form of ossified periosteal bleedings on the external lamina of the skullcap. Some of these proliferating periosteal reactions have been interpreted as the result of increased bone apposition during the forced bending of the infant calotte (Fig. 3.7) (Mendonça et al. 2008). These growth zones are generally aligned with the sutures but do not cover them. The tissues that are trapped directly between the compression device and the bone substrate may also react by bleedings or suffer from the lack of blood circulation, air exposure, and usually lack of hygiene. These conditions may be expressed in the underlying bone in the form of inflammatory responses, such as infection and/or ischemic ulceration of the bone (and obviously also of the overlying skin in the living).

Local resorptive or appositional responses to compression devices are most noticeably described in the compacted occipital eminence above the inion crest. Here, the bone changes are described as resorptive, porotic remodeling, retaining a nodular sclerotic appearance when healed. In infant skeletons, these changes are often associated with signs of endocranial inflammation and increased impressions of meningeal

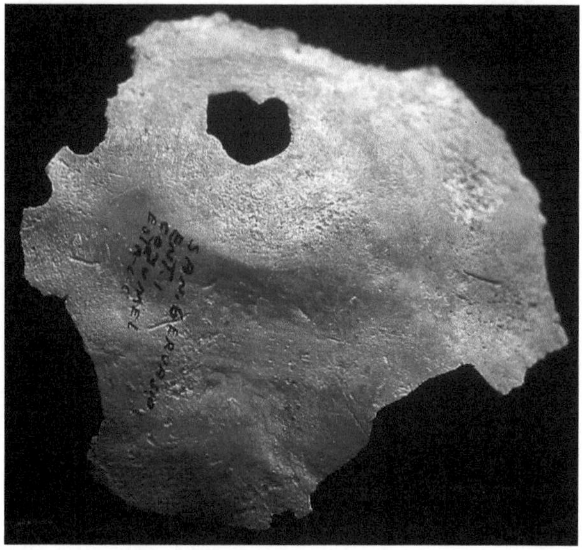

Fig. 3.8 External view of suprainiac thinning, which led to the penetration of the occipital bone in infant cranium (San Gervasio, Quintana Roo, Mexico; Proyecto Auqeológico San Gervasio, INAH; photo V. Tiesler)

vessels (Dean 1995; Mendonça et al. 2008). Most scholars associate these occipital changes, called suprainiac depressions, to circulatory bone necrosis, infectious gangrenes associated with cradleboard use or head modeling in general. Trephining has been cited as a possible cause when the occipital bun is penetrated (Fig. 3.8; see also Holliday 1993; Lagunas 1972; Romano 1975; Stewart 1976; Tiesler 2006; Torres-Rouff 2003, pp. 103; Weiss 1981; for a discussion on this topic). In this book, we will return to suprainiac lesions in Chap 4.

Also the tissues that are enclosed by the skullcap may suffer changes. Blood and cerebrospinal fluid circulation may overcompensate, leading to complications such as hypertension, herniation, or hemorrhaging, although research of the neurological or cerebral impacts is far from conclusive and in any case tends to argue against any harmful effects (Enchev et al. 2010; Gerszten 1993; Schijman 2005). However, Moss (1958) notes that vector destabilization of the cranial base's internal growth (and the concomitant expansion of encapsulated cephalic structures) can lead to health problems. He also infers that tabular erect modifications engender greater risks due to the pressure that vectors opposed physiological vectors, whereas oblique modifications allow to reallocate cephalic structures toward the posterior fossa. Also, vessel changes in the interior of the skull vault bear potential health risks for the molded individual. Studies of endocranial casts show that the pattern, shape, and depth of vessel impressions are altered. The latter appear flattened beneath the areas of growth restraint. By contrast, vessels in the surrounding areas show compensatory enlargement. Expansion of occipital and marginal sinuses were observed in one part of the series of endocasts (Dean 1995).

An alternative approach to assessing cases of fatal head compression is the comparison of age at death profiles of artificially modified vs. unmodified skulls. For ancient Maya skeletal samples, this association has been attempted by comparing the rates of the ages-at-death among individuals with moderate to severe deformations from the large and well-preserved collection from the coastal site of Classic period Jaina with its varied head formations. It was noted that infant mortality appeared to have been significantly higher for those individuals with severe head elongation and reclination, when compared with more subtle expressions of the practice. However, any health correlation may be biased because it does not take into consideration postmodeling growth recompensation (Tiesler 1998, p. 206).

For now, scholarship still awaits systematic scrutiny of the potentially harmful side effects of head modeling (but see Dean 1995; Mendonça et al. 2008; Moss 1958), and we may only speculate on what the actual health risks, involved in the head's artificial modeling, should have been in the Americas and beyond. With these ideas in mind, I posit parsimoniously that daily modeling practices, which at least in pre-Hispanic thought were deemed a means to protect the infant's health and integrity, sought to avoid any such side effects, although some nuisances may have been accepted by the practitioning caretakers for the sales of the expected benefits. Other side effects, such as asymmetric head morphology secondary to cradling, may be an indication of the experience (or lack of experience for that sake) of the kin in charge of the modeling procedures. Still other side effects, such as signs of infection and necrosis, should express the consequences of procedural problems, such as the lack of hygiene, experience, or more categorically—child care.

References

Anton, Susan. 1989. Intentional cranial vault deformation and induced changes of the cranial base and face. American Journal of Physical Anthropology 79:253–267.

Bautista, Josefina, Emma Limón, and Alberto Brown. 2003. La aplicación de la osteología antropológica a la oftalmología. In Antropología física. Disciplina plural, ed. Josefina Mansilla and Xabier Lizarraga, 229–238. Mexico: Consejo Nacional para la Cultura y las Artes, Instituto Nacional de Antropología e Historia.

Björk, Arne. 1968. Roentgencephalometric growth analysis. In Congenital abnormalities of the face and associated structures, ed. Sandra Pruzansky, 237–250. Springfield: Charles C. Thomas Publisher.

Björk, Arne, and Lise Björk. 1964. Artificial deformation and craniofacial asymmetry in ancient Peruvians. Journal of Dental Research 43(3):353–362.

Björk, Arne, and Toshio Kuroda. 1968. Congenital bilateral hypoplasia of the mandibular condyles associated with congenital bilateral palpebral ptosis. American Journal of Orthodontics 54:584–600.

Brailowsky, Simón, Donald G. Stein, and Bruno Will. 1992. El cerebro averiado, plasticidad cerebral y recuperación funcional. Mexico: Fondo de Cultura Económica.

Campillo, Domingo. 1997. Problema antropológico de las craneostenosis y patologías afines. Estudios de Antropología Biológica 7:93–133.

Carlson, Bruce M. 2004. Human embryology and developmental biology. Philadelphia: Mosby, Elsevier.

Cheverud, James M., and James E. Midkiff. 1992. Effects of fronto-occipital cranial reshaping on mandibular form. American Journal of Physical Anthropology 87(2):167–171.

Cheverud, James M., Luci A. P. Kohn, Lyle W. Konigsberg, and Steven R. Leigh. 1992. Effects of fronto-occipital artificial cranial vault modification on the cranial base and face. American Journal of Physical Anthropology 88(3):323–345.

Creutzfeld, Otto D. 1988. Modelle de Gehirns, Modelle de Geistes. Mannheimer Forum 87/88:9–58.

Dávalos, Eusebio. 1965. La deformación craneana entre los tlatelolca. In Temas de antropología física, ed. Eusebio Dávalos, 9–55. Mexico: Instituto Nacional de Antropología e Historia.

Dean, Valerie L. 1995. Sinus and meningeal vessel pattern changes induced by artificial cranial deformation: A pilot study. International Journal of Osteoarchaeology 5:1–14.

Delashaw, Johnny B., John A. Persing, William C. Broaddus, and John A. Jane. 1989. Cranial vault growth and craniosynostosis. Journal of Neurosurgery 70(2):159–165.

Dembo, Adolfo, and José Imbelloni. 1938. Deformaciones intencionales del cuerpo humano de caracter étnico. Buenos Aires: Biblioteca Humanior.

Dingwall, Eric J. 1931. Artificial cranial deformation; a contribution to the study of ethnic mutilations. London: Bale & Sons & Danielsons.

Dufier, J. L., M. C. Vinurel, D. Renier, and D. Marchac. 1986. Les complications ophtalmologiques des craniofaciosténoses. Journal Francais d' Ophtalmologie 9(4):273–280.

El-Najjar, Mahmoud Y., and George L. Dawson. 1977. The effect of artificial cranial deformation on the incidence of wormian bones in the lambdoidal suture. American Journal of Physical Anthropology 46:155–160.

Enchev, Yavor, Grigoriy Nedelkov, Nadezhda Atanassova-Timeva, and Jordan Jordanov. 2010. Paleoneurosurgical aspects of proto-bulgarian artificial skull deformation. Neurosurg Focus 29(6):1–7.

Enlow, Donald H., and William Roger. 1990. Facial Growth. Philadelphia: Saunders.

Fabié, Michel. 1968. Le crâ ne toulousain. Master diss., Universidad de Toulouse.

Farmer, T. 1983. Pediatric Neurology. New York: Harper & Row.

Fehlow, Paul. 1987. Sozialpsychiatrische und klinische Aspekte der Dyszephalie mit besonderer Berücksichtigung des Turmschädels. Ph.D. diss., University of Greifswald.

Fehlow, Paul. 1988a. Neuropsychologische Aspekte der Ovon Kraniosyostose und-Atenose. Zentralblatt fuer Neurochirurgie 49:224–289.

Fehlow, Paul. 1988b. Turmschädel und Kraniosynostosesyndrom. Zentralblatt für klinische Medizin 43(19):1667–1669.

Fehlow, Paul. 1988c. Pseudo-Crouzon-Syndrom mit psychischer Fehlentwicklung. Pädiatrische Grenzgebiete 27(4):227–230.

Fehlow, Paul. 1988d. Epilepsie und Spitzenpotentiale bei Dyszephalie. Zeitung für ärztliche Fortbildung 82:827–828.

Fehlow, Paul. 1990. Historische, sozialpsychiatrische, genetische und anthropologische Aspekte des Turmschädels. Pädiatrische Grenzgebiete 29(4):305–317.

Galaburda, Albert M., Marjorie LeMay, Thomas L. Kemper, and Norman Geschwind. 1978. Right-left asymmetries in the brain. Science 199:852–856.

Gerszten, Peter G. 1993. An investigation into the practice of cranial deformation among the pre-Columbian peoples of northern Chile. International Journal of Osteoarchaeology 3:87–98.

Gervais, Veronique. 1989. Déformations artificielles de crâ nes préhispaniques au Guatemala et au Mexique. Ph.D. diss., Universidad de Caen.

Geschwind, Norman, and Walter Levitsky. 1968. Human brain: Left-right asymmetries in temporal speech regions. Science 161(3837):186–187.

Geschwind, Norman, Albert M. Galaburda, Marjorie LeMay, and Thomas Kemper. 1978. Left-right asymmetries in the human cerebral hemispheres. Science 199:852–856.

Graham, John M. Jr, Jeannie Kreutzman, Dawn Earl, Andy Halberg, Carlos Samayoa, and Xiuqing Guo. 2005. Deformational brachycephaly in supine-sleeping infants. The Journal of Pediatrics 146:253–257.

Hanihara, Tsunehiko, and Hajime Ishida. 2001. Os incae: Variation in frequency in major human population groups. Journal of Anatomy 108:137–152.

Helmuth, Hermann. 1970. Über de Bau des menschlichen Schädels bei künstlicher Deformierung. Zeitung für Morphologie und Anthropologie 62:30–49.
Herring, Susan. 1972. Sutures-a tool in functional cranial analysis. Acta Anatómica 83:222–247.
Holliday, Diane. 1993. Occipital lesions: A possible cost of cradleboards. American Journal of Physical Anthropology 90(3):283–290.
Jiménez, Publio, Arturo Martínez-Insua, Jaime Franco-Vazquez, Xose Luis Otero-Cepeda, and Urbano Santana. 2012. Maxillary changes and occlusal traits in crania with artificial fronto-occipital deformation. American Journal of Physical Anthropology 147(1):40–51.
Kohn, Luci Ann. 1991. The role of genetics in craniofacial morphology and growth. Annual Review of Anthropology 20:261–278.
Kohn, Luci Ann, Steven R. Leigh, Susan C. Jacobs, and James M. Cheverud. 1993. Effects of annular cranial vault modification on the cranial base and face. American Journal of Physical Anthropology 90:147–168.
Kohn, Luci Ann, Steven R. Leigh, and James M. Cheverud. 1995. Asymmetric vault modification in Hopi crania. American Journal of Physical Anthropology 98:173–195.
Kräußel, W. 1979. Ueber das postnatale Wachstum des Schädels in der Sagittalen mit besonderer Berücksichtigung von Rechts-links-Abweichungen und des Aditus Orbitae. Ph.D. diss., Wuerzburg, Germany.
Lagercrantz, Hugo, and Theodore A. Slotkin. 1986. Las tensiones del nacimiento. Investigación y Ciencia 117:80–89.
Lagunas, Zaid. 1972. La trepanación suprainiana en cráneos de Cholula, Puebla. Comunicaciones Proyecto Puebla Tlaxcala 8:47–50.
Larsen, William J. 2003. Embriología humana. 3rd edition, Madrid: El Sevier Science.
Lieberman, Daniel E., Osbjorn M. Pearson, and Kenneth M. Mowbray. 2000. Basicranial influence on overall cranial shape. Journal of Human Evolution 38:291–315.
Lippert, Hanne. 1990. Anatomie. Munich: Urban & Schwarzenberg.
Littlefield, Timothy R. 2004. Cranial remodeling devices: Treatment of deformational plagiocephaly and postsurgical applications. Seminars in Pediatric Neurology 11:268–277.
McNeill, R. William, and George N. Newton. 1965. Cranial base morphology in association with intentional cranial vault deformation. The American Journal of Physical Anthropology 23(3):241–253.
Mendonça de Souza, Sheila M. F., Karl J. Reinhard, and Andrea Lessa. 2008. Deformación craneana como causa de muerte de un niño del Valle de Chillón, Perú. Chungara 40(1):41–53.
Morax, Serge. 1984. Ocul-motor disorders in craniofacial malformations. Journal of Maxilar-Facial Surgery 12:1–10.
Moss, Melvin. 1958. The pathogenesis of artificial cranial deformation. American Journal of Physical Anthropology 16(3):269–286.
Moss, Melvin L., and Richard W. Young. 1960. A functional approach to craniology. American Journal of Physical Anthropolgy 18:281–292.
National Center for Health Statistics, official webpage: http://www.cdc.gov/growthcharts/html_charts/hcageinf.htm. Accessed 20 Feb 2013.
O'Loughlin, Valerie Dean. 2004. Effects of different kinds of cranial deformation on the incidence of wormian bones. American Journal of Physical Anthropology 123:146–155.
Oetteking, Bruno. 1924. Declinations of the pars basilaris in normal and in artificially deformed skulls. Indian Notes and Monographs 27:1–25.
Ogura, Mikinori, Abdullah Al-Kalaly, Reiko Sakashita, Tetsuya Kamegai, and Shouichi Miyawaki. 2006. Relationship between anteroposterior cranial vault deformation and mandibular morphology in a pre-Columbian population. American Journal of Orthodontics and Dentofacial Orthopedics 130(4):535–539.
Ossenberg, Nancy S. 1970. The influence of artificial cranial deformation on discontinuous morphological traits. American Journal of Physical Anthropology 33:357–371.
Palacios, Jaime L., and Juan Games espacio. 1988. Introducción a la pediatría. Mexico: Fco. Méndez Oteo.

Patten, Bradley. 1960. Embriología humana. Buenos Aires: El Ateneo.
Planas, Pedro. 2008. Rehabilitación neuro-oclusal (RNO). Caracas: AMOLCA.
Pucciarelli, Héctor M. 1974. The influence of experimental deformation on neurocranial wormian bones in rats. American Journal of Physical Anthropology 41:29–38.
Pucciarelli, Héctor M. 1978. The influence of experimental deformation on craniofacial development in rats. American Journal of Physical Anthropology 48:455–462.
Renier, Dominique, Christian Sainte-Rose, Daniel Marchac, and Jean-François Hirsch. 1982. Intracranial pressure in craniostenosis. Journal of Neurosurgery 57:370–377.
Rhode, Matthew, and Bernardo T. Arriaza. 2006. Influence of cranial deformation on facial morphology among prehistoric south central Andean populations. American Journal of Physical Anthropology 130:462–470.
Richards, Gary D., and Susan C. Anton. 1991. Craniofacial configuration and postcranial development of a hydrocephalic child (ca. 2500 B.C.–500 A.D.): With a review of cases and comment on diagnostic criteria. American Journal of Physical Anthropology 85:185–200.
Romano, Arturo. 1965. Estudio morfológico de la deformación craneana en Tamuín, S.L.P., y en la Isla del Idolo, Veracruz. Serie de Investigaciones 10. México: Instituto Nacional de Antropología e Historia.
Romano, Arturo. 1974. Deformación cefálica intencional. In Antropología física, época prehispánica, ed. Juan Comas, 197–227. Mexico: Instituto Nacional de Antropología e Historia.
Romano, Arturo. 1975. Dolicráneos prehispánicos en Teotenango, Méx. In Teotenango: El antiguo lugar de la muralla. Memorias de las excavaciones arqueológicas, ed. Román Pia Chán, 417–421, Mexico: Dirección de Turismo, Gobierno del Estado de México.
Rougerie, J., and P. Tessier. 1981. Craniosynostosis and craniofacial dysmorphia. Zeitung für Kinderchirurgie 34(2):179–183.
Sala, Gabriel. 1897. Apuntes de viaje. Exploración de los ríos Pichis, Pachitea y Alto Ucayali y de la región del Gran Pajonal. Lima.
Sánchez-Lara, Pedro, John M. Graham Jr, Anne V. Hing, John Lee, and Michael Cunningham. 2007. The morphogenesis of wormian bones: A study of craniosynostosis and purposeful cranial deformation. American Journal of Medical Genetics 143A:3243–3251.
Sánchez, Mirna. 2008. Emblema visual de identidad y distinción. Reconstrucción de rostros mayas (fisionomía y rasgos bioculturales). Master diss., Universidad Autónoma de Yucatán.
Scheuer, Louise, and Sue Black. 2000. Developmental juvenile osteology. San Diego: Academic Press.
Schijman, Edgardo. 2005. Artificial cranial deformation in newborns in the pre-Columbian Andes. Classics in Pediatric Neurosurgery 21:945–950.
Solow, Beni, and Antje Tallgren. 1976. Head posture and craniofacial morphology. American Journal of Physical Anthropology 44:417–436.
Starck, Dietrich. 1965. Embryologie. Ein Lehrbuch auf allgemein biologischer Grundlage. Second edition. Stuttgart: Georg Thieme Verlag.
Stewart, T. Dale. 1948. The true form of the cranial deformity originally described under the name "Tete Trilobée". Journal of the Washington Academy of Sciences 38:66–72.
Stewart, T. Dale. 1976. Are supra-inion depressions evidence of prophylactic trephinations? Bulletin of the History of Medicine 50:414–434.
Sullivan, Patrick K., Birte Melsen, and John B. Mulliken. 1990. Calvarial sutural abnormalities: Metopic synostosis and coronal deformation—an anatomic, 3d radiographic, and pathologic study. Plastic and Reconstructive Surgery 86(6):1072–1077.
Testut, Leo, and Michel Latarjet. 1985. Tratado de anatomía humana. Barcelona: Salvat.
Tiesler, Vera. 1998. La costumbre de la deformación cefálica entre los antiguos mayas: aspectos morfológicos y culturales. Mexico: Colección Científica/Instituto Nacional de Antropología e Historia.
Tiesler, Vera. 1999. Rasgos bioculturales entre los antiguos mayas: Aspectos arqueológicos y sociales. Ph.D. diss., Universidad Nacional Autónoma de México.

Tiesler, Vera. 2006. ¿Hubo trepanación en la antigua sociedad maya? Una apreciación regional. Antropología Física Latinoamericana 4:169–218.

Tiesler, Vera. 2012. Transformarse en maya. El modelado cefálico entre los mayas prehispánicos y coloniales. Mexico: Universidad Nacional Autónoma de México, Universidad Autónoma de Yucatán.

Tischler, W. 1979. Zur Problematik der Kraniosynostose und Kraneostenose. Pädiatrische Grenzgebiete 18:335–348.

Torres-Rouff, Christina. 2003. Shaping identity: Cranial vault modification in the pre-Columbian Andes. Ph.D. diss., University of California.

Trias de Bes, Luis, Pedro Bosch Gimpera, Jose de C. Serra-Ràfols, D. José Batista y Roca, and D. Alberto del Castillo. 1927. Las razas humanas. Su vida, sus costumbres, su arte, Vol. 1. Barcelona: Instituto Gallach.

Tubbs, R. Silane, George Salter, and Jerry Oakes. 2006. Artificial cranial deformation of the human skull. A review. Clinical Anatomy 19:372–377.

Vyslozil, Otto Von, and Rudolf Slavicek. 2001. Vergleichsuntersuchung an künstlich deformierten und undeformierten Schädeln. Annuario des Naturhistorischen Museums 102 A:245–274. Vienna.

Weiss, Pedro. 1981. La trepanación ritual suprainiana. Boletín de la Sociedad Mexicana de Historia y Filosofía de la Medicina 5(36):193–211.

White, Christine D. 1996. Sutural effects of fronto-occipital cranial modification. American Journal of Physical Anthropology 100:397–410.

White, Timothy, and P.A. Folkens. 2000. Human osteology. San Diego: Academic Press.

Chapter 4
Reconstructing Ancient Head-Shaping Traditions from the Skeletal Record

Now they are gone with all their songs and sins, Women and men, to dust; their copper penny, Of livings, spent, among these dusty inns; The glittering one made level with the many.

Their speech is gone, none speaks it, none can read The pictured writing of their conqueror's march; The dropping plaster of fading screedCeils with its mildred the decaying arch.

John Masefield (2005 [1920–1923], p. 96)

Past performance of head-shaping traditions is prone to leave tangible expressions in archaeologically retrieved skulls. Their systematic examination allows scholarship to recreate the ancient modeling techniques and their morphological results, even in those populations of the remote past that have left no written record. This chapter examines the different natural and cultural origins of human cranial shape. A brief review describes those typologies that have been employed in classifying artificially modified crania, techniques and compression implements in the Americas and specifically Mesoamerica. The cranial typology adapted from the taxonomy of the Italo-Argentinian anthropologist José Imbelloni (1885–1967) is described at length here (Comas 1970a; Dembo and Imbelloni 1938), as are some complementary metric criteria, originally proposed by Frédérique Falkenburger (1890–1965; Falkenburger 1938).

Today, this classification system, proposed 80 years ago, is still employed in most studies on Mesoamerican and Andean cranial modifications. Its standardized use benefits comparisons within and between areas and warrants inferences of Mesoamerican shaping techniques and implements, which are explored subsequently in this chapter (Dembo and Imbelloni 1938; Dávalos 1951, 1965; Falkenburger 1938; Romano 1965; Romano 1974; Tiesler 1998, 2012a). From here, I outline skeletal and contextual criteria that provide useful points of departure for deducing the social dimensions and temporal trends of past head practices, such as skeletal sex and age estimates and the evaluation of associated mortuary attributes, shown in the form of graves' construction and offerings, architectural associations and orientation.

4.1 Sorting Out the Different Origins of Human Skull Shape

Given the multifold origins of cranial alterations in archaeologically retrieved human skeletons, it is problematic to surmise a priori any cultural origin, as Table 4.1

V. Tiesler, *The Bioarchaeology of Artificial Cranial Modifications*,
Interdisciplinary Contributions to Archaeology 7, DOI 10.1007/978-1-4614-8760-9_4,

Table 4.1 Different origins of morphological change in the human skull vault

1	Postmortem modification
	1.1 *Taphonomic damage*
	1.1.1 Mechanical pressure
	1.1.2 Biochemical substitution
	1.2 *Faulty restoration of fragmented skulls*
	1.2.1 Faulty fitting of cranial fragments
	1.2.2 Faulty restitution of missing cranial areas
2	Antemortem modifications
	2.1 *Pathological conditions*
	2.1.1 Congenital defects
	2.1.2 Metabolic disease
	2.1.3 Unknown etiology (primary premature surture closure, etc.)
	2.2 *Populational morphological diversity of skull shapes*
	2.3 *Cultural modifications of the infant skull*
	2.3.1 Occupational causes (tumpline use, postural flattening)
	2.3.2 Modern therapeutical measures
	2.3.3 Cultural manipulation of the infant head (head shaping)

illustrates. Changes in dimensions of the skull vault are often part of posthumous degradation processes. Extrinsic pressure, caused by the weight of the overlying soil and earth compaction, may lead to mechanical suture separation and the displacement, fragmentation, and re-shaping of bone. Likewise, biochemical disintegration and extrinsic substitution may de-naturalize the organic tissue and cause changes in contour (Crist et al. 1997; Gervais 1989, pp. 79–80; Hansen 1919). Tensile and compression stress may result in cranial bending during the early stages of decomposition, as do moisture and changes in humidity and temperature, as actualistic research has shown (Crist et al. 1997, pp. 324–232). Under these circumstances, the thin ductile skulls of newborns show especially pronounced warping effects due to their original flexibility.

Postexcavational deformation, which may also resemble antemortem cranial modification, commonly happens during the curation process. Nonalignment of edges between two adjoining segments occurs when the surfaces to be united are covered with earth or are eroded. Also the restorer's restitution of missing cranial areas may result in unintended changes of its original vault contour (Roberts 2009, p. 95). Distinguishing these artifacts from antemortem modeling is sometimes difficult and requires experience by the observer, although the appearance of postmortem changes differs from those originating in the organic, i.e., living substrate. Posthumous changes of dimension in neurocranial contour and surface tend to be isolated and lead to discontinuities when the skull is assembled, the lines of fracture can no longer be united.

Also antemortem modifications in head form are not always cultural (see Chap. 3). Pathological changes in head morphology sometimes mimic artificial modeling, such as systemic birth defects (Down syndrome, achondroplasia, microcephaly, etc.). Also synostoses, caused by the premature fusion of skull sutures, or some metabolic diseases that intervene in bone formation (such as acromegaly or rachitis), influence the expansion of the cranial vault. In healthy individuals, skull shape is strongly controlled by biological factors which relate to their populational affiliation.

Among those head alterations that are extrinsically induced, some are the circumstantial side-products of occupational stress, adornment, or head posture. These changes are cultural indeed, but are not customs per se, as they stem from unrelated habitual daily activities, such as crib positioning, fitting tight hair ribbons on baby heads, or tumpline use. As mentioned in Chap. 3, posturally induced anteroposterior shortening or plagiocephaly occur when the infant habitually sleeps in the identical supine position on a rigid surface (Graham et al. 2005). This position can lead to the visible flattening of the baby's occipital bun. This sort of restraint is distinguished from anteroposterior compression by the absence of any anterior flattening by definition and, as I follow, of any secondary effects from extrinsic splinting, such as postcoronary grooving.

A more active form of occupational head shaping relates to the use of tumplines (carrying bands), as has been argued for some of the very early cases of frontal flattening (Trinkaus 1982, 1983). Also, some Mesoamerican cases of irregular parietal and frontal flattening may suggest the carrying of heavy loads initiated at a very young age (Gervais 1989; Tiesler 1999). In each of the cases documented from Mesoamerica, the flattening affects both the frontal bone and the anterior portion of the parietals. Conversely, there are no bipolar constrictions or any morphological changes in the occipital counterpart, which appear rounded. The anterosuperior flattening is convex but leads to a generally reclined appearance of the overall skull shape. The rounded occipital silhouette and the joint fronto-parietal flattening, which are not seen in skull modification practices, suggest that the flattening might be the inadvertent effect of tumpline use at an early age, although the growth physiology raises questions on the exact age tumpline use must have started to leave permanent traces in the forehead.

Recent actualistic research by Veronique Gervais (2001) on seven habitual tumpline carriers from Guatemala confirms our ideas and adds information on the side effects of customarily suspending weight from the forehead. Her radiographic observations suggest that tumpline use led indead to forehead flattening. She found also vertebral shortening, shifts in spine curvature and agnesia in some of the habitual carriers she study. Tumplines, called *mecapal* in Nahuatl, were the common forms of transportation in pre-Hispanic Mesoamerica, where no wheels or animals eased the burdens of humans. Boys and girls were initiated early in life in the matters of physical transport. Fraile Gerónimo de Mendieta (1997, pp. 227–228) specifies, for instance, that Aztec parents initiated their children well before the age of five in carrying burdens suspended from a *mecapal*.

Apart from tumpline use, there are still more forms of changing head morphology inadvertedly by habitual use. This may result from the quotidian adjustment of hair bands, hair circlets, or other tight head gear on the baby's head. Also, these may lead to permanent changes in the bone, as it has been documented in historical studies on French folklore (Foville 1834). A bone relic of this type will appear on the head as an isolated groove that tranverses the calvarium at the height of habitual collocation. In France, this hairdo stands at one end of a much broader range of local traditions, most of which were indeed directed to the shape of the infant head and therefore are classified as head shaping (Dingwall 1931, pp. 46–61).[1] This distinction, which is never clearcut, introduces the last category to be discussed here: those head modifications that were the object of cultural practices (Table 4.1).

So far in this section, we have characterized different types and origins of skull changes that are unrelated to infant head-shaping practices. Also some of the antemortem cranial imprints of human behavior are unrelated. They may be indicative of day-to-day practices, but their morphological imprints on the neurocranium hardly communicate any choices by their practitioners that are directed to the head itself and its shape; their object is not the body. Separating them from the marks likely left from infant modeling is not always easy or categorical. Careful case-by-case examination of relevant morphological evidence, contextualization, and patterning are required and only after excluding all possible alternative causes of skull modifications, may artificial infant shaping be surmised by exclusion.

Confirmed infant head treatments by themselves also denote diversity and complexity. These are prone to lead to conflation, bias and confusion upon examination. It is possible that some of this confusion is generated by the artificial distinction according to the criterion of assigned intentionality, which some authors support by distinguishing between what they consider purposeful head modifications from those cultural head alterations that are considered "unintentional," as for example artificial flattening due to cradleboard use (Neumann 1942; Saul 1972). As I have argued in Chap. 2, it may be less problematic to decide whether the objective of infant head manipulation is the head, thereby qualifying as body practice, than to assume cultural purposes *a priori.* What is more, it may be misleading to comprehend the morphological changes in the head as the expression of one single practice. This may be especially problematic in those cultural frames where more than one technique, implement or practitioner is involved, such as is the case in Mesoamerica. In this cultural frame, in particular, the dichotomy of "intentional" vs. "unintentional" oversimplifies the nuanced and often multi-layered meanings and roles that the body practice(s) once held in the native ideology (Chap. 6). Here, infant head manipulations express multifaceted goals and meanings, some of them protective, others preventive and phenomenological, most of them being unrelated to the visible end result in the head (Tiesler 2011, 2012a, b).

4.2 A Review of the Anthropological Literature on Cranial Classification

Over the last centuries, a host of criteria has evolved to classify culturally induced head shapes of the past. Most classifications rely on the formal qualities of the modification, some have attempted to correlate specific head shapes to the ethnicity,

[1] During the nineteenth century, in most French regions or *departments*, infant heads were still molded with hair ribbons or bandages (*bandeau* or *crémé*). In some areas, tight caps and preformed head-dresses (*serre tê tes*, *béguins*, *fromages*) were put on daily. Head restrictions of this kind were widely distributed in rural France and many of the resulting forms were recognized as characteristic of certain areas, confirming their cultural quality as head modification (Delisle 1880, 1902; Dingwall 1931, pp. 46–61; Pereira da Silva and Miya 1994).

area of use, or social distinction of their human carriers (Delisle 1902; Dingwall 1931; Weiss 1967) or to their intentional or unintentional nature (Neumann 1942). Some taxonomies have aimed at assuming specific compression techniques and apparatuses by combining metric and nonmetric criteria (Dembo and Imbelloni 1938; Falkenburger 1938). More recent quantitative classification parameters have recurred to landmark studies and three dimensional statistical models (Anton 1989; Arnold et al. 2008; Cheverud and Midkiff 1992; Cheverud et al. 1992; Gómez-Valdés et al. 2007; McNeil and Newton 1965; Stojanowski and Euber 2011). Still other classifications are founded on detailed descriptions of overall skull form or examine the appearance of constriction grooves or compression planes (for example Buikstra and Ubelaker 1994, pp. 160–163).

As regards the study of New World cranial-vault modifications, it was not before the nineteenth century that interest in native head-shaping practices arose. The new attention is evident in some early reports by naturalists, anatomists, and curious travel reporters, such as Stephens and Catherwood (1963). These descriptions were founded mainly on the analysis of skull collections, because head shaping had long disappeared from most native repertoires by the nineteenth century (Armas 1885; Boas 1890; Morton 1839; Morton 1841; but see also Comas (1958) on modern Conibo Shipibo from Ucayali, Perú; see Chaps. 5 and 11). The early accounts largely reflect the Zeitgeist of the times when antiquarianism thrived and filled the magazines of natural history museums, anatomy departments, and hospitals of the US and overseas. Soon, the New World became known as the main territory of head modifications (Flower 1881; Imbelloni 1933) thanks to its near omnipresence in the Americas and the puzzling diversity of artificial head forms herein produced (Fig. 4.1).

The pioneering volume Crania Americana (1839), *by* Samuel G. Morton, isolates four formal types in the Americas: cylindrical and conical shapes and frontal and occipital flattening. Alternative taxonomies, valid also outside the Americas, were established by Magitot, who distinguished ten skull types in a paper presented in 1880 at the Congress for Anthropology and Prehistoric Archaeology in Lisbon, Portugal (Magitot 1884). Hrdlička would cut these down to two (Hrdlička and Lumholtz 1912). Additional taxonomic systems were established by Gosse (16 types and 2 varieties), Lunier (10 types), Tschudi (3 types), Wyman (2 types), Topinard (5 types), Virchow (3 types), Lehnossek (6 types), and Sergio Sergi (4 types) (Gervais 1989). Naturally, this inflationary number of classifications of head modifications, each of which established to cover specific regions and respond to diverse anthropological interests, was destined to cause unreconcilable confusion on broader scales of comparison. Evident contradictions among different classification systems were decried by Rudolf Virchow (1988, 1890) and by Dembo and Imbelloni (1938, p. 251; see also Imbelloni 1925) who specified that:

> *... until recently, craneologists had access to a growing number of classifications. The disparity of criteria that was taken into account by the taxonomists was such that only confusion could result out of the coexistence of so many systems and nomenclatures. Each author has deemed correct to ignore the foundations of previous classifications in order to pronounce the basis of a new one, obliviating that for this purpose, the anterior classifications need to be deconstructed first by way of constructive criticism.* (Dembo and Imbelloni 1938, p. 249; translation from the Spanish by the author)

Fig. 4.1 Cultural areas which staged the most visible cranial modifications in the Americas; shaded in dark grey (adapted from Imbelloni 1933, p. 218; drawing by B. Ceballos)

As skull shapes were being meticulously described and measured during the nineteenth century, their geographic distributions were examined in the light of diffusionist theory. Lamarckian and Darwinistic ideas on evolutionary mechanisms inspired some authors to address the question of the heredity of acquired traits from the focus of cranial modeling. While Gosse (1855) still considered the possibility of hereditary transmission of artificially produced shapes, Delisle (1880, pp. 18–22) concluded his blant parental analysis of French families, who still practiced head modulation, with the statement that artificial head form was not inherited. Additional questions regarding possible neurological side effects were examined, especially with the advent of neurological sciences in the second half of the nineteenth century. These were correlated to phenomenological ideas by attributing a specific function to each area of the brain. In this frame, also artificially modeled skulls were the object of speculations with racist undercurrents and interpretations that now appear obsolete and unfounded, some of them unbearable transgressions.

At least methodologically, most of the taxonomic foundations in modern Latin American research on head shaping were established during the 1920s and the 1930s. These were the years during which cranial research grew more systematical and breached different cultural spheres. Geographically overarching research on the topic was published by Dingwall (1931), Imbelloni (1925, 1930, 1933, 1938; Dembo and Imbelloni 1938) and Falkenburger (1938). While Dingwall assembled

Fig. 4.2 Portrait of Italo-Argentinian anthropologist José Imbelloni, who developed a systematic taxonomy for classifying artificially modified cranial vaults in the Americas. Wikipedia website: http://commons.wikimedia.org/wiki/File%3AJos%C3%A9_Imbelloni.jpg

over one thousand publications to synthesize ethnological work on artificial head shaping around the globe (with important input also on the New Continent), José Imbelloni meticulously gathered metrical, osteological, and ethnic criteria to establish suitable foundations for studies on modified skulls in the Americas, consigning importance, distribution patterns, and changes through time (Fig. 4.2). In tandem with his Argentinian colleague, Frédéric Falkenburger (1938) recorded metric findings from a detailed analysis of 302 South American skulls. He correlated cranial indices and angles with different modification techniques and provided metric ranges for each. His parameters and those originally established by Imbelloni are detailed in the following section (Sect. 4.3).

North of the Mexican border, Neumann (1942) published a different classification system, designed to put order among the different cranial shapes that characterize the eastern United Sates, while for ancient Perú, Pedro Weiss (1961, 1962) proposed a classification that is still being applied in Andean research and therefore deserves mention. Weiss (1893–1985) was a distinguished physician trained in Perú. During his career, he developed a passion for pre-Hispanic Andean archaeology and osteology and developed an approach to skeletal studies known as "Cultural Osteology"

(*Osteología Cultural*). For his taxonomy of artificial cranial shapes, Weiss chose a more *emic* classification than Imbelloni and names each head type according to the native group, arguing that Imbelloni's categorical distinction between annular and tabular modification types would not fit well with ancient Peruvian practices. Here, many cultures combined rigid, semirigid, and soft head implements that resulted in complex head shapes that are characteristic for each cultural area (Weiss 1961, pp. 10–12). In fact, the emblematic function of Andean cephalic models would turn into a recourse for classifying different skull shapes in South American "cultural osteology" (Weiss 1962). Here, the taxonomies of cranial vault shaping borrowed heavily from ethnic divisions and horizons of Andean cultural evolution. These are still being adopted by more recent scholarship to examine ethnic composition, residence, mobility, and migration in the pre-Columbian Andes (Allison et al. 1981; Lozada 2011; Torres-Rouff 2002, 2003; Weiss 1962). Weiss' system starts out by establishing a broad distinction between those peoples who employed cradleboards and others who used *llautus* (head apparatuses). From here, and based on the observations of preserved head implements and skull shapes, each head type is characterized according to the cultural area it represents. Although, this classification adapts well to the Andean cultural sphere, it is regionally too specific to be transferable to other areas of the world, such as Mesoamerica.

Post-WWII research on Old and New World cranial modifications has addressed also an increasing number of specific morphological topics, most of them related to anatomy and physiopathological cranial growth (Moss 1958; Pardal 1938). Conducted mostly by medical practitioners, the studies generally draw on measurements or morphological observations obtained from series of artificially modified skulls, which are compared to the craneometric impact of pathological growth induced by premature suture closure for instance, as was explained in Chap. 3.

The second half of the twentieth century saw also more anthropologically motivated general studies on New World head modifications, which discussed various general and regionally relevant mechanical and formal aspects and their cultural distribution. There is a diverse orientation in the anthropological work by Rogers (1975), and Stewart (1941, 1958, 1963, 1975) for North and Central America, by Romano (1965, 1974), Weiss (1967), Saul (1972) and Stewart (1943b, 1975) specifically for Mesoamerica, and Pedro Weiss (1961, 1962, 1981), Allison et al. (1981), Munizaga (1974) and Stewart (1943a, 1963), among others, for the Andean Region. Note that in the last two decades, anthropological work on head shaping in the New World has relied increasingly on archaeological and sociological theory in the interpretation of skulls as part of the mortuary record and has reappeared in new conceptual frames anchored in native schemes of cosmology, semiotic and in general cognitive approaches, body theory, and embodiment (Blom 2005; Geller 2006; Lozada 2011; Tiesler 2007, 2010, 2011; Torres-Rouff 2002; Yépez 2006, 2009; see Chap. 2).

In Mexico, specifically, it was Eusebio Dávalos (1909–1968), who gave academic recognition to pre-Hispanic cranial vault modifications. Dávalos started out in the mid-twentieth century on this subject. He first conducted research on a skull collection from Aztec Tlatelolco (Dávalos 1951; Dávalos and Romano 1965) and later in

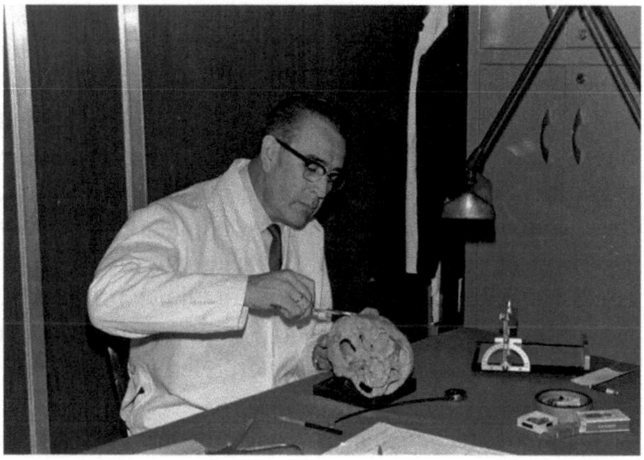

Fig. 4.3 Portrait of Mexican anthropologist Arturo Romano Pacheco, who has promoted the anthropological study on pre-Hispanic head-shaping practices in Mesoamerica and specifically in Mexico. (Photo: Archives of the Dirección de Antropología Física/INAH; courtesy of M. T. Jaén Esquivel)

his career actively promoted the study of Mesoamerican cranial modeling in general. Also, Juan Comas (1900–1979) covered pre-Hispanic head-shaping practices in different parts of Mexico, mainly collections from the Mexican state of Veracruz (Comas 1960, 1970b; Comas and Marquer 1969). Also, noteworthy is his engaged case study of head flattening among recent Peruvian Amazon communities (Comas 1958).

After 1960, Mexican research on Mesoamerican skull modifications was consolidated and has been dominated since by Arturo Romano Pacheco (born in 1921) of the National Institute of Anthropology and History (Fig. 4.3). Trained as a physical anthropologist at the Mexican National School of Anthropology during the 1940s, Romano specialized in osteology from the beginning of his career and showed a profound academic passion in Mesoamerican craniometry, for what he establised an impressive database during his long professional career. For this purpose, Romano designed his own approach to measuring skulls, which he called "craneotrigonometry." His approach is founded mainly on detailed morphological descriptions of the skull and complemented by a series of measurements, oriented in quadrilateral skull polygons, as established by the German physician Hermann Klaatsch. The subject of cranial vault modifications soon became one of Romano's preferred lines of research and teaching. His profound grasp of the subtleties involved in cranial morphology and variation, not only permeates his own research but has left a lasting imprint of many generations of Mexican colleagues and students who over half a century have received Romano's training on cultural craniology (Bautista 2004; Jaén 1962; Martínez 2009; Sánchez 1971; Tiesler 1998, 1999; Yépez 2001, 2006).

Since the seminal volume titled Estudio morfológico de la deformación craneana en Tamuín, S.L.P., y en la Isla del Idolo, Veracruz of 1965, Romano himself has published prolifically on the subject of Mesoamerican cranial vault modifications, although most of the distribution of his work is limited to Hispanic media. For studying Mesoamerican skull modifications, Romano adopted Imbelloni's classification and adapted it to the regional skeletal record of native practices. Here, he combined its parameters with a set of specific craniometric distinctions, which he derived from the work of Falkenburger (1938), Moss (1958) and his own detailed craniometric work (Romano 1965, 1996). This combined approach allows him to generalize on types, forms, and techniques in the skull series he reports. By combining the craniological approaches with iconography the work by Romano already observes some of the complex cultural and specifically cosmological associations that different Mesoamerican head forms once held. Romano establishes some of these connotations in a Preclassic "olmecoid" skull from Pampa el Pajón for example (Romano 1977, 1980), which he interprets as manifestations of ideologically conferred beauty ideals expressed in Olmec imagery. Also, the strong superior flattenings in skulls from El Zapotal (Romano 1977) and the conical head shapes of the Huasteca (Romano 1987) are interpreted in terms of visible assimilations of native sacred entities.

On a more technical note, Romano affirms that Mesoamerican compression gears correspond almost exclusively to rigid devices and it follows that cradleboards and head splints could be combined to produce "mimetic" shapes, their compression effect sometimes being enhanced with horizontal (pseudocircular) or sagittal constriction, the latter being designated "bilobular variety" (Romano 1965, 1973). His distribution map of cranial shapes within the confines of native Mexico, published in 1974, is supported by a review of the literature and his own studies of the voluminous skeletal collections curated in the National Museum of Anthropology.

Also, scholarship—north of the Mexican border—has contributed importantly to the study of Mesoamerican cranial modifications. Seminal work has been conducted, for instance by T. Dale Stewart. Among his work on the topic counts his early treaty, titled Human Skeletal Remains from Dzibilchaltún, Yucatán, Mexico, with a Review of Cranial Deformity Types in the Maya Region (Stewart 1943b, 1953, 1975), published in 1975. In this work, the author compares artificial head forms between different time periods and between regions and concludes that the distribution of head forms varies significantly between the times. Comparing the scholarly approaches from both sides of the border, the lack of integration between local work and international studies becomes apparent. This separation is underlined by the irreconciliable classifications employed in documenting head shapes. Still today, the international community has only reluctantly adopted the prominent Mexican classification system, based on Imbelloni's taxonomy (Duncan 2009; Duncan and Hofling 2011; Saul 1972). A noticeable lack of cross references between Hispanic, French, and Anglo-Saxon publications on Mesoamerican head shaping has come to limit the amount of comparable data, a restriction already voiced by Stewart (1975; see also Gervais 1989; Tiesler 2012b).

Apart from language barriers or national boundaries, the bioarchaeological study of Mesoamerican head shaping is surely also restrained by the degree of deterioration, which usually affects organic substrate, such as bone much more than most inorganic remains and translates into a lack of analytical possibilities and reduced sample sizes available for study. Also the topic's interdisciplinary quality challenges any academic coverage. Placed at the point between physical anthropology and archaeology, biocultural or bioarchaeological studies on head shapes depend both on contextual information and the anatomical knowledge of the observer. In research practice, this disjunction has resulted in the separation or omission of findings on head-shaping practices from those derived from mainstream archaeological research. Many archaeological site reports do not mention head-shaping practices or distinguish them only in the dichotomic terms of presence vs. absence even if quantifiable.

4.3 Classifying Ancient Mesoamerican Cranial Forms

The following paragraphs summarize the classification scheme currently used in Mexico, which goes back to the work of Argentinian anthropologist José Imbelloni (Dembo and Imbelloni 1938, pp. 258–259) and whose taxonomy is presented in an adapted version in Table 4.2. Although not suitable for describing the most complex annular shapes, such as those common in Europe, Melanesia, or Africa, it benefits the classification of most tabular modifications in the Americas and specifically in Mesoamerica. The classification system facilitates assumptions of modeling techniques from the formal properties. Tabular compression is distinguished categorically from annular forms, accomplished by constriction bands, single strings, bandaging, or tightly fitted hats. Conversely, tabular modifications are associated to rigid compression devices (Dembo and Imbelloni 1938, pp. 289–303). According to Imbelloni's scheme, head splints result in tabular oblique shapes with their characteristic backwards inclination showing parallel compression planes (Fig. 4.4a). Cradleboards (or body kits) determine the tabular erect type and its corresponding shortened head forms (Fig. 4.4b). This association does not exclude the possibility of obtaining oblique forms by cradling or, inversely, erect shapes from head splinting (Romano 1974, p. 204). Sometimes, more than one compression implement was used, and therefore the formational effects on the skull vary accordingly. Also, the duration of compression and the applied pressure determined the severity of formal changes in the skull cap, ranging from invisible to extreme.

4.3.1 Visual Recording and Craniometrics

To characterize and comprehend the artificially produced changes in skulls, both formal (visual and craniometric) and procedural (technical) criteria are relevant. Useful points of departure in studying cranial vault modifications are standardized descriptions of the presence, extent, and anatomical relationship of each flattening. In this endeavor, the definition of contours (straight, convex, or concave) and the degrees of compression (ranked in each plane and assigned to the overall alteration) facilitate distinctions of cultural flattening and their formal qualities. Apart from the flattened areas, associated attributes, such as bilateral bulging vs. reduction of bilateral width, suprainiac impressions, assymetry, and postbregmatic changes, provide complementary indications on ancient head-shaping procedures. Graphic representations of the overall cranial silhouette in sagittal, vertical, and dorsal schematic drawings are enormously helpful to visually comprehend the head shape. This careful assessment facilitates the assumption of technical procedures and classification

Table 4.2 Taxonomic criteria, adapted from Dembo and Imbelloni (1938, p. 275)

Type	Distinctive compression technique	Degrees (0–4)	Distribution of compression (in degrees ≤ 3)	Circular wraps (0–3)	Sagittal constriction (0–3)
Tabular oblique modification	Fronto-occipital compression with help of head splints	Extreme (> 3)	Intermediate	Absence	Absence
			Occipitally curved	Presence of horizontal constriction (pseudocircular)	Presence of band (> 0–2)
			Frontally curved		Bipolar separation (> 2)
		Intermediate (≤ 3)	Parallelepided (or obelionic)		
			Mimetic		
Tabular erect modification	Posterior compression in supine position	Extreme (> 3)	Intermediate	Absence	Absence
		Intermediate (≤ 3)	Occipitally flattened	Presence of horizontal constriction (pseudocircular)	Presence of band (> 0–2)
			Frontally flattened		Bipolar separation (> 2)
			Parallelepided (or obelionic)		
			Mimetic		
			Conical		
Annular modification	Symmetric annular constriction through elastic bands or wraps	No information	No information	No information	No information

To avoid conflation, all varieties in this scheme are subdivided in four columns according to their criteria of distinction (severity, form, presence of circular constriction, presence of sagittal constriction)

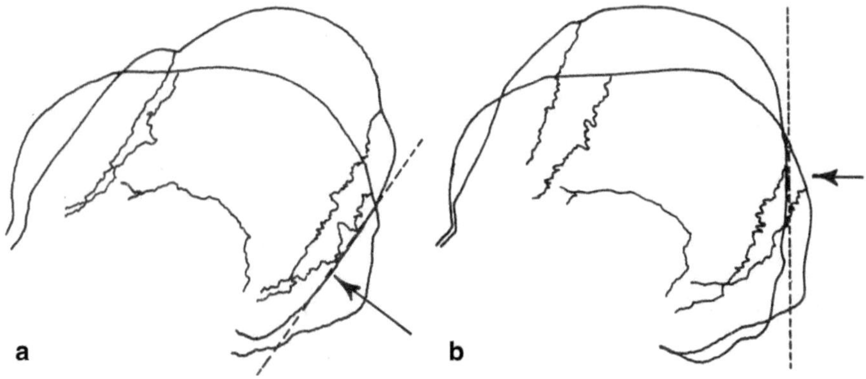

Fig. 4.4 Compression vectors in the **a** tabular oblique modification and **b** tabular erect modification (adapted from Dembo and Imbelloni 1938, p. 250; drawings by V. Tiesler)

assignment. Given the complexity involved in reconstructing living practices from organic substate, it is clear that all classification efforts cannot be more than encouraging approximations toward understanding broader patterns and behaviors sustained by the ancient head-molding practitioners. No categorical distinction, however well-founded, will be enough to account for the real complexity of the processes involved in head shaping and their expression in the bony substrate. With this cautionary note in mind, the following paragraphs make a set of general criteria available for distinguishing head shapes and for surmising modeling techniques from the skeletal record.

4.3.1.1 Basic Modification Types

Apart from the *annular* (artificial orbicular) type, the taxonomy developed by José Imbelloni identifies two basic types of modification: *tabular oblique* and *tabular erect shapes* (Fig. 4.5a, b). Annular head forms produced exclusively by infant head constriction constitute one of these groups, which are either erect or oblique and always lead to a round cross section of the cranial vault and to its reduction in bipolar width. In Mesoamerica, the oblique annular form has been documented in Western Mexico (Gómez-Valdés et al. 2007, pp. 121–122). However, bandaging must also have clearly dominated the practitioners' constriction routines among those groups who practiced cradling or head splinting. These cases are called "pseudocircular" in Imbelloni's taxonomy and will be described in Sect. 2.3.1.3.

Regarding the two tabular types, they are differentiated by their final appearance in the cranium: in tabular erect shapes, the opposing compression planes are usually not parallel to one another due to the position of the posterior plane which tends to cover *lambda*. Here, front-and-back flattening results in a high and broad configuration of the cranium (brachycephalic) when not combined with lateral constriction. In *tabular oblique* shapes, the posterior compression plane centers on the occipital bun or inion crest, situated below *lambda*. The back compression vector opposes directly

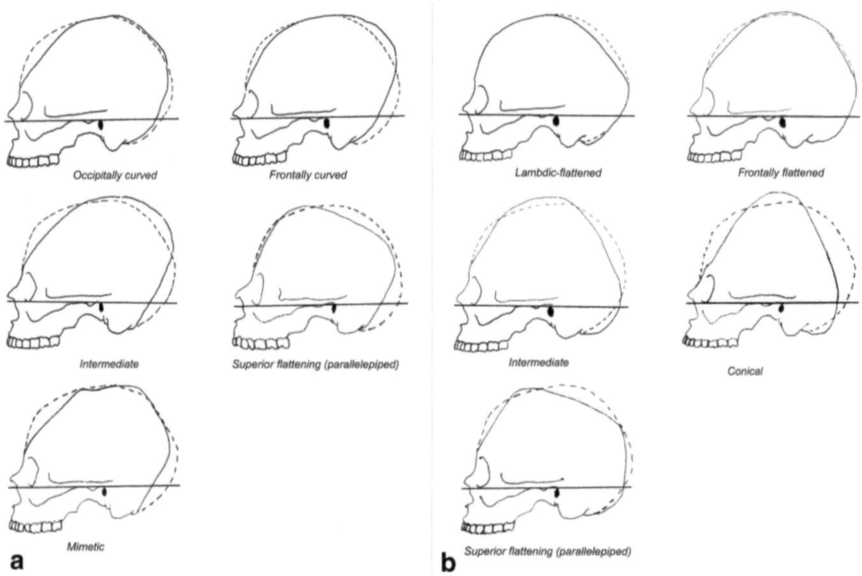

Fig. 4.5 Different formal varieties of tabular oblique modifications (a) and tabular erect modifications (b) (drawings by V. Tiesler)

the frontal vector of mechanical force, commonly resulting in the parallel orientation of both planes and a backward inclination of the cranial vault.

Typical expressions of tabular oblique and erect types were coined by Imbelloni *intermediate variants* of anteroposterior compression. In these classic forms of tabular oblique and erect modeling, both compression planes (front and back) are visible and are shown to a similar degree. In Mesoamerica, most intermediate forms of tabular oblique and erect head shapes describe moderate to severe forms of artificial modeling. Imbelloni's category of *extreme* modifications also matches this touchstone of balance. In this case, both compression planes are expressed in their highest possible degree of modification, being the final result of the daily pressure applied and the yearlong span of compression (see Chap. 3).

4.3.1.2 Craniometry and Landmark Studies

In complete skulls, the morphological findings and their classification and technical interpretation are confirmed and complemented metrically. To this end, cranial landmarks are commonly recorded and now digitalized prior to statistic processing (Barrientos and L'Heureux 2009; Béguelin et al. 2006; Clark et al. 2007; Gómez-Valdés et al. 2007; Manríquez et al. 1884; Pomeroy et al. 2010; Pérez et al. 2009; Rhode and Arriaza 2006). Most of these measurements, which are recorded using 3D digitizers, craniophores and/or callipers, are designed to distinguish presence from absence of cultural flattening and to differentiate general head types. In Latin America, systematic metric evaluations of artificially modeled skulls go back to the efforts by Falkenburger (1938) and Imbelloni (Dembo and Imbelloni 1938). Early work proposed the vertical and horizontal foraminal *clivus* angles as convenient classificatory parameters to distinguish different forms of tabular and annular modification. Also, the angle between basion-nasion and its cord toward *vertex* (highest point on the cranial vault) is considered a useful metric parameter to establish the vault's degree of backward inclination and therefore the type of head modification (Romano 1965; Fig. 4.6a, b; Table 4.3)

Other distinctive morphometric traits used in current research include the anatomical location of the apex and the distance of the opistocranium from lambda and opistion (Tiesler 1999). Proportional measurements of bilateral width provide indications on bipolar constriction vs. bilateral bulging. The angle between the Frankfurt Plane and orientation of the occipital aperture provides an indication of basicranial bending (see Chap. 3), while the angle established between the Frankfurt plane and the cord between *porion* and *vertex* expresses the level of cranial inclination in relation to its highest point. In tabular erect types, this point is located above or below the compression plane, while in oblique and mimetic specimens it tends to be located in the area of compression.

Recent approaches have also successfully used multivariate calculations of cranial landmarks and distance measurements to evaluate either continuous change in skull form or to examine and compare the morphological features of previously assigned modeling types in subscribed areas in Mesoamerica and South America (Gómez-Valdés et al. 2007; Pérez et al. 2009; Pomeroy et al. 2010; Rhode and Arriaza 2006). Also on a more general level the sophisticated digital measure meuts of coordinates, angles and cords, aligned according to the culturally flattened areas of the cranium, offer useful objective tools to distinguish the vault's shifts, bulges and reductions, reclination and heightening (see Fig. 8.2).

4.3.1.3 Varieties of Tabular Oblique and Tabular Erect Shapes

Not all anteroposterior compressions fall clearly into the two basic tabular categories described above. Some specimens may instead show characteristics that identify both erect and oblique attributes. Imbelloni describes these hybrid forms as *tabular mimetic* or "cranial specimens that share their formal characteristics with the family to which they belong, but also manifest accessory attributes that make them visibly similar to another type of modification" (Dembo and Imbelloni 1938, p. 277). Such is the case when more than one compression plane is observed in the back of the skull.

In Mesoamerica, most mimetic modifications are expressed in this form in fact. In these specimens, the superior area of posterior flattening typically suppresses the surface around lambda, whereas a second, inferior compression zone flattens the central and lower portions of the back of the skull. When both occipital compression planes are combined in a sagittal drawing to generate one single composed line of posterior compression, this is almost always reclined and finds itself roughly parallel to the anterior compression plane, as described for the tabular oblique category (see Fig. 4.5a). In fact, the contour of the mimetic cranial vault, modified in the

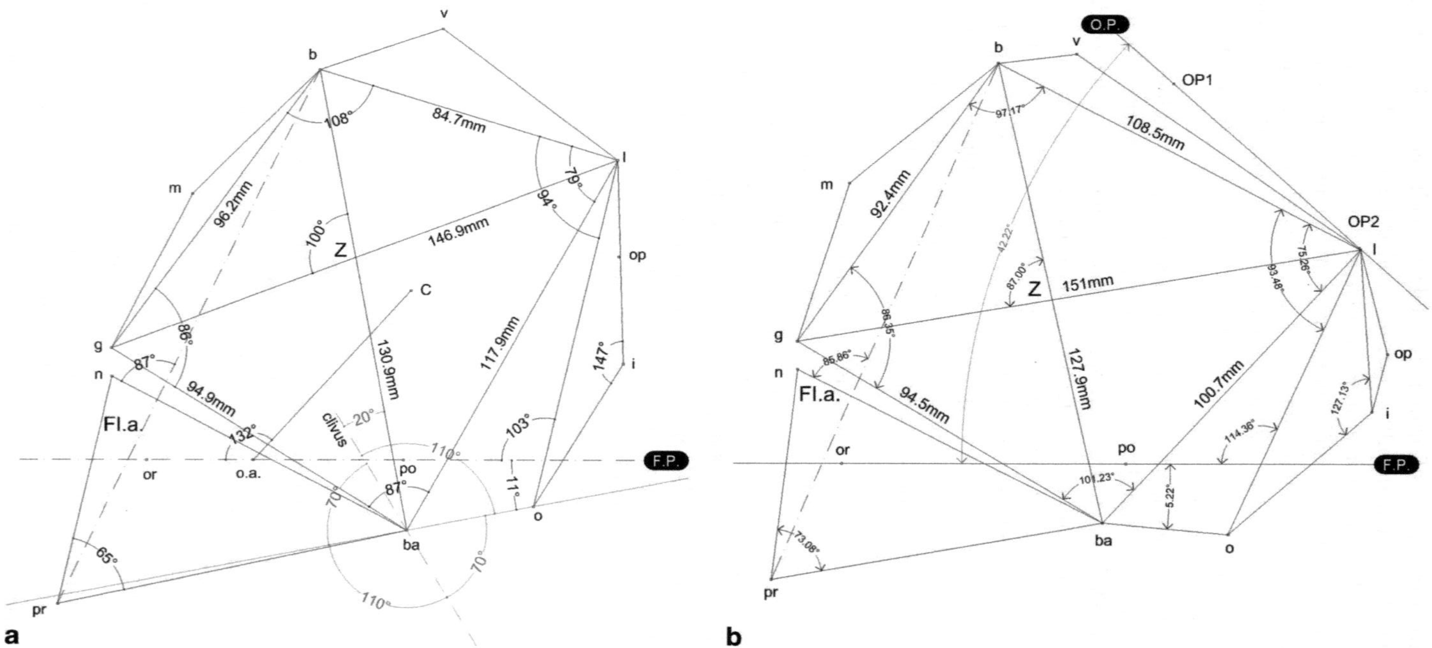

Fig. 4.6 Klaatsch quadritateral on skull with corresponding measurements, obtained from digital 3D landmark recording of different artificial head shapes **a** Tabular oblique, **b** Tabular erect with superior flattening (measurements and digital drawings by J. Gómez-Valdés)

Table 4.3 Metric parameters in distinguishing different types of pre-Hispanic Maya cranial modifications

Parameters	Tabular erect[a] (N: arithmetic average ± standard deviation)	Tabular oblique[a] (N: arithmetic average ± standard deviation)	Total[a] (N: arithmetic average ± standard deviation)
Foraminal *clivus*	26: 52.71 ± 11.15	5: 60.75 ± 12.09	36: 152.67 ± 9.50
Horizontal *clivus*	9: 56.61 ± 6.11	2: 52 ± 0	14: 54.61 ± 6.10
Vertical *clivus*	9: 22.17 ± 4.34	2: 33.50 ± 9.19	14: 25.68 ± 7.40
Angle of reclination with *vertex* as reference point	10: 109.75 ± 5.25	2: 123.25 ± 7.42	15: 113.47 ± 7.78
Angle of reclination with *opistocranium* as reference point	9: 173.22 ± 7.79	2: 161.25 ± 3.89	14: 168.5 ± 10.23

[a]Only adult crania were included

described form, achieves a posterior inclination that is characteristic of all oblique modifications. Craniometry appears to confirm this association on most mimetic skull silhouettes, the proportions and angles of which approach or even exceed the arithmetic averages assigned to the tabular oblique category (Tiesler 1999, pp. 212–213). It follows that mimetic modifications of this sort should be comprehended as a formal variety of tabular oblique shapes rather than a separate type.

This oblique mimetic configuration, which dominates in many Classic Maya collections, is different from other mimetic skull contours, which appear broader and less inclined and therefore seem to stem from cradleboard use instead of free head compressors (Fig. 4.5b). In the latter case, both posterior planes are centered around, or even above, lambda, although they do not encompass much of the parietal surfaces of the upper skull vault, which distinguishes them from skulls with obelical or superior flattenings. It is significant that there are still other *mimetic* head forms, although there are few that do not fall within the oblique or the erect categories. Maybe, the latter express the combination of separate rigid devices on the infant head. Naturally, for the observer, these undefined "tabular" specimens also signal the practical limitations in determining modification techniques from head morphology alone. From all of the above, it seems wise to consider "mimetic" skull shapes not as a separate entity but rather as variants of either tabular oblique and tabular erect types, depending on the inclination and technique used (Dembo and Imbelloni 1938, p. 275).

A variety that is similar to mimetic forms equally implies the implementation of a third compression plane; however, this additional area of complression is not situated on the back of the head but flattens its top (Fig. 4.6b). In this case, the superior plane usually covers the majority of the sagittal cord and often leads to postbregmatic bulging in front. In its severe expression, the entire upper parietal area (from bregma to lambda) appears flat, acquiring the contours of a round, flat plate. This specific shape has been coined *obelical, obelionic* "Zapotal" type, parietal or *superior flattening* by scholarship (Dembo and Imbelloni 1938, p. 273; Martínez 2009; Nelson and Mandimenos 2010; Romano 1973, p. 59; Stewart 1939; Tiesler et al. 2013, p. 55). The description of this morphology closely resembles the tabular erect *parallelepiped* variety that Imbelloni's system characterizes as a cubic (or parallelepiped) appearance of the head (Dembo and Imbelloni 1938, p. 273). Unfortunately, Imbelloni's original definition leaves doubts on whether this form (apart from its front-and-back reduction) is limited to a single superior plane or if additional bilateral compression was used in order to ensure a more cubic appearance. This ambiguity and the discrepancies with Falkenburger's (1938) diagnostic angles led Arturo Romano (1973, p. 59) to substitute the term "parallelepiped" in favor of "superior flattening" in his study on skulls with strong superior flattening from the site of El Zapotal, in the state of Veracruz. More recent work has returned to the original term (i.e., parallelepid) to assign superior head flattening, because these resemble the cubic formation that results from the combined application of anterior–posterior and superior pressure (Tiesler 2012a; Yépez 2001).

Similar to mimetic skull shapes, some forms of superior flattening appear on reclined skulls which show approximately parallel compression planes, associated with free head-compressor devices and the tabular oblique type. Other specimens that show the superior compression plane appear to stem from cradling instead, as is likely the case of the skulls from El Zapotal. Their foreheads do not show any backward inclination and, when the cranial polygon is drawn and craniometric landmarks measured, their discriminant values align with erect, not with oblique head types (Fig. 8.2; Tiesler et al. 2013, pp. 59–60). However, also in the case of the "Zapotal" like obelionic flattenings in the Mixtequilla area and as a generic form, there is controversy in the literature as to the device responsible, which some authors hold to be head splints (Martínez 2009; Nelson and Mandimenos 2010), and others to be cradleboards (Stewart 1939; Tiesler et al. 2013). An enlightened opinion on the devices responsible for parietal flattening is provided by T. Dale Stewart (1939, p. 464–465), who comments on crania from Southern Florida, which is worth transcribing at length here.

> "Except where [head modification] is caused simply by the weight of the head upon a hard surface, some intentional pressure must be exerted to hold the head against the [compression] surface. In the present case the point of counter pressure appears to be the chin or some part of the anterior trunk. Perhaps, therefore, the child was bound to the cradleboard in such a way that the head was pressed against an inclined endpiece. This view seems more logical than that which envisions a bandage passed under the chin, for the simple reason that the latter mechanism would have deformed the jaw also," and, as the author is quick to point out "there is no evidence of this in the present collections" (Stewart 1939, pp. 464–465).

So, to the point, given the diverse formal criteria of top flattened heads, it is most likely that superior flattening was reproduced both by tabular oblique and erect techniques, leading to different degrees of head reclination and distinctive anatomical distributions of compression planes (Fig. 4.5). Grégory Pereira (1999) comes to a similar conclusion after examining cranial collections from pre-Hispanic Western Mexico and separates top-flattened reclined skulls from unreclined ones on the basis of technical differences in compression devices. Both top-flattened forms appear to have made their appearance in the Mesoamerican cultural repertoire in the latter part of the Classic period (Pereira 1999; Tiesler 2012a, b; Tiesler et al. 2013).

Apart from mimetic and obelical varieties, there are other tabular head formations. These evaluate the expression of each front-and-back compression plane with each other. According to Imbelloni's system, *occipitally curved* and *frontally curved tabular oblique* varieties imply that the cranium is round and convex in this part, while a more pronounced, usually straight of concave compression plane and compare them is observed on the opposite side. Similarly, *frontal* and *lambdic flattenings* are distinguished in *tabular erect morphologies*. These imply that the plane in question is noticeably more defined or more pronounced than the opposing posterior or anterior compression plane. The four variants described here usually relate to slight or moderate degrees of cultural modeling, although they also may occur in severe forms of shaping. The four varieties have in common that, when compared, the front-and-back planes are expressed to a dissimilar degree. This becomes apparent when the extension of both compression planes and their contours are compared with each other (straight, convex, or concave). Unfortunately, there are no established specific measures in the literature that distinguish each of the above varieties metrically.

Although not described in the classification system by Imbelloni, a particular expression of tabular erect head shapes gains relevance in the Mesoamerican cultural sphere and deserves mention. In this form, both compression planes include an important parietal component. Frontal flattening reaches bregma or may extend beyond this point, while posterior flattening is centered over lambda and expands across at least half of the parietal chord. Converging in the postbregmatic area, both planes produce a *conic form* when observed in profile, which recalls Romano's (1980) characterization of Huastec sculptured portraits from Tamuin. The "roof-like" appearance, described here, should not be confused with the conical shape assigned to annular variants by Imbelloni, which implies bipolar constriction and therefore reduction (Dembo and Imbelloni 1938, p. 276; also see Yépez (2006) for the conic shapes identified in the Andean region).

4.3.1.4 Annular and Sagittal Grooves

Annular wraps and sagittal grooves appear abundantly in the Mesoamerican skeletal record. In contrast to Imbelloni's original taxonomy, I prefer to treat these as a separate third classification category to avoid confusion with other variants with which they may concur. Regarding narrowed head forms, these make up one group of artificial head forms in the taxonomy established by Imbelloni (Table 4.2) and have already been briefly described above. In Mesoamerica, pure annular forms are very rare, although flexible devices clearly combine with rigid head compressors in subscribed areas of practice, such as the Olmec-dominated Isthmus areas, the Maya Usumacinta Basin or parts of the Gulf coast during the Classic period (Tiesler 2012a, b; see Gómez-Valdés et al. 2007). Here, many of the severely elongated skulls acquire a tubular form and are clearly reduced bilaterally, a point to be taken up in the next section. But also where rigid compressors were used, bands and ribbons, semirigid pads or cushions—once adjusted on the head—would tighten and mediate the pressure of compression boards or complement their compression effects on the skull. When expressed on the skull, this complementary constriction defines the *pseudocircular* variety of tabular types.

A second form of constriction secondary to tabular compression is referred to as *bilobed* or *trilobed* in Imbelloni's taxonomy and is produced when a tight band connects the front-and-back planes over the top of the skull. Its basic expression consists in an isolated sagittal sulcus that divides the posterior portion of the parietal bones visibly into two lobes. When this sort of cranial adjustment is severe and prolonged past fontanelle closure, a postcoronal furrow may ensue as a physiological consequence of altered growth (see Sect. 3.2.2), giving the final shape a trilobed appearance when observed from above. The so-called trilobed skulls have been a preferred subject of academic interest during much of last century and have lead to several case studies on Mesoamerican skull colections, such as the Mexican Gulf Coast area (Comas and Marquer 1969; Dávalos 1965; Stewart 1948). One head shaper possibly responsible for the sagittal ridges is shown in a Maya figurine from El Salvador (Cavatrunci et al. 1992; Fig. 4.7). The tablet on the forehead, which is similar to that used recently by Peruvian *shipibo-concibo,* appears to be held in place by an occipital band and a sagittal strip that runs over the top of the head.

4.4 Head-Shaping Devices in Ancient (Meso) America

The taxonomy of Imbelloni associates each morphological type to a specific head or body implement. Head splints produce tabular oblique formations, while tabular erect forms are interpreted as the result of cradling (Dembo and Imbelloni 1938, pp. 289–303). Naturally, in practice, this correlation is not a categorical one, as oblique head forms may be obtained also by use of cradleboards depending on the support beneath the occiput and collocation of the frontal plane. Inversely, erect shapes may also be the product of head splinting, as has been argued for the case of the narrow and high head forms of Olmec traditions (Romano 1974; Tiesler 2010). Even within each type, variation in the apparatuses, or additional constriction with strings translate into diverse appearances. When an occipital head board is held in place by strings around the forehead, a frontally curved tabular oblique form will result. When the posterior plane is part of a cradleboard device and tied in front, the tabular erect modification is identified with its lambdic variety (Dembo and Imbelloni 1938).

On a more general note, cushions inserted beneath the compression board produce depressions in the flattened surface, while the joint application of free boards with circular bands results in what is called "pseudoannular" or "pseudocircular" tabular shapes (Romano 1965). In many cases, it appears that more than one compression implement was used, as was already argued for mimetic head shapes. Also, the duration of compression forces and the pressure applied introduce diversity as they condition the visibility of formal change, ranging from invisible to extreme.

The following paragraphs will complement the information on artificial head forms by discussing different implements used in the Americas and specifically in Mesoamerica, where numerous figurines from different times and cultural regions decode for us the technical specificities of this practice. Each of these implements

Fig. 4.7 Head splint displayed by Classic period Maya figurine (The Museo Popol Vuh, Universidad Francisco Marroquín, Guatemala; adapted from Cavatrunci et al. 1992; drawing by B. Ceballos)

will be cautiously related to the anatomical location and morphology of flattened or constricted surfaces and put in context with the classification of head formations that these implements presumably produced.

4.4.1 Cradleboards

In most parts of the American continent, cradleboards or cribs once carried babies during their first few months of lives (Dembo and Imbelloni 1938; Dingwall 1931; Mason and Porter 1889; Romano 1974; Weiss 1961). These carriages basically consisted of a framed backboard, which could be padded with soft plant fibers. Cradleboards could also include foot- or buttock-rests. Hoods over the head of the infant would provide protection against the cold, sun, wind, or rain and provide a candy frame for cloth and other shields. Surrounded by lateral contentions, rims, or head pieces, the crib would protect the little also from the exterior toward each side (Fig. 4.8).

Safeguarded this way, the infant would rest in an extended supine or prone position, firmly strapped to the support with cords or strips of cloth. Thus, cradleboards

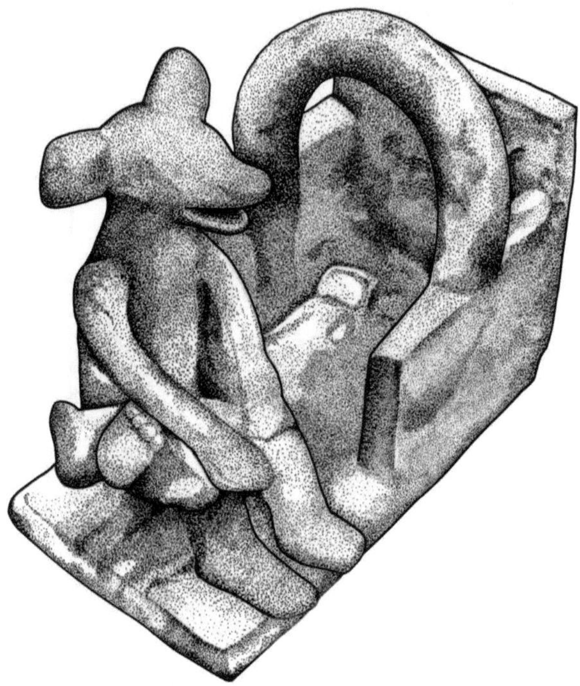

Fig. 4.8 Figurine representations of crib device with handles and side protection, Cempoala, Veracruz (adapted from figurine exhibit of the National Museum of Anthropology/INAH, Mexico City; drawing by B. Ceballos)

usually confined the mobility of babies and at the same time protected them without impeding day-to-day feeding and cleaning by their caretakers. As multipurpose kits, cradleboards were employed also as head compressors. To serve as a modeling device, the headpiece of the cradleboard needed to incorporate two juxtaposed compression planes. Often, the backboard itself or an inserted ring-shaped roll or pad squeezed the head from beneath. The opposite upper plane could be either flexible or rigid and was usually tied directly to the backboard. This compression mechanism usually produced lambdoid flattenings or visible front-and-back reductions of the infant head, expressing the classic tabular erect category developed by Imbelloni. Here, the baby could be tied to the cradleboard in such a way that the head was pressed against the bottom plank, sometimes also against an inclined endpiece, potentially producing the top-flattened head forms known from parts of Veracruz, from Florida and the Southwestern USA and Panama and Chichén Itzá (Nelson and Mandimenos 2010; Stewart 1937, 1939, 1958; Romano 1977; Tiesler 2012a; Tiesler et al. 2013),

Unlike further north or south, concretely in the Andes, no secured body compressors are known for pre-Hispanic Mesoamerica. Instead, its many ceramic sculptures bear witness of cradleboard lore and at the same time inform about head-shaping procedures. Many cradleboard scenes are depicted in the form of naturalistically sculptured pairs of infants and their female caretaker. The cradleboard may rest on the lap of the woman, in her arms or over her shoulder. Some children are being nursed, others are attended by mystical beings while resting inside the cradleboard (Fig. 4.8). Their shape and equipment identify many of them more as steadfast pieces of furniture than mobile carriages (Fig. 4.9). Most appear to be small but bulky cribs, some of which stand on legs. Stiff footrests, hoods and body straps, lateral restraints, handles, or protective hoods over the head and leg area of the infant complement the baby kits. It is assumed that those cradleboards in the figurative record that appear more austere and compact were employed for daily transportation during daytime errands. It is also noteworthy that many of the cradleboard sculptures originally functioned as whistles or flutes, suggesting their uses either as children's toys or as instruments that accompanied infant transition rituals, as argued elsewhere (Tiesler 1999, 2012a).

The above descriptions characterize ancient cradleboards as multifunctional care units beyond doubt, most of which facilitated child protection, immobilization and care. The Mesoamerican folklore images also illustrate some of the compression procedures involved. Tablets or thick strings of cloth cover the forehead of the little one and appear to squeeze the baby's head (Fig. 4.9). Some heads rest on cushions; others are wrapped or carry head compressors that are not directly attached to the cradleboard itself, as several scenes from Teotihuacán demonstrate (Fig. 4.10). In still other figurative settings, a female caretaker hovers over her infant and appears to manipulate its head with her hands. Hence, the depicted settings show a broad range of compression mechanisms, ranging from hand pressure to free head wraps and splints, and to proper cradling, when the compression was conditioned directly by the backboard. Lambdic flattening likely stems from tying the forehead back to the body board. More severe intermediate or extreme erect shapes must have relied on a rigid frontal tablet. Bilobal reduction is conditioned in this process by lateral constraint in the form of complete head wrapping and its side stringing, manifesting pseudocircular shapes. Cushions on the infant forehead are sometimes depicted and may have led to depressions within the flattened area.

Still other questions concern the orientation of the child's head within the crib, which is shown to look sideways and up and backward in some scenes. Surely, the habitual head posture during the modeling procedure left an imprint on the head: some compressors flattened the back of the head in its upper section, others in its lower part. As already explained previously (Sect. 3.3), the slipping of the head toward one side may have caused the compression boards to shift and produce secondary cranial asymmetry (Björk and Björk 1964; Dembo and Imbelloni 1938; Kohn et al. 1995). More than the head apparatuses, cradleboard straps are at risk of becoming loose and side-slipping by opposing the natural movements of the baby's head and body. Their crib's additional uses in infant transportation, cleaning and nursing, in all probablility still increased the chances of slackening (Dingwall 1931; Gervais 1989; Romano 1974).

4.4.2 Head Splints

Free head splints with a clamp-like effect usually bear different results from those obtained by cradling. In contrast to compression cradleboards, head apparatuses do

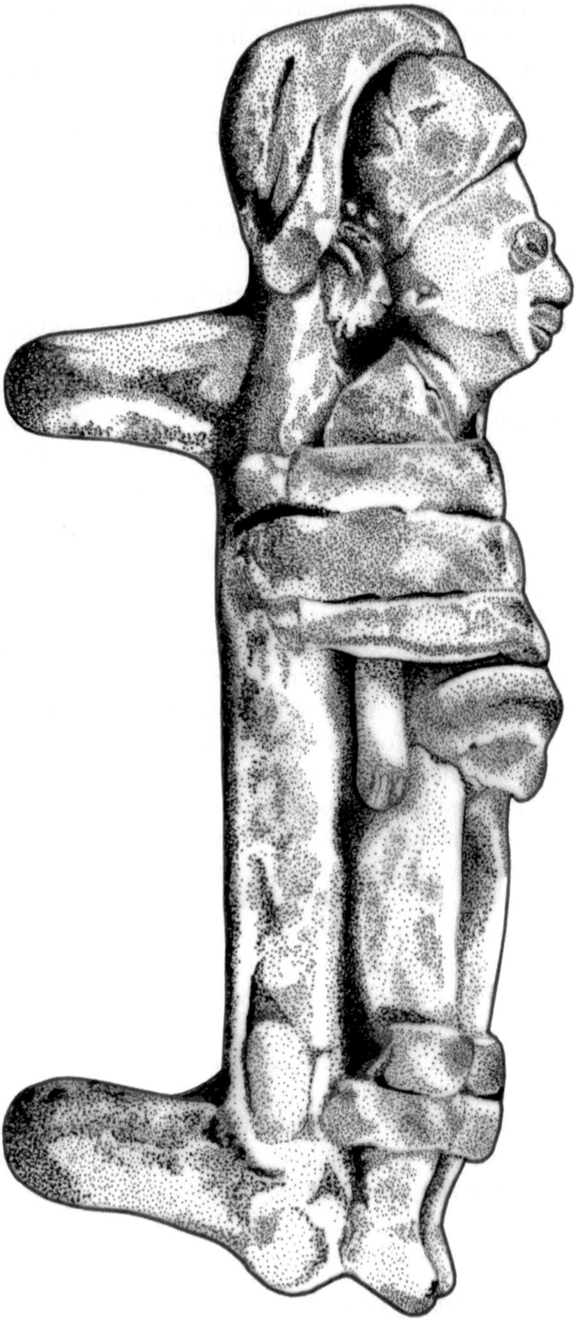

Fig. 4.9 Figurine representations of crib device with legs, Western Mexico (MNA/INAH; redrawn by B. Ceballos from courtesy photograph by A. Romano)

not affect infant mobility during use. Here, the fundamental compression mechanism consists in compressing the skull between two free boards that are allocated on opposite sides of the head and tied together firmly, often resulting in two flat planes that

Fig. 4.10 Figurine representations of infant cradle held by adult female. Note that the baby is compressed by head splints inside the cradle (Teotihuacán, Veracruz; adapted from Musée de Quai exhibit, Paris, France; drawing by B. Ceballos)

are approximately parallel to each other (Fig. 4.4a). The produced overall formation of the vault is typically broad and inclined backwards. That is why the classification system by Imbelloni identifies this technique with the tabular oblique type.

Head splinting was constrained to certain areas of Mesoamerica and on the whole was practiced much less in native lore than cradleboarding. Accordingly, head compressors are much less represented in the pre-Hispanic imagery than compression cribs. The former appear mostly in Classic period figurines from Teotihuacán, the Gulf Coast and the Maya area, where different forms of head elongation were in vogue (Fig. 4.11). Just like the ceramic sculptures of cradleboards, also many of the figurines of children with head compressors originally served as whistles, attesting to potentially similar uses as for cradleboard figurines (Tiesler 1999, 2012a).

In the Maya area, the few portraits of rigid head shapers appear mostly in scenes of female caretakers interacting with their infant. Sometimes, it is a narrow plank, in other cases a broad tablet that is strapped tightly to the forehead by the use of different arrangements of straight or crossed fastenings. It remains unclear in most arrangements, whether there is a rigid implement in the back of the head or if the fastening of the frontal head compressor actually rested on the occipital plane.

The fastening of the front board by occipital constriction, perhaps mediated by cushions, would lead to curvo-occipital forms of oblique shapes, similar to the device still recently employed by Shipibo mothers in Peru. Conversely, inclined mimetic shapes are the likely result when two pairs of upper and lower strings appear to cross over the head and cause a double restraint of the head's back. This arrangement seems to be the most feasible pressure mechanism that is represented in a number

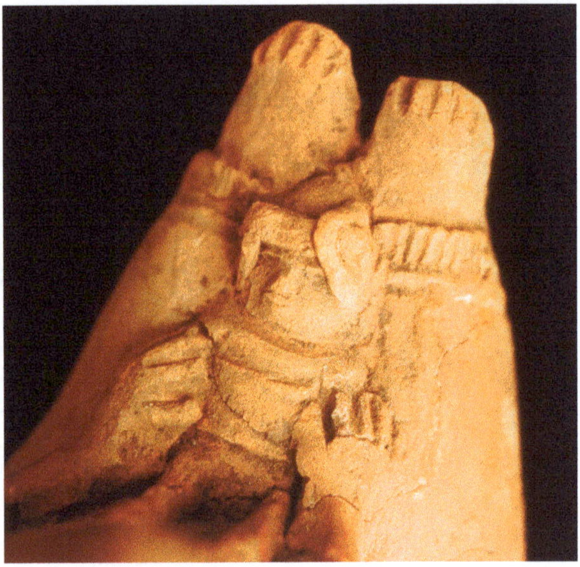

Fig. 4.11 Head board in place on the forehead of an infant that rests in the lap of an eldery woman (The Museo Popol Vuh, Universidad Francisco Marroquín, Guatemala; photo by V. Tiesler)

of Maya figurines (Figs. 4.11 and 4.12; see also Tiesler 1999). In these cases, the resulting head proportions are prone to be restricted in bilateral width when compared to classic all-rigid head compressors, such as the ones from Teotihuacán (Fig. 4.10). These front-and-back squeezers resemble the head supports described by Imbelloni (1933) and Weiss (1961, p. 4) for the Andean Humahuaca or Lambayeque.

A further question concerns the role of head splinting not in elongated oblique head shapes but in high and narrow head forms, assigned by Imbelloni as pseudocircular tabular erect shapes (Dembo and Imbelloni 1938, p. 292). These play a role in ancient Lambayeque culture, where erect head splints have been represented as a large occipital board that is fastened to the forehead by slings that cross over the sides and the top of the head. This device is shown to result in a shortened cranial vault. In the Mesoamerican pictorial record, similar headwear appears frequently in Olmec and Teotihuacán subadult portraits, although, it cannot be distinguished beyond doubt if the headgear was functional or purely ornamental. We remind that head compressors often appear similar in style and shape to the headgear worn by grownups in the Mesoamerican figurative repertoire. This is obvious in Teotihuacan, for example, where long flat pieces of board are carried by many female adults the same way as they appear adjusted on their infants' heads. If we believe the figurative record, infant head splinting was performed in Teotihuacan, while the babies rested in their crib. Also, the artists of the Maya Usumacinta basin would sometimes copy the mother's head gear in her infant's outfit, as smallscale figurines appear to communicate.

Still unique is the ceramic Mesoamerican head compressor that has been published recently by Carlos Jácome et al. (2013, p. 8) for of El Tropél, Colima. This is an archaeological site in Western Mexico that had been settled during the Pre-Classic or Classic period. Among the recovered burias lay a one and a half old infant in a

Fig. 4.12 Infant head board on infant sitting on the hip of a female caretaker (The Museo Popol Vuh, Universidad Francisco Marroquín, Guatemala; photo by V. Tiesler)

lateral prone position. Still in place rested what the authors hold to be a ceramic head compressor. Indeed suggestive is the peculiar form of the artifact, which adapts to the artificial curvature of the neck. Upon discovery, it was still in touch with the little one's neck. In the living baby and toddler, this ceramic valve should have been suitably pressed against the occiput by help of straps or bandages (see also Fig. 2.13).

4.4.3 Wraps and Strings

When constriction complements front-and-back rigid constraint, Imbelloni (Dembo and Imbelloni 1938) speaks of *pseudocircular* varieties. The complementary annular compression of the skull is accomplished with bands, belts or strings of cloth that are either attached directly to the compression boards, as described above, or may have been applied separately during the shaping procedure with the infant. Possibly, tight wraps maintained the visible result of the modeling process and avoided physiological growth re-compensation after the compression boards had been removed (Sect. 3.2.7). This is suggested at least by the Classic period Usumacinta iconography around Bonampak and Palenque, which shows toddlers who frequently wear a head wrap sitting on the hip of their female caretaker (Miller 1995, pp. 52–53; Fig. 4.12). In the Maya figurative repertoire, these horizontal wraps appear again similar in style and shape to the headdresses worn by adults and therefore we do not know if they are adornments or had a constricting role in infants. Imbelloni describes pseudocircular forms in tabular erect and oblique formations (Dembo and Imbelloni 1938, pp. 271–272). In Mesoamerica, the pseudocircular erect shapes characterize for example the pear-shaped "Olmec" head form, which visually divides the vault into an upper and lower part. The mechanism for producing this pear-shaped head formation is still under discussion and both body and head compressors are cited as Olmec devices (see Chap. 7). Lastly, also secondary sagittal bands can create bone furrows or grooves, which visually divides the vault into two hemispheres on the level of the sagittal and/or coronal sutures and has been described as part both of cradling and head splinting (Comas and Marquer 1969; Imbelloni 1933; Weiss 1961; Yépez 2006).

4.4.4 The Role of Suprainiac Depressions

Related to head modeling also are the round depressions and grooves that are frequently observed in the central part of the occipital squama of pre-Hispanic cranial series. They have been described in the literature either as suprainiac depressions, occipital lytic lesions or trephination marks (Curtin 2007; Holliday 1993; Kato et al. 2007; Lagunas 1974; Stewart 1976; Tiesler 2006; Weiss 1981). In Mesoamerica, where these marks are common, they are closely related to cranial modeling, although some few specimens with suprainiac lesions have been documented also in naturally curved crania (Tiesler 1999, 2006). Usually, these indentations appear right above the occipital eminence and vary widely in depth, contours and extensions. Some are shallow and hardly visible on the bony surface, whereas others are deep, their borders being clearly delineated.

There are also documented cases of symmetrically paired, geometric lateral depressions that appear on both lateral wings of occiput and the parietals right above (Fig. 4.13). Their geometric outline and paired appearance suggests they were imprints of the protruding margins of a posterior compression tablet. Other more centrally located occipital depressions appear either directly beneath the occipital flattening (Lagunas 1974; Tiesler 2006; Weiss 1981; see also Romano 1975 on a different case of vault thinning fom Teotenango, Mexico State). Interestingly, the position of the indentation also may vary according to the vector of the flattening. It is probably no coincidence that the depressions of the external bone plate appear not in the suprainiac area but in the obelical region above lambda in many superior flattenings from Veracruz, implying that the depressions are, in fact, associated to the compression plane by whatever mechanism. Although, there seems to be no direct

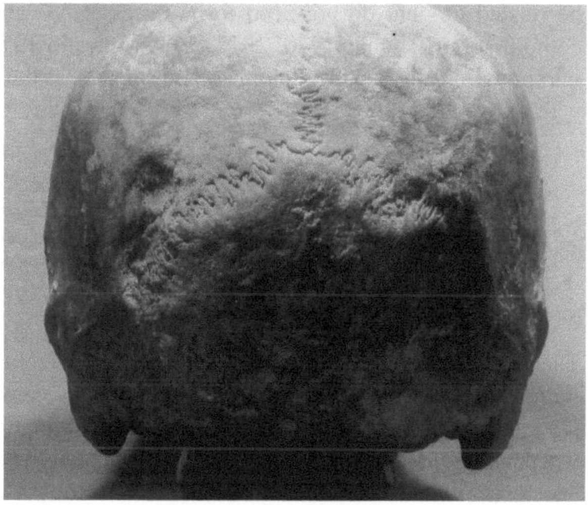

Fig. 4.13 Bilaterally situated supra-inial depressions in an elongated skull from Jaina, Campeche (DAF/IANH; photo, V.Tiesler)

correlation, *per se,* between their presence and the type or degree of vault modification by itself, suprainiac indentations tend to be deeper and larger when the occipital bun is severely flattened. Their characteristics also appear to be more noticeable in certain pre-Hispanic areas and sites, suggesting that specific local shaping traditions played a role.

Several suprainiac indentations are pathologically altered, showing healed or unhealed lithic resorptions or periosteal reactions (see Sect. 3.3.2; Fig. 3.8). Some suprainiac lesions penetrate the bone plate above inion completely, which has led to interpreting some of these extreme expressions as the result of occipital trephining in newborns. Decades ago, Pedro Weiss (1967) found similarities between these Mesoamerican suprainiac lesions and probable suprainiac trephinations in several mummified heads from Peruvian Chancay and Chimu, including subadults, in which Weiss observed "scars over the scalp" (Weiss 1981, p. 206, 1967, p. 24–25). He concludes his cultural comparison by assuming that trephining in the form of suprainiac scraping during first infancy was known both among the Andean and Mesoamerican cultures.

This interpretation sparked controversy in the academic community. Authors such as T. Dale Stewart (1967) and Diane Holliday (1993), prefer to regard suprainiac lesions more as the inadvertent results of modeling apparatuses instead. These instruments may damage the compressed soft tissue by generating ischemic ulcers and infections, which are transmitted to the underlying bone, leading to localized necrosis. More recent studies have suggested that extrinsic intervention (although rare), intrinsic reactions, or purely biomechanical forces may condition suprainiac depressions (Curtin 2007; Kato et al. 2007; Tiesler 2006).

Another question concerns the role of head compression mechanisms in the origin of occipital depressions, especially in view of those skulls that bear these marks but are un-modeled. Some authors have argued that suprainiac indentations might be related to—but do not need to be necessarily—a direct expression of cranial compressors (Lagunas 1970, 1972, 1989; Serrano 1973; Tiesler 2006). The latter interpretation aligns with ethnohistorical sources, as the statement by Francisco Paso y Troncoso (1926, p. 25):

> *Se caracterizan por su modesta presencia física, por el color pardo, por los grandes ojos, por la frente amplia, por la nariz, por la nuca plana aunque esta se debe a la acción de los padres [...] (ellos) consideran de hecho que sea un indicador de belleza las frentes pequeñas y ricas de cabellos y la nuca prácticamente inexistente que viene comprimida por el obstetra (las parteras) por medio de la aplicación de un peso desde cuando ven la luz, cuando el cráneo es tierno y mantiene esa forma cuando el niño viene depositado supino en la cuna" [...].*[2]

The chronicler suggests that midwives applied weights on the occipital eminence immediately after birth and prior to initiating its compression in cradleboard devices. This last scenario would be functional in commencing occipital compression, when taking into consideration the convex shape of the occipital emminence and the resistence of the basicranium which unites with the softer desmal vault on the height of the occipital emminence (Chap. 3). It would make sense, therefore, to prepare a centrally located flattened spot to facilitate the ensuing symmetrical compression of the back in the spot of increased anatomical resistence. I will take up this last idea again in connection with the practices, roles, and meanings in Chap. 6.

In synthesis, it appears that the host of manifestations of suprainiac depressions in different head shapes, sites, and regions, expresses different possible origins, compression techniques used by the practitioners, and day-to-day circumstances. These identify organic biomechanical reactions, as may be imprints of the rigid head compressor itself, or any hard object that may underly it. Others should relate to extrinsic action in the form of conservative cure or trephining; still others should express possible local side effects of the compression process itself.

4.5 Contextualizing Head-Shaping from Archaeological Records

Bioarchaeology is the most suitable approach to deducing broader patterns of past human cultural behavior from artificial skull modifications. This line of research is inherently interdisciplinary in that it examines skeletal remains as part of archaeological contexts (Buikstra 2006). The integration of the human biological substrate from the material record with other sources of information allows bioarchaeological studies to conveniently combine different data-sets, with lenses of approximations according to time and geography. These approximations are facilitated in most cultural

[2] The original Latin version reads: "medioci constant corpais habitudine colore frisco ocuis magnis fronte navibus, plano occiputio, quaquarium parentens hos fiat indishia... ad pularitudianaum apectare puntuant frontes parvas et refetas capillis ac fire nulum occipitum quod eneris quo que ferendi causa deprimitus unión cal < varia et teenrrina servatuga en figura sipinis lascatibus in canis..."

frames where human vestiges constitute central elements of mortuary assemblages and express the individual of the past as a central analytical unit for exploring collective behavior. Thus, regional bioarchaeological reviews of head shaping embrace wider social structures and dynamic social change by combining and patterning representative cranial series from culturally defined areas of the past. On a smaller scale, cities or communities with documented mortuary records decode local customs and their assimilation or rejection of regional ways of enactment or looks. This idea, in turn, introduces notions of social tensions and change. Still more focused are excavated house units or patio groups, which ideally provide punctual glimpses on head shaping as a family practice and in the cycling of individual lives in their sometimes capricious motions. Specifically eloquent is the decoding of gender expression and their role in child rearing and social integration.

On an analytical level, the variables that distinguish individual and collective qualities related to the head-shaping practices (as materialized in the bony skull), confer importance on vital information from the skeleton, such as sex and age-at-death. To carry out these examinations, bioarchaeology combines techniques, methodologies and concepts derived from physical anthropology and archaeology. The latter tends to assign value to the mortal depositories of the dead and their contents, their location and chronology. In some cultural ambits, he funerary attributes may warrant cautiously drawn inferences regarding the social roles played by the individual before death. In Mesoamerican burial contexts, for example, correlates for vertical and horizontal ascription have been prominently applied to tha study of the mortuary record, either scored by individual contexts or processed as part of a multivariate matrix (Welsh 1988; Wright 2006). As for the Classic Maya, relevant status markers include the presence of tomb architecture, container burials or inclusion of exotic materials, such as ambar, cinnabar, jadeite or obsidian (Krejci and Culbert 1995; Tiesler 1999, p. 106). Naturally, these static attributes cannot be more than simplistic approximations to the complexity involved in the multi-tiered social spheres that characterize most of Mesoamerica.

In conclusion, scrutinizing head-shaping practices from the perspective of bioarchaeology shows even greater potential when examined jointly with additional sources of non-archaeological information, gleaned from health sciences and cultural studies. The first acknowledges cranial plasticity, neurology, growth and morphology while the latter warrants a closer look at ideology and the body, gender, ritual performance, social structure, and even at ancient aesthetics and constructed beauty (Chaps. 2, 3). The feasibility of including non-archaeological cultural data-sets in this type of research, naturally depends on the cultural expressions of the past society under examination, their written record and permanence. In the case of the data-rich Mesoamerican study environment, interdisciplinary cultural generalizations are facilitated by a host of additional fields, like, ethnography, iconography, glyphic, ethnohistorical, and semiotic studies. Here, the combination of the bioarchaeology of head-shaping traditions with other sources of information, are suitable points of departure not only to understand the deeper underlying cultural meanings of skull modifications by themselves. Beyond that, they are capable of tracing the underlying long-lasting social dynamics and cultural and ethnic undercurrents that head shapes may express in the form of variety vs. homogeneity, trends, and continuity, frequency vs. scarcity and social distinction. It is this double goal of comprehending the roles of head-modification practices *per se* and exploring the underlying age, guides, social and ideological trends they express in Mesoamerica (and Latinamerica in general), which sets the stage and tempo of the second part of this volume.

References

Allison, Marvin J., Enrique Gerszten, Juan Munizaga, Calogero Santoro, and Guillermo Focacci. 1981. La práctica de la deformación craneana entre los pueblos andinos precolombinos. Chungara 7:238–260.

Anton, Susan C. 1989. Intentional cranial vault deformation and induced changes of the cranial base and face. American Journal of Physical Anthropology 79:253–267.

Armas, Juan Ignacio de. 1885. Les crânes dits déformés; memoire lu en espagnol a la Societé Anthropologique de la Havane. Cuba: El Fénix.

Arnold, Wolfgang H., Victoria A. Fedorischeva, Ella A. Naumova, and Nicolay I. Yabluchansky. 2008. Craniometric measurement of artificial cranial deformation in eastern European skulls. Anthropologischer Anzeiger 66:139–146.

Barrientos, G., and L. L'Heureux. 2009. Una aproximación bioarqueológica al estudio del poblamiento prehispánico tardío del sudeste de la región Pampeana. Revista Intersecciones en Antropología 2:3–18. Buenos Aires: UNCPBA.

Bautista, Josefina. 2004. Evidencias de deformación cefálica intencional en figurillas prehispánicas de México. Ph.D. diss., Universidad Nacional Autónoma de México.

Béguelin, Marien, Valeria Bernal, Mariano Del Papa, Paula Novellino y Gustavo Barrientos. 2006. El poblamiento humano tardío del sur de Mendoza y su relación con el norte de Patagonia: una discusión bioarqueológica. Anales de Arqueología y Etnología 61:5–25.

Björk, Arne, and Lise Björk. 1964. Artificial deformation and craniofacial asymmetry in ancient Peruvians. Journal of Dental Research 43(3):353–362.

Blom, Deborah E. 2005. Embodying borders: Human body modification and diversity in Tiwanaku society. Journal of Anthropological Archaeology 24:1–24.

Boas, Franz. 1890. Cranium from Progreso, Yucatan. Proceedings of the American Antiquarian Society 6:350–357.

Buikstra, Jane E. 2006. A historical introduction. In Bioarchaeology. The contextual analysis of human remains, ed. Jane E. Buikstra and Lane A. Beck, 7–25. Burlington: Academic Press.

Buikstra, Jane E., and Douglas Ubelaker, (ed.). 1994. Standards for data collection from human skeletal remains. Fayetteville: Arkansas Archaeological Survey Research Series No. 44.

Cavatrunci, Claudio, Giuseppe Orefici, and Claudia Terenzi. 1992. Centro America. Tesori d'Arte delle Civiltá Precolombiana. Milán: Fabbri.

Cheverud, James M., and James E. Midkiff. 1992. Effects of fronto-occipital cranial reshaping on mandibular form. American Journal of Physical Anthropology 87:167–171.

Cheverud, James M., Luci A. P. Kohn, Lyle W. Konigsberg, and Steven R. Leigh. 1992. Effects of fronto-occipital artificial cranial vault modification on the cranial base and face. American Journal of Physical Anthropology 88:323–345.

Clark, J.L., S. D. Dobson, S. C. Antón, J. Hawks, K. L. Hunley, and M. H. Wolpoff. 2007. Identifying artificially deformed crania. International Journal of Osteoarchaeology 17:596–607.

Comas, Juan. 1958. La deformación cefálica intencional en la región del Ucayali, Perú. In Miscelánea Paul Rivet, 109–119. México.

Comas, Juan. 1960. Datos para la historia de la deformación craneal en México. Historia Mexicana 36:509–520.

Comas, Juan. 1970a. José Imbelloni (obituario). Anales de antropología 7:290–291.

Comas, Juan. 1970b. History of physical anthropology in Middle America. In Physical Anthropology. Handbook of Middle American Indians, Vol. 9, ed. R. Wauchope and T. Dale Stewart, 3–21. Austin: University of Texas.

Comas, Juan, and Paulette Marquer. 1969. Crânes déformés de l'île de Sacrificios (État de Veracruz, Mexique). Bulletins et Mémoires de la Société d'anthropologie de Paris 4(3):209–257.

Crist, Thomas A. J., Arthur Washburn, Hydrow Park, Ian Hood, and Molly A. Hickey. 1997. Cranial bone displacement as a taphonomic process in potential child abuse cases. In Forensic taphonomy. The postmortem fate of human remains, ed. William D. Haglund and Marcella H. Sorg, 319–336. Florida: CRC Press.

Curtin, Joane. 2007. On the etiology of occipital lytic lesions in artificially deformed crania. Supplement to the paleopathology newsletter: Annual meeting 2007, 17–18. Philadelphia: Paleopathology Association.

Dávalos, Eusebio. 1951. La deformación craneana entre los tlatelolca. B.A. and Masters diss., Escuela Nacional de Antropología e Historia, Instituto Nacional de Antropología e Historia.

Dávalos, Eusebio. 1965. La osteología mexicana en el Museo del Hombre. In Temas de antropología física, ed. Eusebio Dávalos, 103–116. Mexico: Instituto Nacional de Antropología e Historia.

Dávalos, Eusebio, and Arturo Romano. 1965. Las deformaciones corporales entre los mexicas. In Temas de Antropología Física, ed. Eusebio Dávalos, 75–102. Mexico: Instituto Nacional de Antropología e Historia.

Delisle, Fernand. 1880. Contribution a l'étude des déformations artificielles du crâne. Ph.D. diss., Paris.

Delisle, Fernand. 1902. Les déformations artificielles du crâne en France, carte de leur distribution. Bulletins et Mémoires de la Société d'Anthropologie de Paris 3(serie V):111–166.

Dembo Adolfo, and José Imbelloni. 1938. Deformaciones intencionales del cuerpo humano de carácter étnico. Buenos Aires: Biblioteca Humanior.

Dingwall, Eric J. 1931. Artificial cranial deformation. A contribution to the study of ethnic mutilations. London: Bale & Sons & Danielsson.

Duncan, William N. 2009. Cranial modification among the maya: Absence of evidence or evidence of absence?. In Bioarchaeology and identity in the Americas, ed. Kelly Knudson and Christopher Stojanowski, 177–193. Gainesville: University Press of Florida.

Duncan, William, and Charles Hofling. 2011. Why the head? Cranial modification as protection and ensoulment among the maya. Ancient Mesoamerica 22:199–210.

Falkenburger, Fédéric. 1938. Récherches anthropologiques sur la déformation artificielle du crâne. Revista de Antropología de la Universidad de Tucumán 1:1–70.

Flower, William H. 1881. Fashion in deformity. London: Nature Series.

Foville, Achille. 1834. Influence des vê tements sur nos organes. Déformation du crâ ne résultant de la méthode la plus générale de couvrir la tê te des enfantes. Paris.

Francisco Paso y Troncoso del. 1926. Códice de la Real Academia de la Historia (facsimile edition, Museo Nacional de México, fols. 46, 47 de Anticuarios Nova Hispania).

Geller, Pamela. 2006. Altering identities: Body modifications and the pre-Columbian maya. In Social archaeology of funerary remains, ed. Rebecca Gowland and Christopher Knüsel, 279–291. Oxbow: Oxford.

Gervais, Véronique. 1989. Déformations artificielles de crânes préhispaniques au Guatemala et au Mexique. Ph.D. diss., Universidad de Caen.

Gervais, Véronique. 2001. El uso del mecapal en Guatemala y sus consecuencias sobre la columna cervical (primeras observaciones). Estudios de Antropología Biológica 10: 496–501.

Gómez-Valdés, Jorge, Josefina Bautista, and Arturo Romano. 2007. Morfometría geométrica aplicada al estudio de la deformación cefálica intencional. Estudios de antropología biológica 13:117–134.

Gosse. Louis. 1855. Déformations artificielles du crâne. Bailliere: Paris.

Graham, John M. Jr, Jeannie Kreutzman, Dawn Earl, Andy Halberg, Carlos Samayoa, and Xiuqing Guo. 2005. Deformational brachycephaly in supine-sleeping infants. Journal of Pediatrics 146(2):253–257.

Hansen Soren. 1919. On posthumous deformation of fossil skulls. Man 19:121–124.

Holliday, Diane Y. 1993. Occipital lesions: A possible cost of cradleboards. American Journal of Physical Anthropology 90(3):283–290.
Hrdlička, Aleš, and Carl Lumholtz. 1912. Artificial deformations of the human skull with special reference to America. Acts of the 17 International Americanistic Congress, 147–149. Buenos Aires.
Imbelloni, José. 1925. Sobre el número de los tipos fundamentales a los que deben referirse las deformaciones craneanas de los pueblos indígenas de Sudamérica. Anales de la Sociedad Argentina de Estudios Geográficos GAEA 1:183–197.
Imbelloni, José. 1930. Die Arten der künstlichen Schädeldeformationen. Zusammenfassender Bericht. Anthropos 25:801–830.
Imbelloni, José. 1933. Los pueblos deformadores de los Andes. Anales del Museo de Historia Natural 37:209–254.
Imbelloni, José. 1938. Formas, esencia y metódica de las deformaciones cefálicas intencionales. Revista del Instituto de Antropología de la Universidad de Tucumán 1(1):5–37.
Jácome, Carlos, Fernando González, and Laura Almendros. 2013. El Tropel, Colima. Hallazgo de un deformador craneal de cerámica. Arqueología Mexicana 121:8.
Jaén Esquivel, María T. 1962. Comparación de los métodos para estimar la capacidad craneana. B.A. and Master diss., Escuela Nacional de Antropología e Historia, Instituto Nacional de Antropología e Historia.
Kato, Katsutomo, Ken-ichi Shinoda, Yoshikazu Kitagawa, Yoshitaka Manabe, Joichi Oyamada, Kazunari Igawa, Hild Vidal, and Atsushi Rokutanda. 2007. A possible case of prophylactic supra-inion trepanation in a child cranium with an auditory deformity (pre-Columbian Ancon site, Perú). Anthropological Science 115:227–232.
Kohn, Luci Ann, Steven Leigh, and James Cheverud. 1995. Asymmetric vault modification in Hopi crania. American Journal of Physical Anthropology 98:173–195.
Krejci, Estella, and Patrick Culbert. 1995. Preclassic and Classic burials and caches in the Maya Lowlands. In The emergence of Lowland Maya civilization, ed. Nikolai Grube, 103–116. Acta Mesoamericana 8, Möckmühl: Verlag Anton Saurwein.
Lagunas, Zaid. 1970. Nota sobre el hallazgo de cráneos con lesión supra-iniana en Cholula. Boletín del Instituto Nacional de Antropología e Historia 39:1–3.
Lagunas, Zaid. 1972. La trepanación suprainiana en cráneos de Cholula, Puebla. Comunicaciones Proyecto Puebla Tlaxcala 8:47–50.
Lagunas, Zaid. 1974. Observaciones recientes sobre la lesión suprainiana. Boletín del Instituto Nacional de Antropología e Historia 11(2nd series):47–54.
Lagunas, Zaid. 1989. Los antiguos habitantes de Cholula: Prácticas osteoculturales. Notas Americanas 11:8–50.
Lozada, María Cecilia. 2011. Marking ethnicity through premortem cranial modification among the Pre-Inca Chiribaya, Perú. In The bioarchaeology of the human head. Decapitation, decoration, and deformation, ed. Michelle Bonogofsky, 228–240. Gainesville: University Press of Florida.
Magitot, Émile. 1884. Essai sur les mutilations ethniques. CRAA, 9th Session, 549–622. Lisbonne.
Manríquez, Germán, Fermín E. González, Juan Carlos Salinas, and Oscar Espoueys. 2006. Deformación intencional del cráneo en poblaciones arqueológicas de Arica, Chile: Análisis preliminar de morfometría geométrica con uso de radiografías craneofaciales. Chungara 38(1):13–34.
Martínez, Blanca Lilia. 2009. La deformación cefálica intencional tipo tabular, variante superior, en el Zapotal, Veracruz. Estudios de Antropología Biológica 14:489–502.
Masefield, John. 2005 [1920–1923]. Enslaved and other poems by John Masefield. Montana: Kessinger Publishing Reprints.
Mason, Otis T., and J. H. Porter. 1889. Cradles of the American aborigines with notes on the artificial deformation of children among savage and civilized peoples. Washington: Government Printing Office.
McNeill, William, and George Newton. 1965. Cranial base morphology in association with intentional cranial vault deformation. American Journal of Physical Anthropology 23(3):241–253.
Mendieta, Gerónimo de. 1997. Historia eclesiástica Indiana, Vol. 1. México: Consejo Nacional para la Cultura y las Artes, Mexico.

Miller, Mary E. 1995. Bonampak. Arqueología Mexicana 16:48–55.
Morton, Samuel G. 1839. Crania americana. Philadelphia: John Pennington.
Morton, Samuel G. 1841. Mexican crania: Otomí, Chichimec, Tlascalan, Aztec. Proceedings of the National Academy Sciences 1:50.
Moss, Melvin L. 1958. The pathogenesis of artificial cranial deformation. American Journal of Physical Anthropology 16(3):269–286.
Munizaga, Juan R. 1974. Deformación craneal y momificación en Chile. Anales de Antropología 11:329–336.
Nelson, Greg C., and Felicia C. Mandimenos. 2010. Obelionic cranial deformation in the Puebloan Southwest. American Journal of Physical Anthropology 143:465–472.
Neumann, George K. 1942. Types of artificial cranial deformation in the eastern United States. American Antiquity 3:307–318.
Pardal, Ramón. 1938. La deformación intencional del cráneo por los indios de América. Actas Ciba 3:67–81.
Pereira, Grégory. 1999. Potrero de Guadalupe: anthropologie funéraire d'une communauté prétarasque du nord du Michoacán, Mexique. Oxford: British Archaeological Research International Series 816.
Pereira da Silva, Miya A. 1994. Du bandeau aux couvre-chefs ou les origines de la 'déformation toulousiaine'. Musée de'l Homme 4:51–55.
Pérez, Sergio Iván, Claudia Della, Paula Novellino, Paula N. González, Valeria Bernal, Estela Cuneo, and Adam Hajduk 2009. Deformaciones artificiales del cráneo en cazadores-recolectores del Holoceno medio-tardío del noroeste de Patagonia. Magallana 37:77–90.
Pomeroy, Emma, Jay T. Stock, Sonia R. Zakrzewski, and Marta Mirazón. 2010. A metric study of three types of artificial cranial modification from North-Central Perú. International Journal of Osteoarchaeology 20:317–334.
Rhode, Matthew, and Bernardo T. Arriaza. 2006. Influence of cranial deformation on facial morphology among prehistoric south central Andean populations. American Journal of Physical Anthropology 130:462–470.
Roberts, Charlotte A. 2009. Human remains in archaeology: A handbook. New York: Cambridge University Press.
Rogers, Spencer. 1975. Artificial deformation of the head. New World examples of ethnic mutilation and notes on its consequences. San Diego: San Diego Museum Papers 8, Museum of Balbao Park.
Romano, Arturo. 1965. Estudio morfológico de la deformación craneana en Tamuín, S.L.P., y en la Isla del Idolo, Veracruz. Serie de Investigaciones 10. Mexico: Instituto Nacional de Antropología e Historia.
Romano, Arturo. 1973. Deformación cefálica intencional en la población prehispánica de Cholula, Puebla. Comunicaciones Proyecto Puebla Tlaxcala 8:49–51.
Romano, Arturo. 1974. Deformación cefálica intencional. In Antropología Física, Época Prehispánica, ed. Juan Comas, 197–227. Mexico: Instituto Nacional de Antropología e Historia.
Romano, Arturo. 1975. Dólicocéfalos prehispánicos de Teotenango, México. In Memorias del XLI Congreso Internacional de Americanistas, 1974, 417–463. Mexico.
Romano, Arturo. 1977. Los cráncos dcformados dc El Zapotal. In Las Esculturas en Terracota de El Zapotal, Veracruz, ed. Nelly Gutierrez Solana and Susan K. Hamilton, 31. Mexico: Instituto de Investigaciones Estéticas, Universidad Nacional Autónoma de México.
Romano, Arturo. 1980. The skull from El Pajon, Chiapas. Papers of the New World Archaeological Foundation 44: 95–114.
Romano, Arturo. 1987. Iconografía cefálica maya. In Memorias del Primer Coloquio Internacional de Mayistas, 1413–1474. Mexico: Universidad Nacional Autónoma de México.
Romano, Arturo. 1996. La craneología antropológica en México. In La antropología física en México. Estudio sobre la población antigua y contemporánea, ed. Sergio López, Carlos Serrano, and Lourdes Márquez, 35–54. Mexico: Instituto de Investigaciones Antropológicas, Universidad Nacional Autónoma de México.

Sánchez Saldaña, Patricia. 1971. Cuicuilco: Estudio osteológico de la población prehispánica. B.A. diss., Escuela Nacional de Antropología e Historia, Instituto Nacional de Antropología e Historia.
Saul, Frank. 1972. The human skeletal remains of Altar de Sacrificios. An osteobiographic analysis. Cambridge: Papers of the Peabody Museum of Archaeology and Ethnology, Vol. 63, no. 2, Harvard University.
Serrano, Carlos. 1973. La lesión suprainiana en Mesoamérica: Implicaciones arqueológicas. Estudios de la Cultura Maya 9:29–45.
Stephens, John, and Frederick Catherwood. 1963 [1843]. Incidents of travel in Yucatan, 2 vols. New York: Dover.
Stewart, T. Dale. 1937. Discussion and correspondence: Different types of cranial deformity in the Pueblo Area. American Anthropologist 39:169–171.
Stewart, T. Dale. 1939. A new type of artificial cranial deformation from Florida. Journal of the Washington Academy of Sciences 29(10):460–465.
Stewart, T. Dale. 1941. The circular type of cranial deformity in the United States. American Journal of Physical Anthropology 28:343–351.
Stewart, T. Dale. 1943a. Skeletal remains with cultural associations from the Chicama, Moche, and Virú Valleys, Perú. Proceedings of the United States National Museum. Smithsonian Institution 93:153–185.
Stewart, T. Dale. 1943b. Skeletal remains from Tajamulco, Guatemala. In Excavations at Tajamulco, Guatemala, ed. Bertha Dutton and Hilda R. Hobbs:111–114. Albuquerque: Monographs of the School of American Research 9, University of New Mexico.
Stewart, T. Dale. 1948. The true form of the cranial deformity originally described under the name "T ete Trilobée". Journal of the Washington Academy of Sciences 38(2):66–72.
Stewart, T. Dale. 1953. Skeletal remains from Zaculeu, Guatemala. In The ruins of Zaculeu, Guatemala, Vol. 1, ed. Richard B. Woodbury and Aubrey S. Trik, 295–311. Richmond: William Byrd.
Stewart, T. Dale. 1958. Skeletal remains from Venado Beach, Panama: Cranial deformity. Congreso Internacional de Americanistas 33(3):45–54.
Stewart, T. Dale. 1963. Deformity, trephining, and mutilation in South American indian skeletal remains. Handbook of South American Indians 6:43–48.
Stewart, T. Dale. 1975. Human remains from Dzibilchaltun, Yucatan, Mexico, with a review of cranial deformity types in the Maya region. Papers of the Middle American Research Institute, Vol. 31. New Orleans: Tulane University.
Steward, T. Dale. 1976. Are supra-inion depressions evidence of prophylactic trephination?. Bulletin of the History of Medicine 50(3):414–434.
Stojanowski, Christopher M. and Julie Euber. 2011. Technical note: Comparability of Hrdlička's catalog of crania data based on measurement landmark definitions. American Journal of Physical Anthropology 146(1):143–149.
Tiesler, Vera. 1998. La costumbre de la deformación cefálica entre los antiguos mayas: Aspectos morfológicos y culturales. Mexico: Instituto Nacional de Antropología e Historia.
Tiesler, Vera. 1999. Rasgos bioculturales entre los antiguos mayas: Aspectos arqueológicos y sociales. Ph.D. diss., Universidad Nacional Autónoma de México.
Tiesler, Vera. 2006. ¿Hubo trepanación en la antigua sociedad maya? Una apreciación regional. Antropología Física Latinoamericana 4:169–218.
Tiesler, Vera. 2007. Bases conceptuales para la evaluación de restos humanos en arqueología. Mexico: Universidad Autónoma de Yucatán.
Tiesler, Vera. 2010. "Olmec" head shapes among the Preclassic period maya and cultural meanings. Latin American Antiquity 21(3):290–311.
Tiesler, Vera. 2011. Becoming Maya: Infancy and upbringing through the lens of pre-hispanic head shaping. Childhood in the Past 4:117–132.
Tiesler, Vera. 2012a. Transformarse en maya. El modelado cefálico entre los mayas prehispánicos y coloniales. Mexico: Universidad Autónoma de Yucatán, Universidad Nacional Autónoma de México.

Tiesler, Vera. 2012b. Studying cranial vault modifications in ancient Mesoamerica. Journal of Anthropological Sciences 90:1–26.

Tiesler, Vera, Arturo Romano, Jorge Gómez-Valdés, and Annick Daneels. 2013. Posthumous body manipulation in the Classic period Mixtequilla: Reevaluating the human remains of Ossuary I from El Zapotal, Veracruz. Latin American Antiquity 24:47–71.

Torres-Rouff, Christina. 2002. Cranial vault modification and ethnicity in Middle Horizon San Pedro de Atacama, North Chile. Current Anthropology 43(1):163–171.

Torres-Rouff, Christina. 2003. Shaping identity: Cranial vault modification in the Pre-Columbian Andes. Ph.D. diss., University of California.

Trinkaus, Erik. 1982. Artificial cranial deformation in the Shanidar 1 and 5 Neandertals. Current Anthropology 23(2):198–199.

Trinkaus, Erik. 1983. The Shanidar Neanderthals. New York: Academic Press.

Virchow, Rodolphe. 1888. La craneologie américaine. Congrés International des Americanistes 251–262.

Virchow, Rodolphe. 1890. Sur la craneologie américaine. 7 Congrés International des Americanistes, 251–262, Berlín.

Weiss, Pedro. 1961. Osteología cultural. Prácticas cefálicas. Lima: Universidad Nacional de San Marcos.

Weiss, Pedro. 1962. Tipología de las deformaciones cefálicas de los antiguos peruanos, según la osteología cultural. Revista del Museo Nacional 31:15–42.

Weiss, Pedro. 1967. Ensayo de osteología cultural en Guatemala. Antropología e Historia de Guatemala 19:14–26.

Weiss, Pedro. 1981. La trepanación ritual suprainiana. Boletín de la Sociedad Mexicana de Historia y Filosofía de la Medicina 36:193–211.

Welsh, W. Bruce. 1988. An Analysis of Classic Lowland Maya Burials. London: International Series 40, British Archaeological Reports.

Winning, Hasso von. 1968. Process of head deformation shown by Mesoamerican figurines. The Masterkey 42(2):53–58.

Wright, Lori. 2006. Diet, health, and status among the Pasión maya. A reappraisal of the collapse, Vol. 2. Nashville: Vanderbilt Institute of Mesoamerica, Archaeological Series 2, Vanderbilt University.

Yépez, Rosaura. 2001. El Modelado cefálico intencional en los pobladores prehispánicos del barrio teotihuacano de la Ventilla. Exploración 1992–1994. Master diss., Universidad Nacional Autónoma de México.

Yépez, Rosaura. 2006. La práctica cultural de modelar la cabeza en dos culturas andinas del Antiguo Perú: Paracas y Chancay. Un estudio de los procesos de significación de la cabeza modelada intencionalmente. Ph.D. diss., Universidad Nacional Autónoma de México.

Yépez, Rosaura. 2009. El simbolismo de la modificación cultural de la cabeza en la cultura andina de Paracas del antiguo Perú. Estudios de Antropología Biológica 14(2):526–543.

Chapter 5
Source Compilation on Head-Shaping Practices in Hispanic America, by Pilar Zabala

This chapter transcribes and comments on some 100 individual text citations on Hispanic America, an anthology that covers five centuries (fifteenth to twentieth) of directly witnessed head-shaping practices (Tiesler and Aguirre 2011; see also Chap. 10 of this book). This extensive body of written testimonies comes from the geographical sphere first controlled by the Viceroyalties of New Spain and Perú, including the Caribbean region, and provides primary source material for further research on head practices after European contact. We may divide this area roughly into four sections according to the European colonization powers, strategies and timing: the Antilles, Portuguese Brazil, Mesoamerica, and Perú (Fig. 5.1). While the colonization strategies of the last two jurisdictions are similarly characterized by a despotic claim to streamline culture and transform the social fabric to the needs of the Spanish crown (which included the forced assimilation of culture and, more so, religion), the impenetrable forested areas east of the Peruvian Highlands remained largely isolated and unexplored for decades and centuries to come. From the Atlantic side of the continent, Brazilian establishments were mainly restricted to economic exploitation and limited to the coastal areas. Deeper incursions into the Amazon Basin did not occur until the seventeenth and eighteenth centuries (Fournier 1999; Lucena 2005, p. 102). More complex is the Caribbean history, where indigenous traditions either faced extinction along with their human carriers, or persisted on the more isolated islands.

As for ancient head-shaping practices, the historical extracts to be presented in the following sections, give a voice to native body practitioners, Iberian eyewitnesses, government analysts, adventurers, and naturalist bystanders, and lastly, to modern anthropologists who describe waning head traditions still being applied in the Amazon Basin and parts of Mesoamerica during the twentieth century. Most colonial documents come from the *Archivo Histórico Nacional* in Madrid and the *Archivo General de las Indias* in Seville, while the majority of transcriptions on head shaping, dated to postcolonial times, derive from the admirably thorough recompilation of historical sources by T. Eric Dingwall (1931). Care was taken to examine each transcription critically within its particular historical context to avoid oversimplification and interpretational biases.

V. Tiesler, *The Bioarchaeology of Artificial Cranial Modifications*,
Interdisciplinary Contributions to Archaeology 7, DOI 10.1007/978-1-4614-8760-9_5,

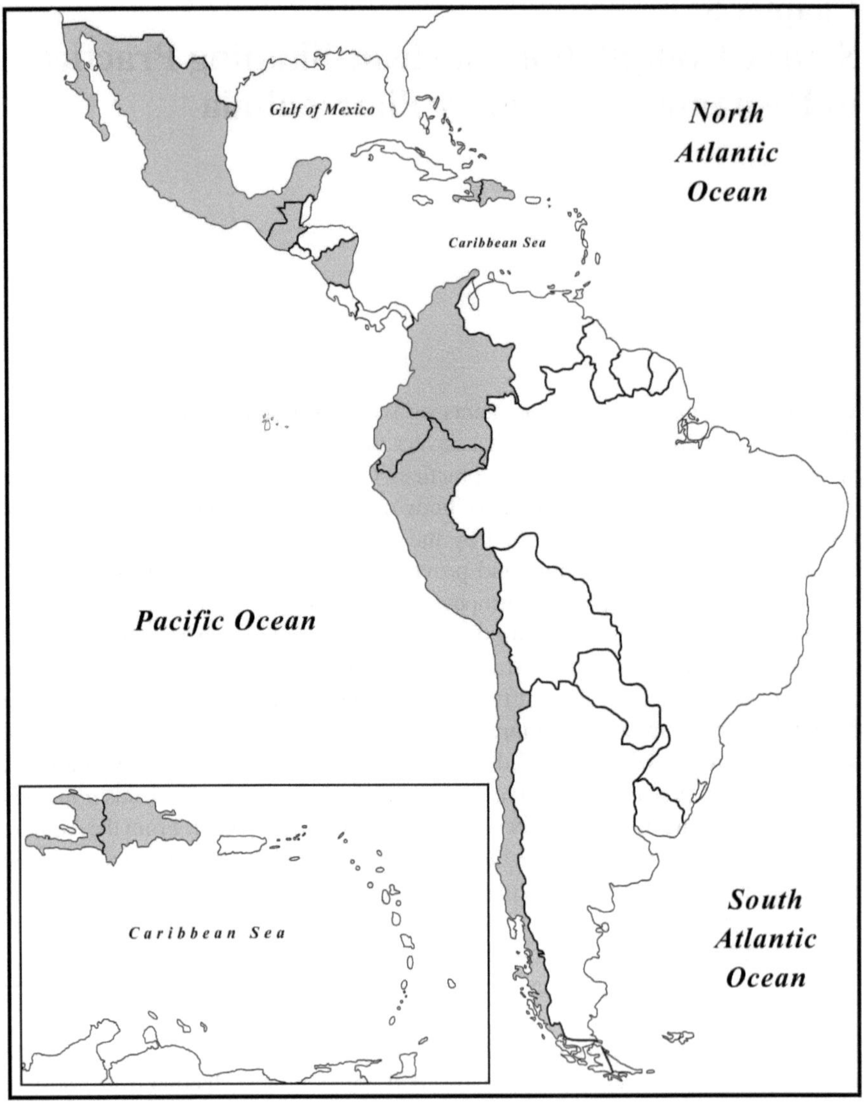

Fig. 5.1 The countries of Latin America (highlighted) staged cranial modification practices as described by early colonial references. (Drawing by V. Tiesler)

5.1 Cranial Modification in Colonial Sources

The custom of artificial cranial flattening constituted a cultural practice among the majority of native peoples that populated the American continent still at the time of European contact. Subsequently, as new civil and ecclesiastical powers were imposed by the colonizers from afar, head practices began to disappear. Especially, in the Iberian colonies, the new laws imposed on the natives were intended to eradicate ancestral customs, including the modification of head shapes on their newborns. Notwithstanding the acculturation efforts, these were not always effective, as shown here by their continued practice and historical mentions. Some Ibero-American historical accounts witness a gradual abandonment of the custom whereas others noted a rapid change. It appears that the custom disappeared earlier in the territory that was dominated by the Viceroyalty of New Spain; at least, this can be argued from the little information that we were able to gather from this region. However, given the uniformity of the cephalic shapes practiced among Mesoamerican natives at that time, there is also a real possibility that the custom may have passed unnoticed by most incoming Europeans, who might have considered the shortened and broadened head forms as a natural aspect of native physiognomy (see Chap. 10). This contrasts with the diversity of forms and the longer survival of the custom of head shaping among the Caribbean and South American groups under both Spanish and Portuguese rule (Fig. 5.1).

5.1.1 *The Caribbean Region*

Mentions of cranial modification are not numerous; however, it can be deduced from the timespan of chronicles and travel accounts that the tradition was enacted among native islanders and survived for a long time after Conquest.

5.1.1.1 La Hispañola

The first mention of cranial transformation practice among the indigenous populations of La Hispañola is attributable to Christopher Columbus himself, who includes it in a description of the physiognomy of the natives he encounters during his first voyage. He describes the form in which the mothers performed this operation on their newborns, affirming nonchalantly that the natives practiced it because they liked the visible aspect.

> Estos isleños aunque no muy altos eran de una estatura regular y proporcionada la cabeza aplanada porque desde niños, sus madres, se la ponían muy apretada entre las manos ó entre dos planchas de madera como en una prensa, de donde provenía que doblado el cráneo y criándose más espeso con este artificio se volvía el casco tan duro que los españoles hicieron pedazos más de una vez sus espaldas, queriendo descargar el golpe de tajo sobre la cabeza de estos infelices. Esta mala inconformidad de la cabeza y frente los agradaba mucho, y si se junta a eso que tenían las narices muy abiertas (...) todo este conjunto de facciones contribuía mucho á este aire salvage y feroz que se observaba en aquellos pueblos (Christopher Columbus in Vega 1826, p. 172).

Other references to the custom of head shaping come from Santo Domingo, also in La Hispañola. Eric Dingwall cites Thamara, whose work was published in the mid-sixteenth century, some 60 years after Columbus' earlier mention:

> In S. Domingo, Thamara noted the practice of deforming the head in 1556. He states that certain of the population have their foreheads narrowed artificially by lateral compression, and like other authors he comments upon the fact that under the constriction the eyes protrude from the head (saltar los ojos) (Dingwall 1931, p. 158).

Likewise, we found mentions of the custom among the Carib "archers" of San Juan, around the Xaragua lake of Santo Domingo. In this description, Fernández de Oviedo focused exclusively on the physical description of the "Caribs" as he reports the artificial manipulation of their heads, but offers no interpretation of the ritual motifs of this practice.

> Tornando a nuestro propósito digo que la color desta gente es lora: son de menor estatura que la gente de España comúnmente, pero son bien hechos e proporcionados, salvo que tiene las frentes anchas e las ventanas de las narices muy abiertas e lo blanco de los ojos algo turbio. Esta manera de frentes se hace artificialmente porque al tiempo que nasçen los niños, les aprietan las cabezas de tal manera en la frente y en el colodrillo, que como son las criaturas tiernas, las hacen quedar de aquel talle, anchas las cabeças delante e detrás e quedan de mala graçia (Fernández de Oviedo 1945 [1557], p. 137).

5.1.1.2 Island of Martinique

A few centuries later, Father Labat, was still able to observe more carefully, the custom among the native groups living in the Lesser Antilles of the eastern Caribbean and relates on his trip to the Antilles during the eighteenth century:

> todos eran bien hechos, bien proporcionados y de agradable fisonomía, aunque sus frentes parecían un tanto extraordinarias, debido a que se presentaban aplanadas y como hundidas. No nacen así, sino que obligan a las cabezas de sus niños a tomar esta figura, colocando sobre la frente de los recien nacidos una tablilla amarrada fuertemente por detrás de la cabeza, que dejan allí hasta que la expresada frente haya tomado consistencia, permaneciendo entonces aplastada, de modo que desde lo alto de la cabeza ven casi perpendicularmente por encima de ellos, pero la figura o la desproporción de la frente, hace que parezcan de tamaño regular (Montalvo 1884, pp. 17–18).

5.1.1.3 Guayana

We also found a mention from eighteenth century Guyana on this modification process among the children from the lands surrounding the Caribbean region. It was confirmed by a German soldier who participated in the revolt in Surinam and was published by Stedman:

> Most of these people esteeming a flat forehead as a mark of beauty, they compress the heads of their children, it is said, immediately after their birth like the Chactaws of North America (Stedman 1729, p. 398).

5.1.2 Viceroyalty of New Spain

Data from documentary sources and chronicles on the Mesoamerican native practice of cranial modification, are relatively few in number and most of them are very brief. Many are limited to a simple description of the technique used to transform the shape of the skull of newborns; others attempt a more culturally sensitive explanation. These interpretations differ according to author, as well as the ethnic groups to which they refer.

5.1.2.1 Mexica (Aztecs)

López de Gómara (1987 [1552]) conceives to the practice among the Nahua populations of New Spain in a way similar to how Christopher Columbus mentions it in his physical description of Hispañola natives, i.e., as an almost given part of their physiognomy. López de Gómara adds that the first phase of this practice was carried out on the newborn child by the midwife who attended the birth, after which the mother controlled the child's care until the head had acquired the desired shape. Gómara likewise made no observations of importance regarding the cranial modification process, simply declaring that it was a custom that was deemed desirable:

> Crían largo el cabello, lo ponen negro con tierra por gentileza y para que los mate los piojos. Las casadas se lo rodean a la cabeza con un nudo a la frente, las vírgenes y por casar lo llevan suelto y echado y atrás y adelante. Se pelan y untan todas, para no tener pelo sino en la cabeza y cejas y así tienen por hermosura tener la frente pequeña y llena de cabello y no tener colodrillo (. . .). Las parteras hacen que las criaturas no tengan colodrillo, y las madres las tienen echadas en cunas de tal suerte que no les crezca, porque se precian sin él (López de Gómara 1987 [1552], p. 451).

In the same way as López de Gómara, Paso y Troncoso describes the physical characteristics of the Aztecs, interpreting infant cranial transformations as an adornment, deemed beautiful in the eyes of native Nahua:

> Se caracterizan por su modesta presencia física, por el color pardo, por los grandes ojos, por la frente amplia, por la nariz, por la nuca plana aunque esta se debe a la acción de los padres (. . .) consideran de hecho que sea un indicador de belleza las frentes pequeñas y ricas de cabellos y la nuca prácticamente inexistente que viene comprimida por el obstetra por medio de la aplicación de un peso desde cuando ven la luz, cuando el cráneo es tierno y mantiene esa forma cuando el niño viene depositado supino en la cuna (Paso y Troncoso 1926, p. 25).

Some other references, although somewhat later in time, go beyond plain customary testimonies. Toward the close of the sixteenth century, indigenous chronicler Cristobal del Castillo transcribes an assertion written in Nahuatl, which relates the practice to military hierarchy, deeming that cranial transformation was related to aspects of courage, and high military rank.

> los guerreros, los varones, "los tehihuaque" (los que tienen cargo), los "quahiqueque" (los raspados de la cabeza) que se consideran altos grados en la jerarquía militar (Castillo 2001 [late sixteenth century], p. 99).

5.1.2.2 Purepecha/Chichimec

From the vast stretches of New Spain's western provinces comes the *Relación de Michoacán* (1977 [1541]). It provides a compilation of several accounts by natives living in the Michoacán province. Franciscan Friar Jerónimo de Alcalá probably transcribed these from the Purepecha language just 20 years after the Spanish Conquest. These first and third person testimonies relate dramatic conflicts and massive migrations in the territories around Lake Patzcuaro. There, the Tarascan settlers at that time participated in broader population movements between the Great Chichimeca to the north (whose populations did not have the habit of modifying the head) and local Purepecha stock (Pereira 1999, pp. 167–168). Also, artificially shortened heads are mentioned in these contexts and referred to as desirable attributes of bravery and gallantry of the local aristocracy ("los señores"). A man from the Patzcuaro area, called Hirípan, sadly exclaims in his monologue that he is ashamed of his rounded head and small stature:

> ¡Oh Hirípan: aunque soy de tal estatura y tan pequeño, y aunque tengo la cabeza redonda, que no es de valientes hombres, nunca me tengo que olvidar de aquella injuria de Hiuacha! (Relación de Michoacán 1977 [1541], p. 145).

Some sentences further down, the story's protagonist is described again:

> de pequeña estatura, y tenía la cabeza redonda. Que los que tenían de tal manera, no los tenían por valientes hombres, y por eso a los señores les allanaban las cabezas, y se las asentaban y hacían como tortas (Relación de Michoacán 1977 [1541], p. 145).

5.1.2.3 Totonac and Huaxtec

An additional reference is offered by Francisco Hernández for the eastern mountains of New Spain, toward the Gulf Coast of present-day Tamaulipas and Veracruz. He states that the Indians of the Panuco area thought that the head practice enabled them to carry greater weights and to transport larger loads during their workday. This assertion is clearly related to tumpline (*mecapal*) use, observed still among today's native population. It consists in a broad band that passes over the head is fastened to the load on their backs:

> Creen que es cosa bella tener las frentes pequeñas y cubiertas con los cabellos y casi ninguna nuca, la cual, para que puedan llevar carga, se le aplanan por las parteras en cuanto ven la primera luz, porque entonces la calavera es muy tierna y esa figura se conserva por los recién nacidos acostado en sus cunas (Hernández 2001 [1576], p. 111).

Another sixteenth century chronicler, Friar Bernardino de Sahagún, characterizes the artificial head form among natives from northern Veracruz, who "call themselves Huastec," as shortened, without mentioning whether these shapes stem from artificial modeling. This sketchy appraisal stands in surprising contrast to with the minute detail in which his chronicle, *Historia General de las cosas de la Nueva España* narrates other ethnically specific adornments, like hairstyles and headdresses.

> Estos "totonaques" están poblados a la parte del norte, y estos se dicen ser "guastecas". Tienen la cara larga y las cabezas chatas; y en su tierra hace grandísimos calores (Sahagún 2005 [sixteenth century], Vol. 3, p. 201).

We find a more nuanced explanation in the work of Walter Krickeberg, who attempts to explain the differences in head shapes between both groups, by translating the native terminologies used in mentioning this practice:

> Los totonacas ejecutaban el aplanamiento de la cabeza y en eso imitaban a los huaxteca. La expresión "Quavacaltic" (de uacalli, una canasta que se echa a la espalda) se usa aquí, como en el párrafo de Sahagún que trata de los huaxteca, para describir la deformación de la cabeza, ... A las cabezas de los huaxteca se les llama anchas y aplanadas (patlachtic), aquí se acentúa lo largo de la cara. Molina traduce "melatic" como "cosa derecha y luenga" (de Melaua, "enderezar alguna cosa tuerta"), y "mimiltic", o "cosa rolliza, como rolliza, como pilar" (Krickeberg 1933, p. 47; see also Yépez 2001).

Krickeberg concludes that:

> Eso solamente puede significar que los totonaca ejecutaban otra forma de deformación de la cabeza que los huaxteca, y se tiene la tentación de emplear la antigua distinción de "long-heads" (cabezas largas anulares) "flatheads" con esas dos tribus (Krickeberg 1933, p. 47).

5.1.2.4 Maya

Friar Diego de Landa counts among the few early historiographers to go into detail when recounting cranial modification practices among the Yucatecan Maya of the sixteenth century. Expressing himself in a clearly reproachful attitude, he contemplates the pain that the head compression must have caused the infant and the danger it could have caused to its life.

> The women brought up their little children with all the roughness and nakedness in the world, since four or five days after the infant was born, they placed it stretched out upon a little bed, made of sticks of osier and reeds; and there with the face [down], they put its head between two small boards one on the back of the head and the other on the forehead, between which they compressed it tightly, and here they kept it suffering until at the end of several days, the head remained flat and molded, as was the custom of all of them. There was so great inconvenience and danger to the poor children that some were in danger, and the author of this book saw the head of one perforated behind the ears, and this must have happened to a great number (Landa [sixteenth century], in Tozzer 1941, p. 125; see also Landa 1982, pp. 54–58).

Historian Molina-Solís echoes Landa's view when talking about this custom in Yucatán. He recounts the sources dating to the European contact with the Maya during the years of the conquest of Yucatán. The author recounts that the natives attributed great importance to changing their physical appearance for social and religious reasons. He likewise notes that cranial modifications were present in those people since infancy, emphasizing the role of mothers in the early physical adaptation of the body.

> Pero si era una raza bien dotada por la naturaleza, adolecía de vicios de conformación en un gran número de individuos, que acarreaban las necesidades de la crianza, con las preocupaciones más banales sociales y religiosas. A menudo se encontraban sujetos estavados, bizcos, con la cabeza aplastada, horadadas las orejas y arpada la ternilla de las narices. Todos eran defectos artificiales o adquiridos, ora porque las madres, en la edad de la lactancia, llevaban a sus hijos de un lugar a otro ahorcados sobre sus caderas, ya también porque gustaban de usar zarcillos, o bien se imprimían crueles arpaduras para consagrarse con sus divinidades (Molina-Solís 1943, p. 218).

Similar to the statement by Cristóbal del Castillo on the Mexica, is the characterization of the custom that Bartolomé de las Casas provides for Guatemaltecan. The chronicler, expert in the native ways in New Spain and Peru, states that they flattened their heads in order to appear more ferocious in combat.

> Cuanto a la costumbre de querer parecer fieros en las guerras, ordenaron a los principios hacerse las caras y cabezas, por industria de las parteras o de las mismas madres cuando las criaturas son tiernas y chequitas, empinadas y hacer las frentes anchas (Casas 1967 [sixteenth century], p. 177).

5.1.2.5 Nicaragua

Further east and down the land bridge between North and South America, we find one other invaluable testimony from the cultural fringes of Mesoamerica. Among Nahua-speaking folk of Nicaragua, Friar Francisco de Bobadilla notes different connotations for their head-shaping practices in his description of native customs and beliefs. He emphasizes the notion of "beauty" that this body modification holds for locals and cites other, possibly more practical considerations, such as promoting a docile nature from early age on and to prepare the children to lift heavier burdens later in their working life. Using the recourse of dialogue, Bobadilla recites:

> F. ¿cómo no tenéis vosotros la cabeza de la hechura que los cristianos?
> Y. Cuando los niños nacen, tienen las cabezas tiernas, y hácenselas como veis que las tenemos con dos tolondrones a los lados dividiendo, y queda por medio de la cabeza un gran hoyo de parte a parte; porque nuestros dioses dijeron a nuestros pasados que así quedamos hermosos y gentiles hombres, y las cabezas quedan más recias para las cargas que se llevan en ellas (Dialogue by Fray Francisco de Bobadilla [sixteenth century], in D'Olwer 1963 p. 352).

5.1.3 Viceroyalty of Perú

The limited amount of information available on New Spain contrasts with the wealth of detail that diaractorizes many references collected from the records of the Viceroyalty of Peru. Here the native head practices clearly inspired more numerous and explicit writings.

5.1.3.1 Inca, Perú

As an example of eloquence, we mention Bartolomé de las Casas and Juan de Torquemada. Both bring the cultural practice up only briefly among the natives of New Spain, but describe and justify it at length in the case of the Viceroyalty of Perú. In addition to purely technical aspects, the friars provide various interpretations for this custom among the Inca and the Andean peoples dominated by the Viceroyalty. They regard the enactment of cranial transformation as an exclusive right of the members of the social hierarchies, i.e., the only individuals who held the privilege of practicing and of conferring this right onto others. On a more general note, diverse models of the cranial modification are ascribed to Peruvian natives as a visible body attribute used to identify place, tribe or lineage.

> Dije algunos de los del Perú, porque por la mayor parte, cuasi en cada provincia tenían propia costumbre y diversa de las otras, de formar con industrias las cabezas. Y es cosa de maravilla ver la diligencia e industria que tienen para entallar las cabezas mayormente de los señores; éstas de tal manera las atan y aprietan con lías o vendas de algodón o de lana, por dos y tres años a las criaturas, desde que nacen, que la empinan un palmo grande, las cuales quedan de la hechura y forma de una coroza o de un mortero de barro muy empinado (...). Por privilegio grande concedían los del Perú a algunos señores, y que ellos querían favorecer, que formasen las cabezas de sus hijos de la forma que los reyes y los de su linaje las. No había provincia en toda la tierra, con ser innumerables, que los vecinos de cada una no trujesen su señal en la cabeza, que entrando en la plaza de la ciudad de Cuzco, en la cual entraban por cuatro partes, como en cruz, y viéndolos de lejos, no cognosciesen de qué provincia eran, sin que más del traje viesen; y esto hasta hoy dura.
> A aquesta diligencia destas señales para conocerse las personas de qué provincias eran, parece poderse ayuntar la costumbre antigua, que también tenía cada provincia, de formar las mismas cabezas, porque fuesen cognoscidos los vecinos de cada una dellas, Y así, cuando infantes que acababan de nascer y de allí adelante, mientras tenían las cabezas muy tiernas, les ataban ciertas vendas o paños con que se las amoldaban según la forma que querían que tuviesen las cabezas; así, unos las formaban anchas de frente y angostas de colodrillo; otros, anchas de colodrillo y angostas de frente; otros, altas y anchas, y otros de otras maneras; finalmente, que en la forma de las cabezas tenían muchas invenciones, y ninguna provincia, al menos de las principales, había que no tuviese forma diferente de las otras, de cabezas.
> Los señores tomaron para sí e para todo su linaje, que se llamaba ingas, tres diferencias de cabezas, puesto que después algunas dellas comunicaron a otros señores de algunas provincias, sin que fuesen del linaje de los ingas, por especial privilegio (Casas 1967 [sixteenth century], p. 594).

In Chap. 15 of Book 14 of the *Monarquía Indiana*, Friar Juan de Torquemada speaks about the physical description of different ethnic groups in the Indies, including the "manners that they had of forming their heads." It is noteworthy that native Mexican head shapes are described here omitting that their head shapes were most likely the result of artificial modification, a connotation, which Torquemada does point out for Peruvian head practices.[1] For the latter, he reports different modification techniques and a diversity of forms. He mentions particular head shapes that were a privilege granted only to certain lords, who in turn granted this right to others as a favor so that their heads could take the shape of the "kings".

> Diximos algunos de los del Pirú, porque por la maior parte, casi en cada Provincia, tenían propia costumbre, y diversa de las otras, de formar con industria las cabeças, y era cosa de maravilla verla diligencia, que tenían para entallar, y formar las cabeças, maiormente de los Señores; estas de tal manera las ataban (...) y apretaban con lias, o vendas de algodón u de lana, por tiempo de dos, o tres años, desde que nacian, que las empinaban mas de vna quarta,

[1] Note that Friar Juan de Torquemada writes his *Monarquía Indiana*, written in the early 1600s. It was published in 1615 while the author still lived. His work based on writings left by previous chroniclers such as Hernán Cortés, López de Gómara, Bartolomé de las Casas, Motolinia, Mendieta and other Franciscans as well as the Jesuit Acosta or the royal chronicler Antonio de Herrera. Here, we find again the works by the authors mentioned, but we believe it is important to include the complete text on head shaping offered by Torquemada in this chapter.

> las quales quedaban de la hechura, y forma de vna coroça, u de vn mortero de barro, mui empinado, y alto, y en esto ponían mucha diligencia, y por privilegio grande concedían los del Pirú a algunos Señores a quienes querían favorecer, que formase las cabeças de sus hijos, de la manera, que los Reies, y todos los otros de su linaje (Torquemada 1969 [1615], Vol. 2, p. 583).

An account of Santa Cruz Pachacuti, written many years after the Conquest, recalls that the head shapes of the underprivileged and marginalized were also controlled by the Inca. He states that the Inca, Manco Capac, ordered the heads of his people molded so that they would be obedient:

> el Inca Manco Cápac había ordenado el uso del modelamiento en sus provincias conquistadas para que sean simples y sin ánimo, porque los indios de gran cabeza y redonda suelen ser atrevidos para cualquier cosa: mayormente, son desobedientes (Santa Cruz Pachacuti 1995 [1613], p. 23).

The same information is also voiced in later years of Inca occupation and attributed to the third Inca, Lloque Yupanqui.

> Lloque Yupanqui tercer Inca, también había mandado que todas las naciones a él sujetas los atasen las cabezas de las criaturas para que sean largas y quebrantadas de frente, para que fuesen obedientes (Santa Cruz Pachacuti 1995 [1613], p. 130).

There are no uniform criteria among the different authors regarding the underlying purpose or intention of the cultural practice of cranial modeling; different interpretations are advanced by chroniclers of the first colonial years and later writers, as Dingwall already makes clear, when he states:

> The fact that, generally speaking, the practice seems to have been a privilege of a class suggests, as Wiener has pointed out, that some advantage was supposed to accrue from it, although he is wrong in stating that the heads of men only were deformed. The reasons given by Sacamayua cannot, I think be considered seriously in their literal interpretation; and similarly Angrand's idea that the custom had taken root in the belief of a serpent ancestor cannot be admitted (Dingwall 1931, p. 219).

Other, more recent scholarship has equally attempted to confer deeper meanings for this Peruvian custom from the colonial references. Purizaga analyzes birthrights among the Incas, reporting on the importance of the child's placement in the compression crib and the custom of cranial modification carried out in these areas (Fig. 5.2). Purizaga specifically analyzes the term *ayuscay.* While considered by some authors as referring to the birth, Purizaga argues that the interpretation made by chronicler Cristóbal de Molina in his *Fábulas y ritos de los indios*, tentatively dated 1574, should be correct. Cristóbal de Molina believed that the term *ayuscay* referred to the modification practice.

> Cristóbal de Molina el cusqueño, manifiesta que la palabra "Ayuscay" está más bien relacionada con una ceremonia especial que forma parte del ritual del nacimiento. En efecto, el cronista dice, "El ayuscay, era que cuando paria la mujer, el cuarto día ponían las criaturas en las cuna que llama quirao". Es decir, que ayuscay se refiere al hecho de colocar a la criatura en la cuna. La tradición andina establece que se realiza al cuarto día. Esto no admite dudas (Purizaga 1991, p. 6).

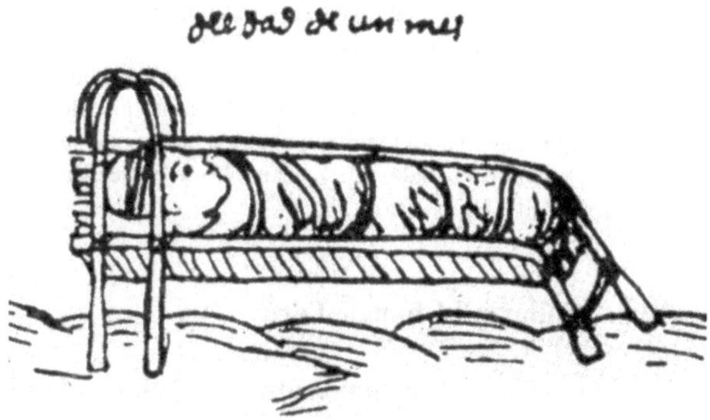

Fig. 5.2 Inca cradle (*k'irawpi kaq*), holding a month-old baby. (Guaman Poma de Ayala 1944 [1615], p. 212; Fig. 79)

After reviewing the various interpretations that different authors propose, Purizaga finds only one of them unsatisfactory: In the 1930s, Frenchman Louis Baudin considered that cranial modification was for political purposes and uses the statements by Bernabé Cobo on los Collas, Manco Capac and other, more recent authors:

> Hasta es posible que Amautas de una inteligencia superior pretendiendo ir más lejos por este camino, hayan probado por dicho medio –la deformación- comprimir ciertas circunvalaciones cerebrales, a fin de crear individuos de mentalidades determinadas. Un cronista indio pretende que el inca utilizó ese procedimiento para hacer a sus súbditos obedientes. Allí estaría el término lógico de una política de racionalización: la fábrica de esclavos (Purizaga 1991, pp. 56–57).

Nevertheless, Purizaga rejects this interpretation, finding that Baudin's argument clearly manipulates the primary information at hand:

> Además, no está demostrado que la deformación craneana lleve consigo ninguna alteración ni mental ni emocional. Lo cierto es que tal costumbre está más bien vinculada con distintivos tribales, totémicos, tal vez sociales y raspados con fines profilácticos de ciertas enfermedades. Este "raspado bautismal" -dice un autor- evitaría males en el adulto y una consiguiente intervención curativa penosa, mediante la trepanación del cráneo (Purizaga 1991, p. 58).

Finally, Dingwall transcribes the technique used for this practice among the tribes of Perú, by citing the report by Cachot, dated 1923:

> The board, as placed in position over the occiput, is much larger than the back of the head. By means of bands, one going round the head and the other over the head in the direction of the sagittal suture, the board is fixed to the head (Dingwall 1931, p. 218).

5.1.3.2 Manta, Ecuador

The Inca, Garcilaso de la Vega, also relates the practice of cranial modification in his writings on northern Andean folk, stating that the custom of cranial-vault modification was performed on both boys and girls in the "Manta" territories, which then occupied the northwestern shores of the Inca Empire. He describes the techniques employed and the motives (in order to appear fiercer). Interestingly, Garcilaso de la Vega's description reflects some of the earlier interpretations in New Spain, where the practice is related to a more ferocious or monstrous physical appearance.

Hombres y mujeres se labraban las caras con puntas de pedernal, deformaban las cabezas a los niños en naciendo ponianles una tablilla en la frente y otra en el colodrillo, y se las apretaban de día en día hasta que eran de cuatro a cinco años, para que la cabeza quedase ancha de un lado al otro y angosta de la frente al colodrillo y no contentos de darles la anchura que habían podido trasquilaban el cabello que hay en la mollera, corona y colodrillo y dejaban los de los lados (...) rizándolos y encrespándolos para su aspecto monstruoso (Vega 1982 [1609], p. 333).

5.1.3.3 Colla and Chauco, Perú

Returning to the Andean highlands, not all chroniclers offer such extensive explanations on the head custom as Bartolomé de las Casas or Torquemada. Pedro Cieza de León simply mentions it among the Colla Indians.

En las cabezas traen puestos unos bonetes a manera de morteros, hechos de su lana, que nombraban chucos; y tiénenlas todos muy largas y sin colodrillo, porque desde niños se las quebrantan y ponen como quieren, según tengo escrito (Cieza de León 1984 [sixteenth century] in D'Olwer 1963, p. 484).

Among the native settlers around the Colca River watershed (Arequipa), cranial modification was known to identify two different natives: Collagua and Cabanas. Corregidor Juan Ulloa-Mogollón testifies on the infant practice in 1586:

Estos Collaguas antes de la visita general que se hizo por mandamiento del excelentísimo virey don Francisco de Toledo, traían en la cabeza unos que llamaban en su lengua Chucos, á manera de sombreros muy altos sin falda ninguna, y para que se pudiesen tener en la cabeza, se la apretaban á los niñosrecien nacidos tan reciamente, que se la ahusaban y adelgazaban alta y prolongada lo más que podían, para memoria que habian las cabezas de tener la forma alta del volcán de donde salieron. Esto les está ya prohebido por ordenanza (...) Estos [los Cavanas] son muy diferentes en la cabeza á los Collaguas, porque recien nacidos los niños é niñas, se la atan y la hacen chata y ancha, muy fea y desproporcionada; la cual se atan con unas cuerdas blancas á manera de mechas, y dando muchas vueltas alrededor, quedan las cabezas ensanchadas (...) Conócense bien en la hechura de las cabezas el ques natural de Cavana y el ques Collagua, que, como está dicho, los Collaguas se ahusan la cabeza larga y estos Cavanas ancha y chata (Ulloa-Mogollón 1885 [1586], p. 40).

In the following, Corregidor Juan Ulloa-Mogollón offers still additional information on these peoples and their specific head shapes:

Los Collaguas usaban una especie de sombrero alto sin falda en forma de cono truncado, imitando la del volcán y para que la cabeza se conformase a ese tocado recibía la correspondiente deformación. Los de la Provincia de Cauana creían proceder de otro cerro nevado que se llama Ualca ualca, del cual salieron sus antepasados, venciendo a los naturales y echándolos fuera de la tierra. Así poblaron Cauana Colla en la sierra alta y Cauana Conde al otro lado. Estos Cauanas se deforman también la cabeza pero no alargada como los Collaguas, sino chata, cubriéndola con unas cuerdas blancas con que se dan varias vueltas. Así quedaban bien diferenciados unos de otros (Ulloa-Mogollón 1885 [1586] in Yépez and Romano 2008, p. 56).

Although mentions of artificial head forms became fewer with the passage of time, they apparently became more detailed, as the authors attempted to understand not just the manner in which the infant skull was altered, but wanted to comprehend more fully the underlying motives of the different peoples who practiced the custom.

Bernabé Cobo in the second half of the seventeenth century describes the peoples of the Andean region, mentioning the "Collas" and "Chaucos" and referring to the array of shapes that differentiated one group from the other, along with their modeling techniques. Cobo associates its purpose with the maintenance of health and to foster productivity. However, I think that this explanation is slightly contradictory, when he asserts that the natives adapted the shape of their heads to the form of certain headdresses (a type of hat) they wore and not the opposite:

> Usaban algunas naciones (andinas) en naciendo la criatura, formarle la cabeza en diferentes figuras con muchas supersticiones y tanto rigor, que algunos niños morían del dolor que padecían, y a no pocos hacían saltar los sesos o quedar siempre enfermos y lisiados. Desproporcionaban de esta suerte la hechura del hombre y no contentos con las cabezas que Dios les dio, querían enmendar la naturaleza humana y dar a sus cabezas el talle que más les agradaba y cuanto con mayor desproporción y disformidad quedaban, tanto lo juzgaban por más gala y gentileza. Unas naciones las lucían anchas de frente apretándoles para darles esta forma con unas tablillas fuertemente liadas. Los collas formaban la cabeza larga y puntiaguda, con tanto extremo que pone admiración ver los viejos que yo alcance con aquel uso de su gentilidad y esto lo hacían porque usaban ellos de unos bonetes de lana llamados chucos a manera de moriscos o de sombreros, sus faldas muy altas y puntiagudas y porque mejor cayesen y ajustasen formaban la cabeza al molde del tocado y no el tocado a proporción de la cabeza y para dar esta figura a las cabezas de los niños, las liaban y apretaban con vendas y las traían así hasta la edad de cuatro o cinco años que ya quedaban endurecidos y amoldados a su tocado, largas escusadas y sin colodrillo. Decían ellos que ponían deste talle las cabezas porque fuesen más sanos y para más trabajo y haciéndoles el primer bonete con muchas ceremonias y supersticiones (Cobo 1893 [1653], pp. 175–176).

Cieza de León mentions the practice among the Chauco near Cali. His description of the natives' physical appearance states vaguely that the head shapes were the readily visible result of an intentional manipulation.

> Los chaucos (...) tan grandes que parecen pequeños gigantes espaldudos, robustos, de grandes fuerzas, los rostros muy largos, tienen cabezas anchas, porque en esta provincia y en la de Quimbaya, y en otras partes destas Indias cuando la criatura nasce le ponen la cabeza del arte que ellos quieren que la tenga; y así, mas quedan sin colodrillo, y otros la frente sumida, y otros hacen que la tenga muy larga, lo cual hacen, cuando son recién nacidos con unas tablitas, y después con sus ligaduras (Cieza de León 1984 [sixteenth century], p. 145).

5.1.3.4 Carangue, Perú

An even more extensive interpretation is provided by Pedro Cieza de León on the Carangue. In addition to the variety of head shapes acquired through the modification process, Cieza de León also advances some interpretations that allude to different physical qualities, such as improving health or increasing strength for work.

> Los carangues y sus comarcanos es otro linaje de gente y no son labrados y eran de menos saber que sus vecinos porque eran behetrías; por causas muy livianas se daban guerras unos a otros. En naciendo la criatura la abajaban la cabeza, y después la ponía entre dos tablas, liadas de tal manera que cuando era de cuatro o cinco años le quedaba ancha o larga y sin colodrillo; y estos muchos lo hacen y no contentándose con las cabezas que Dios les da, quieren ellos darles el talle que más les agrada y así unos la hacen ancha y otros larga. Decían ellos que ponían de estos talles las cabezas por serían más sanos y para más trabajo (Cieza de León 1984 [sixteenth century], p. 227).

5.1.3.5 Chucuito, Perú

A similar interpretation is provided by Garcí Diez de San Miguel, when he discusses the custom of cranial modification practices on newborns during his visit to the Highland Province of Chucuito:

> En toda la dicha provincia generalmente tienen por costumbre las indias cuando paren, apretar con las manos las cabezas de los niños para hacerlas largas y delgadas y se las traen liadas y apretadas más de un año con unas trenzas de lana para que vayan creciendo y adelgazando solo a fin de que cuando sean hombres se les encajen en las cabezas unas caperuzas largas y angostas que entre ellos usan que llaman *chucos* de manera que en lugar de hacer las caperuzas conforme a las cabezas hacen las cabezas al talle de las caperuzas en lo cual ha habido y hay tan gran exceso que ordinariamente vienen a morir de ello muchos niños y los que quedan por la mayor parte se crían enfermos y traen los ojos malos y quedan sordos como lo he visto y entendido en la dicha visita y aún ha acaecido salírsele a alguno los sesos por las orejas y para evitar los dichos daños provei un auto que Vuestra Señoría vera conviene se ejecute (Diez de San Miguel 1964 [1567], p. 224).

5.1.3.6 Uros (Lake Titicaca), Perú and Bolivia

Father Antonio de la Calancha briefly reports the existence of cranial transformation among the Uru natives who during colonial times, settled the shores and islands of Lake Titicaca although he does not offer any deep cultural interpretation:

> Traen en las cabezas como turbantes moros (...) crían a sus hijos atormentándolos, porque traen la cuna en las espaldas, parada la criatura y fajada por toda la cuna, i desde el punto que nacen le van apretando la cabeza para que sea prolongada y no redonda (Calancha 1974 [1638], p. 1467).

We also find other mentions of these folk. Bandelier, in his 1910 work, transcribes information on the old ways of Titicaca islanders, some still based on chroniclers and travelers of the sixteenth and seventeenth century.

> While the women on the Island are usually of the low stature of other female Indians, there among them some middle height and more slender than, for instance, the Pueblo Indian women of New Mexico. Among the men there are some tall and well formed figures, with pleasant faces; many are of low stature and have sinister countenances. It is not unusual to meet an Indian with remarkably low forehead and abnormally elongated skull. It is known that flattening of the forehead was carried on for at least half a century after the Spanish authorities had peremptorily forbidden the practice (Bandelier 1910, p. 67).

5.1.3.7 Amazon Natives of Eastern Perú and Brazil

The Omagua-speaking native inhabitants of the eastern Amazon Valley were once a populous, highly organized native society (Fig. 5.3). Like most of the indigenous populations, they suffered a steep decline during the early years of Iberian colonization. Some early references to their indigenous head practices are reported by Cristóbal de Acuña, although he only describes the techniques used in the practice without offering any kind of cultural explanation:

Fig. 5.3 Engraved portrait of Cambeba native with miter-shaped head and necklace from Brazil's Amazon Valley. (Riou 1867)

> Son todos cabezas chatas, que causa fealdad en los varones, si bien las mujeres mejor lo esconden con el mucho cabello; y está en ellos tan entablado el uso de tener las cabezas aplastadas, que desde que nacen las criaturas se las meten en prensa, cogiéndoles por la frente con una tabla pequeña y por la parte del cerebro con otra grande que sirviendo de cuna recibe todo el cuerpo del recién nacido; el cual puesto de espaldas sobre ésta y apretado fuertemente con la otra, queda con el cerebro y la frente tan llanos como la palma de la mano y como estas apreturas no dan lugar a que la cabeza crezca más que por los lados, viene a desproporcionarse. De manera que más parece mitra de obispo mal formada que cabeza de persona (Cristóbal de Acuña [1641] in D'Olwer 1963, p. 677).

Another reference to these people is provided by Friar Laureano de la Cruz, who in 1653 described the form in which mothers performed the head maneuvers on their children.

> Las mujeres se envuelven en unas mantillas de algodón tan cortas y angostas que les honestan muy poco. El modo que tienen en apastarse las cabezas es el siguiente: Toman la criatura de pocos días de nacida y cíñenle la cabeza por la parte del cerebro con una faja de algodón ancha, y por la frente con una planchuela que hacen de cañas bravas, que les coje desde los ojos hasta el cabello muy bien apretada, y de esta manera lo que la cabeza había de crecer en redonda, crece para arriba y queda larga, chata y muy desproporcionada (Cruz 1900 [1653], p. 99).

Later references to the Omagua appear published in the mid-eighteenth century in a work by La Condamine, as transcribed by Veigl (Fig. 5.3). He states that the Omagua wanted to resemble the full moon by transforming their heads.

> Condamine reached on the 27th the mission station of the Omaguas, formerly a powerful nation, whose dwelling extended along the banks of the Amazon for a distance of 200 leagues below the Napo. Originally strangers in the land, they are supposed to have come down some river rising in Granada, and to have fled from the Spanish yoke. The word Omagua means flat-head in Peruvian, and these people have the singular custom of squeezing the foreheads of new-born babies between two flat pieces of wood, to make them, as they say, resemble the full moon (Veigl 1798, p. 78).

Father Acuña, a Spanish Jesuit, notes the practice of modelling the children's heads in his seventeenth-century work on the exploration of the Amazon River Basin. Among other local customs, he describes the techniques and visible results that were obtained, which simulate the shape of a bishop's miter:

> ... when a child is born its head is placed in a press (en prensa), a small board (tabla) being fastened to the forehead and another against the occiput (...). The larger board on which the child lay, the top board being fastened over it in a manner which must have been almost identical with that adopted by the Chinook of the North West Pacific Coast. The result of this treatment was that the front and back of the head became flat and the general form resembled, according to the old ecclesiastic, an ill-shaped bishop's mitre instead of a human head (Acuña 1641, p. 24; in Dingwall 1931, p. 198).

Late descriptions of native groups from the Brazilian side of the Amazon Basin also go at lenghs to describe this custom. The late dates of these references should not be surprising since the colonization of these territories by the Portuguese began later than their exploration of the western stretches toward the Andes. From the Atlantic, population expansion toward the interior only occurred from the eighteenth century onward. For the topic treated here, it is noteworthy that at that time Ribeiro de Sampaio still describes the visible outcome of infant head flattening among the native folk during an official visit of the pioneer territories in 1774. As others before him, Sampaio compares the artificial head form to a bishop's miter. Apparently, the practice was in the course of being abandoned.

> Formerly children used to have their heads compressed between two boards, the effect of a bishop's miter being produced. He himself compares these head form to those of the Macrocephali mentioned by Hippocrates, but adds that at the time of his visit those natives whom he observed had abandoned the custom (Dingwall 1931, pp. 198–199).

5.1.3.8 Codes Prohibiting the Custom of Cranial Modification

Cranial modeling also figures prominently in the colonial codes of both civil and ecclesiastic law. These announced the prohibitions that were formulated by the Iberian colonizers to eradicate the ancestral head practices still pursued after contact and conquest. In the case of the Viceroyalty of Perú, prohibitions of this type also reveal the techniques and forms used to transform the children's heads. The legal codes of Viceroy Toledo in 1573, state:

> ITEM, mando, que ningun Indio, ni India apriete las cabezas de las criaturas recien nacidas, como lo suelen hazer para hazerlas mas largas, porque de averlo hecho se les a recrecido, y recrece daño, y vienen a morir dello (Bandelier 1910, p. 25).

Thirteen years later, in 1586, we find a similar prohibition in a text mentioned previously on the visit made by Corregidor Juan de Ulloa-Mogollon to the Province of the Collaguas in the Department of Arequipa in Perú. To define what had to be prohibited, the official goes into detail to describe the different shapes displayed by each ethnic group.

> Estos Collaguas ... traían en la cabeza unos que llamaban en su lengua chucos, á manera de sombreros muy altos sin falda ninguna, y para que se pudiesen tener en la cabeza, se la apretaban á los niños tan reciamente, que se la ahusaban y adelgazaban alta y prolongada lo más que podían, para memoria que habian las cabezas de tener la forma alta del volcán de donde salieron. Esto les esta ya prohibido por ordenanza (Ulloa-Mogollón 1885 [1586], pp. 40–41).

Similar prohibition edicts condemn to the natives of Cavana and the Collaguas who were distinguished practitioners of head modeling still during the colonies.

> Estos [los Cavanas] son muy diferentes en la cabeza á los Collaguas, porque recien nacidos los niños é niñas, se las atan muy recio y la hacen chata y ancha... Conócense bien en la hechura de las cabezas el ques natural de Cavana y el ques Collagua, que, como está dicho, los Collaguas se ahusan la cabeza larga y estos Cavanas ancha y chata, muy fea y desproporcionada; la cual se atan con unas cuerdas blancas á manera de mechas, y dando muchas vueltas alrededor, quedan las cabezas ensanchadas (Bandelier 1910, p. 25).

It is further specified that:

> Estáles prohibido ya esto por ordenanza. Conócense bien en la hechura de las cabezas el ques natural de Cavana y el ques Collagua, que, como está dicho, los Collaguas se ahusan la cabeza larga y estos Cavanas ancha y chata (Bandelier 1910, p. 25).

The last prohibition we came upon was issued by the Council of Lima of 1614, years after the above edicts. The Constitution offers one of the most exhaustive references to be found on head modeling among New World natives. The resolution had been generated during the meeting of the Bishops of Lima in 1614. This document includes a series of "constitutions", written in Latin, which list the customs to be eradicated among the different ethnic groups (Fig. 5.4). These also condemn the practice of cranial modeling as a superstitious rite and list the techniques used by the natives. Here, we transcribe Constitution 100:

> In antiquity and in a generalized fashion did the Indians of those provinces engage in the custom of modifying the heads of their children in different manners by flattening it from the back as part of a superstitious rite of the lineages of the different provinces (...) In some provinces, women compressed the heads of their newborns in a pointed fashion with their hands, a custom that in the zaitahoma language is called "pilleo" when it led to an elongated form, generally this horrid custom implied binding the head (...) However, in other provinces, the head without its posterior portion is custom [and accomplished] by flattening the occiput, which among the native language is called "paltahoma", when different effigies are put on the soft head of the children and compressed with a band until they break it. This custom must not be continued in this part, conscience must be gained that of its evil nature that can result in other insanities still worse, when the natural order of the human kind is changed, one has to put an end to the tradition of such an undesirable custom of

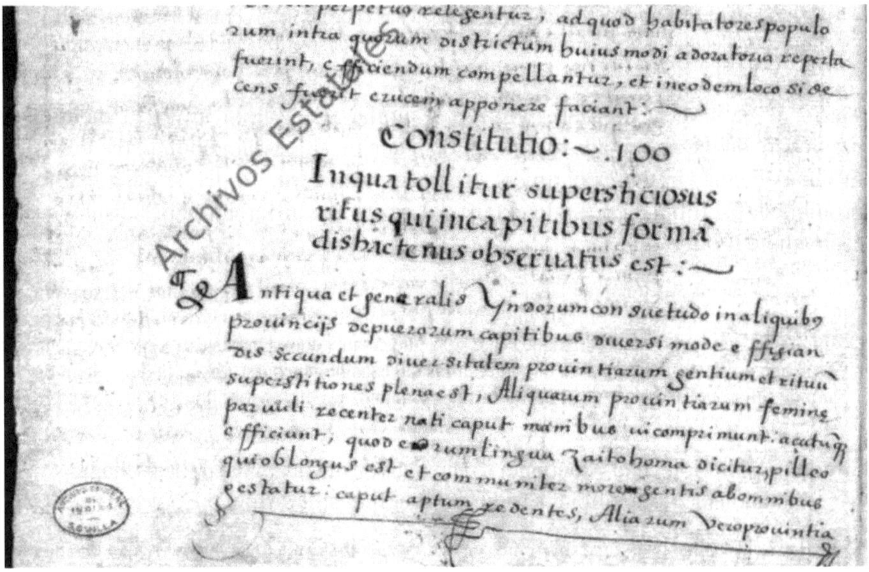

Fig. 5.4 Extract of original copy of Constitution 100 (A.G.I. Patronato, 189-R. 40), prohibiting native head modifications in infants

> some lineages (. . .) The Christian religion (. . .) detests these ceremonies and sacrilegious rites [that] put such harm to the care of the diligent priests of the Indians. (A.G.I., Patronato, p. 189, R. 40; translation from Latin to English by Pilar Zabala and Vera Tiesler).[2]

---

[2] *From the original text in Latin: Inquatollitur supersticiosus ritus qui in capitibus formam disbactenus observatus est: Antiqua et generalis yndorum consuetudo in aliquibus provinciis de puerorum capitibus diversi mode effisiandis sucundum diversitatem provintiarum gentium et rituum superstitiones plena est. Aliquarum provintiarum femine parvuli recenter nati caput manibus vi comprimunt acutumque efficiunt quod eorum lingua zaitohoma dicitur pilleo qui oblongus est et communiter morem gentis abominibus gestatur caput aptum redentes. Aliarum vero provintia cum mos est capita excerebrantes plano occipite deformare quod yndorum lingua dicitur paltahoma dum vero diversas capitum effigiens componunt mollia puerorum capita quarantes et vittes ea comprimentes frangunt ex quo pars eorum non minima sepe obiit vel discomposbitis sensuum cellulis in amentiam decidit est et aliud maius malum quia tempore quo hec tanto pere humano generi nociva fiunt et dum nature ordinem mutare contendunt multi (ut fertur) solis sacra vel peragunt vel phanis de more gentis e mira nota per soluunt que omnia sine gravi Christiane religionis contemptu fieri non possunt. Igitur has ceremonias et ritus sacrílegos abominatur et damnat omnibusque sacerdotibus yndorum curam agentibus districte precipit hoc predictos sacrílegos semel et iterum et pluries admonere ut abhisvamis supersticionibus et nocivis ommine desistant supost predictas admonitiones res sipiscere no luerint iuris pena que supersticiosis statuitur plectantur exhortamus in domino quantum posumus huius provincia guvernatores et presides catholicosque judices et civitatum prestores quod sui muneris est in hac parte exequi iubeant et ut indicibus ecclesiastieis ope et auxilio faventes ecclesiastieis reformationi non desint.*

5.2 Persistence of the Custom of Cranial Modification Between the Eighteenth and Twentieth Centuries

Up to this point, we have transcribed all references available to us on head-modeling practices among the different ethnic groups of the Caribbean region, New Spain, and the Peruvian viceroyalty, as witnessed by the chroniclers from the first years of colonization. However, we also know that the practice persisted in some areas throughout the centuries of the Colonization, including the independent era and in some cases, up to the twentieth century. The following sections incorporate those late colonial and postcolonial references that testify to the continuity of the practice in some regions that were colonized later or among groups that were further from the centers of control, where the tradition also persisted.The survival of Ibero-American head practices to relatively recent dates leads us to question the total success of the different laws, codes, synods, etc., that were issued by civil as well as ecclesiastical authorities for the purpose of eradicating the custom. Naturally, the domination process was not equally intense in all provinces of Hispanoamerica; in some places, such as Portuguese Brazil, colonization was slower. In some regions of South America, native groups remained isolated from Western civilization and closer to their ancestral ways.

5.2.1 Continued Practices in the Caribbean Area

5.2.1.1 Caribbean Area During the Nineteenth Century

In his account of his voyage to the Antilles, published in 1813, Father Leblond refers to the modification process among the native inhabitants of the islands, where he saw children, four or five months of age, wearing head splints (Fig. 5.5):

> That the boards were kept in place for nine days at time: then they were taken off and subsequently soon replaced, until the head seemed to the parents to be satisfactorily formed, when they were finally discarded. That occasionally the device was worn for some time may be seen (Leblond 1813, pp. 197–199).

Father La Borde, in his history of the Caribs, mentions the modification process among local natives and explains that it was practiced for the sake of beauty:

> La Borde declares that the Carib artificially flatten the heads and noses of their children, the mother compressing them at birth and during the time that they are suckled, thinking the result beautiful (Dingwall 1931, p. 197).

5.2.1.2 British Guyana During the Nineteenth and Twentieth Centuries

There are also references to the continued practice of the custom among the Maopityan on the Carib's southern fringes of British Guyana during the advent of the twentieth century. Roth cites testimonies of the existence of the custom and narrates the form in which it was accomplished:

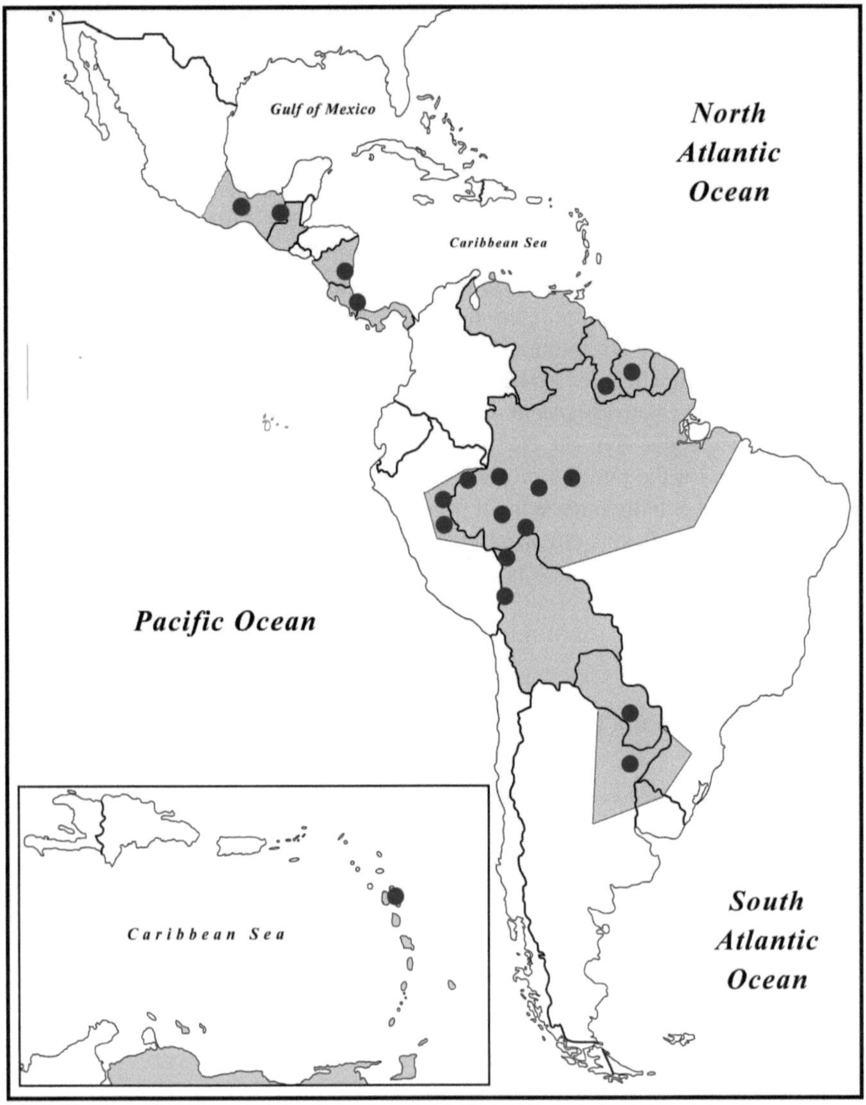

Fig. 5.5 Map of Latin America: highlighted countries denote the areas of sustained head practices as cited by late colonial and post-independence historical sources. Dots indicate the locations where the custom is referred to among the sources transcribed in this chapter. (Drawing by V. Tiesler)

> Among the Taruma, this same traveler speaks of seeing two more flat-headed Maopityan, a man and a woman... It is said that the Maopityan, who called themselves Mawakwa, were so named from the Wapishana word mao, a frog, and pityan, folk or tribe (SR, II, 472). But in connection with the mention of Taruma it is interesting to note that Schomburgk some 30 years previously had recorded a side-to-side flattening of the head among this same people, but was careful to make the statement that the deformity was not artificial (Roth 1924, p. 501).

Another mention from British Guayana dates back to the nineteenth century in a work published in the first decades of the twentieth century. Here the author argues that the practice continued to that date in the Carib region:

> In a remote part of British Guiana or perhaps beyond the frontier near the sources of the Essequibo, there lives a little known people which was in the habit of tying boards to the heads of its children in such a way as to flatten them. Early writers record the fact that the custom formerly prevailed among all the Carib of this region but it has now fallen into disuse (Roth 1924, p. 501).

On a more general note, Gillin discusses recent head-shaping practices in the Suriname area in his contribution for the Handbook of South American Indians and states:

> Permanent ornamentation by deformation of the head is mentioned occasionally among the Cayenne and Suriname coastal Carib, Tarumá, and Maopityan, although the evidence is not clear that the deformation was intentional. Frontal deformation occurs on the coast, and fronto-occipital and side-to-side in the interior (Gillin 1948, p. 834).

5.2.2 Southern Mexico and Central America

5.2.2.1 Maya and Mixe, Mexico

We also came across some recent references which demonstrate that the cultural practice persisted among Mexican Lacandons (Palka 2005). Maudslay describes the physical features of the native Lacandones in his work published in 1899. He concludes that the custom of head modeling was slowly being abandoned, even though it was still evident in the physical features of the elder folk.

> I was much impressed by the striking likeness, which the feature of the elder man, who appeared to be the leader of the village, bore to those carved in stone at Palenque and Menché. The extremely sloping forehead [of elder Lacandon males] was not quite so noticeable in the younger men, and it may be that the custom of binding back the forehead in infancy, which undoubtedly [was] obtained amongst the ancients, is being now abandoned. These people still use bows and stone-tipped arrows, which they carry with them wrapped in a sheet of bark (Maudslay and Maudslay 1899, pp. 236–237).

Further west, native Mixe from Coatlán in the Mexican state of Oaxaca are photographed still at the turn of the twentieth century with a similarly receding forehead and a flattened occiput (Shattuck 1933, p. 29; Trias de Bes et al. 1928, p. 84). The artificial nature of this appearance is only briefly noted by Trias de Bes and colleagues.

5.2.2.2 Nicarao, Sumo and Miskito, Nicaragua

Dingwall also refers to various authors who mention the persistence of the custom of cranial modification, among Nicarao groups of what today is Nicaragua. Here he transcribes Lehmann who writes at the beginning of the twentieth century that the custom was still in use among this group:

> Apparently the custom is still persisting in certain regions. In Managua, Lehmann reported that some of the natives possessed the custom, and there is no doubt that in the more isolated districts there is the possibility of the distortion still being practiced although not to the same extent as formerly (Lehmann 1909, p. 535; in Dingwall 1931, p. 155).

In his work on the archaeology of Central America, Joyce recalls the late practice of this custom in the Sumo region (east of Nicaragua), at the beginning of the twentieth century:

> In the Sumo district in the east of Nicaragua one tribe at least, according to Joyce, practised deformation upon the heads of their children by pressing their infant skulls between pieces of wood and stones (Joyce 1916, p. 35).

Joyce also documents the practice of head shaping by Nicarao groups and other Maya settlers further north, although no defined time frame is provided. They alleged that the gods had told their forebears that in this way they would acquire a more noble aspect, and gain strength to carry heavy loads:

> The Nicarao moulded the heads of infants so as to produce a boss on either side and a depression in the middle. They alleged two reasons for this custom: first that the gods had told their ancestors that it gave a noble appearance to the individual so treated, and second that it made the head harder for carrying burdens. The Maya tribes further to the north also practised head-deformation, but their aim was to produce a flat and receding brow. The ears were pierced for ornaments, and probably the lips also, in Mexican fashion (Joyce 1916, pp. 31–32).

Regarding the Miskito groups that settled the Mosquito Coast of Central America, Pim and Seemann communicate in their work of 1869.

> The Mosquito also flattened the heads of their babies, although when they became adults their tangled hair concealed the signs of the deformation (Pim and Seemann 1869, p. 308).

5.2.2.3 Talamanca Area, Costa Rica

Head practices are also mentioned in some areas of the region of Costa Rica, which induces Mac Curdy, to believe that there must be a certain relationship between this region and Perú:

> In some parts of Costa Rica it appears that the custom was not usually practised although in the Talamancan area it was at one time apparently prevalent. A form of frontal deformation seems to have been the rule, and MacCurdy has suggested that connection with Perú might be established through an examination of the archaeological material, although the details are at present too scanty to be discussed (Dingwall 1931, p. 155).

5.2.3 The Amazon Basin

5.2.3.1 Brazilian Amazon Natives of the Nineteenth Century

A brief reference from 1822 notes the practice of this custom in Brazil. Again, the visible result of the cranial transformation is compared to with a bishop's miter.

> Spix and Martius in 1822 simply mention the canoe-shaped cradles in which children are placed and secured whilst their heads are compressed between boards to give them a mitre shape (Dingwall 1931, p. 199).

5.2.3.2 The Amahuaca and Eastern Perú and Brazil, During the Nineteenth Century

The reference to these people is even briefer, discussing the custom of cranial modification at the end of the nineteenth century:

> Similarly the Amahuaca, who live in the high country near the head waters of the Sepauha and Piedras rivers, artificially flatten the heads of their children by tying boards to their foreheads, and in addition they try to compress their noses by tying bands across them (Dingwall 1931, p. 201).

5.2.3.3 Macheyenga, Eastern Perú

Farabee mentions the cultural practice of cranial modification among groups of Macheyengas, theorizing that the custom satisfied the need for group identity:

> A rather different result is achieved by the Macheyenga, a tribe related to the Campa who lived along the middle course of the Urubamba river. Here heads are deformed by binding a board upon the occiput and a roll of cotton over the forehead, so that a groove is formed on the frontal bone, which, it is said, can be felt distinctly (Dingwall 1931, p. 202).

5.2.3.4 Omagua, Eastern Perú

Another nineteenth-century reference to the practice by Jorge Juan y Antonio Ulloa is transcribed by Dingwall. The "monstrous" appearance of the Omagua is emphasized by the author who considers the practice of cranial modification to be very ancient.

> Jorge Juan and Antonio Ulloa describing the monstrous appearance of the Omagua head state that in proportion as the forehead is compressed it rises upwards to such a height that the space between the bridge of the nose and the beginning of the hair is greater than that between the bridge of the nose and the tip of the beard. They add that the practice is of some antiquity, nevertheless, they rigidly conform to it and deride other tribes amongst whom it is not met, calling them contemptuously "calabash heads" (Dingwall 1931, p. 199).

5.2.3.5 Loreto District, Eastern Perú, During the Nineteenth Century

There is also a visual reference from 1860, showing the practice of cranial modeling in the Loreto region of Perú, although no name of any specific native affiliation is provided. The photo is provided by Raimondi who, in the Sarayaco Mission, observed a boy who showed a cranial modification:

> He says that the child had been brought to the Mission to be baptized, and had its head elongated behind, with a rounded projection on the frontal bone, the remainder of this region being much depressed apart from this protuberance. Upon being questioned, the mother replied that there was a hole of considerable size in the board, a statement, which can be paralleled as we have previously seen in the case of the North American Indians (Raimondi 1862, p. 20).

5.2.3.6 Conibos, Shipibos, Ribera del Ucayali, Perú, of the Nineteenth Century

A fairly detailed description of cranial modeling among the Conibo of Perú dates from the turn of the nineteenth century.

> The same custom has been reported of the Conibo of the Pampa del Sacramento and the Ucayali river. Skinner in 1805 says of the Conibo what La Condamine had said of the Omagua, namely, that they flattened their foreheads and occiputs with the view of resembling the full moon and of becoming the strongest and most valiant people in the world (Dingwall 1931, pp. 199–200).

Condamine specifies on the way the compression boards were adjusted:

> the forehead of the child, . . . , is first of all wrapped in cotton, and then a small square board is laid on it, another being applied to the occiput and the two adjusted with cords. Skinner remarks that this practice cannot fail to alter the functions of the brain, and states that the reproach of stupidity has been levelled against certain Japanese priest whose heads are compressed into the form of sugar-loaves, a statement for which I have not seen any reliable evidence elsewhere (Dingwall 1931, pp. 199–200).

Father Sala goes into further detail in his 1897 work when characterizing the the instrument that served for the head-flattening procedure.

> The first part of the operation consisted in the construction of a sort of straw comb-like arrangement made out of a bunch of reeds and fastened with two lumps of metal. This device measured seven inches in length and two and a half inches broad. Under it was placed a small cotton pad or cushion shaped rather like a money roll, and this was fastened firmly behind the occiput by a bandage or thong. As the child was forced to become used to the device just after birth, the head was flattened without much suffering on its part, and the child developed a wedge-shaped head (Sala 1897, p. 80; in Dingwall 1931, p. 201).

The cleric goes on to declare the custom, which was practiced on boys as well as girls, to be barbaric. The simple interpretation he gives for it resides in the high forehead, which was to prevent the hair from blocking their vision.

Church, in his study on the native tribes of South America, likewise briefly mentions this custom among the Conibo, commenting that the purpose of this operation was for reasons of beauty:

> That two boards were used and children who were still being suckled were to be seen wearing these attachments the people thinking the result "very pretty" (Church 1912, p. 185).

5.2.3.7 East Peruvian Natives of the Twentieth Century

Farabee writes at the beginning of the twentieth century on the Conibos and the Shipibo on the banks of the Uacayali River, and mentions that the Conibo still practice head modeling.

> Soon after birth, he says, the head of the infant is fitted with a board bound upon the forehead, and a pad of cotton upon the occiput, and these are left in place for five or six months. A similar method is followed by the Sipibo, another tribe of the Ucayali, and the high C.I. found amongst them is probably due to the distortion arising from the board method of deformation (Dingwall 1931, p. 201).

Cranial modification among the Conibo and Shipibo also fills Reichlen's work, as he describes a device that was used for this cranial modification and details how the device was fastened to the heads of children for 5–6 months:

> The Conibo admire a flat, broad head, and plump arms and legs. Soon after birth, the child's head is bound with a board on the forehead and a pad of cotton behind. This bandage is kept in place for five or six months, which insures the permanency of the deformation. This method is followed also by the Shipibo, and this accounts for the high cephalic index of these two tribes (Reichlen 1961, p. 61).

Tessmann, speaking of Conibo, Shipibo, *Cashibo, Nokamán,* and other natives of northeast Perú, discusses the persistence of the custom up to the time of his writing:

> The Omagua are no longer addicted to the practice, and similarly the custom is not reported among the Kokama, Panobo, Ssensse, Koto, Pioché, Lamisto, Aguano, Kandoschi, Kichos, Bora, Ssabela, Uitoto, Muinane, Mayoruna, Tschamikuro, Chebero, etc. Amongst the Cashibo, who dwell between the Pachitea and the Cushabatay, the customs both of cranial deformation and of nose perforation have been reported.
> The head press is called wuömidi, and consists of a wooden tablet with woven band (nyumbi), which is secured upon the forehead of infants shortly after birth (Tessmann 1930, 211; in Dingwall 1931, p. 202).

Tessmann further specifies that:

> Again the Nokamán (Pano), who are now settled near the source of the Inuya, a tributary of the lower Urubamba, use a piece of apparatus which they call yewuitsígage, and the Auschiri (Pano-Tukano) try to produce long heads by means of bandages, a custom also practiced by the mixed Kahuarano. Similarly the Zaparo, who lived north of the River Tigre, mould and press the heads of their infants to produce long faces, whereas the Ikito and the Pioché try to produce broad faces by the same means (Tessmann 1930, p. 211; in Dingwall 1931, p. 202).

Tessmann goes on to describe the device used by the Conibo and Sipibo to carry out this cranial modification, known as *vuitá-nete*. This devise was applied on the head of the children for at least three months:

> The Conibo constitute, together with the Sipibo and other tribes, the so-called Tschama peoples who have long been known to practice head deformation. Generally speaking it appears that the newborn infant is forced to wear the head-presser (vuitá-nete) continuously for three months, except when the child is being washed. They say that if the head is not treated thus it becomes ugly and thus care is taken to see that the head is properly flattened when the bones are still plastic (Dingwall 1931, p. 200).

In another work by Reichlen, we find a more detailed description of the technique of *vuitá-nete* mentioned by Tessmann:

> D'après le témoignage de Tessmann, en 1928, le vuita-nete (étymologiquement: bande de front) de Indiens Chama de lÙcayali se compose essentiellement dùne planchette frontale, faite dùn bois souple et doublée dùn épais coussinet (don't il níndique pas la matière et dúne angle sagitto-occipitale en forme de T renversé dont la partie horizontale, reforcée au centre par un rectangle plus large, se prolonge des deux côtés par des cordelettes qui sénroulent dans des encoches pratiquées sur les bords latéraux de la planche frontale (Reichlen 1961, p. 60–61).

Reichlen also affirms that:

> Léxtrémité libre de la bande verticale se noue à líntersection des liens qui partent des quatre coins du coussinet et se croisent sur la face externe de la planchette à laquelle ils le fixent. Garçons et filles portaient láppareil dès la naissance et durant deux à trois mois jour et nuit, sauf pendant le bain (Reichlen 1961, p. 61).

Reichlen mentions an additional device used for this cranial modification. This was described later, in 1954, by Juan Comas (1958) for the Cashibo of the Laguna de Yarinacocha:

> Il parle dúne tablette antérieure en bois et dúne sange postérieure, mais pas de sagital. Ságitil dún oubli? La photo, prise de profil, ne permet pas de láffirmer, mais Tessmann, qui ne la mentionne pas non plus dans sa description de láppareil cashibo (vuömidi, et bande de tissu, nyumbi), la représente cependant sur sa Carte de répartition (Reichlen 1961, p. 62).

The author offers still another description, transcribed from the Franciscan missionary Father Sala, who had observed the custom as practiced by the Cunebo, describing the techniques used and the forms acquired by the heads. Father Sala also stated that the modification process was applied to both sexes, edioing previous interpretations.

> También pude observar en las muchas criaturitas que llevaban, el método que usan para aplastar su frente. Primeramente forman una especie de peine de paja del plumero del carrizo, sujetándolo con hilo con dos rieles por la parte de abajo. Este rastre ó peine tiene unas siete pulgadas de largo por dos y medio de ancho. Debajo de este peine se coloca una almohadita de algodón, como un cartucho de cuarenta soles, y por último, se sujeta con una venda ó cinchito muy fuerte, por detrás del occiput. De modo que poniendo al infante recién nacido este instrumento, se va acostumbrando a ello sin mucho dolor, y por otra parte se le va aplastando el cráneo hacia atrás con mucha suavidad y facilidad, quedando al fin su cabeza en forma de un cono ó de una cuña toda su vida. La única razón que dan de esta barbaridad es que de este modo no les tapa la vista los cabellos y tienen la frente más grande, lo que no sucede con las campas y otros chunchos. Y esto lo practican tanto con los hombres como con las mujeres (Sala 1897, in Reichlen 1961, p. 60).

More insightful the testimony by still another missionary, Father Izaguirre, who spent many years among the Cunebo. He communicates his observations on the customs in a work published in 1922. After describing the techniques used for this practice, he interprets head modeling from the native perspective of animistic worldviews, equating the head to the sun:

> El distintivo de la tribu [Cunebo] es el Pánchaue, manera de achatar la frente de los niños de ambos sexos, por la aplicación de un aparato ad hoc, que consta de las siguientes piezas: una tableta cuadrada en forma de rectángulo (abi), sobre la que se coloca un almohadoncito de arcilla (buitanoti), adaptable a la frente del de niño; este almohadoncito esta forado en tela

> de algodón y atado a una ranura practicada en ambos extremos de la tableta, en la misma que se ata el tuibanoti, liga que abraza la cabeza por el cerebro y la coronilla (Izaguirre, cited by Reichlen 1961, pp. 60–61).

Father Izaguirre goes on to detail on the specific measures taken:

> Después de dos o tres días que un niño ha nacido, se le ata el buitanoti a la frente, al principio tan suavemente que solo el peso de la arcilla ejercerá su acción; pero conforme crece se ajusta la liga progresivamente, hasta dejar la frente del niño muy por debajo del nivel del rostro; lo que se consigue después de diez a doce meses. La cabeza trepanada de tan extraña manera toma entonces una forma semejante a la mitra de un obispo; creen ellos así asemejarse al Sol. Naturalmente no todos los niños achatados viven, pues hay muchos que con la vida pagan su tributo a tan bárbara costumbre (Izaguirre 1922; in Reichlen 1961, pp. 60–61).

5.2.4 Southern Bolivian Highlands During the Nineteenth and Twentieth Centuries

Some authors consider that in the Bolivian Highlands native head-modeling practice existed only during the colonial past and by the twentieth century was no longer known. Yet, Bandelier, in his work on the Islands of Titicaca and Coati, published in 1910, mentions the sustained practice of head modeling in some isolated districts of those regions that were less exposed to mainstream national life:

> It is probable that the practice of deformation is dying out today throughout the whole of this area. Chervin stated in 1912 that in the Bolivian highlands the custom was no longer found(. . .)two years earlier that certain of the Indians on Lake Titicana were still practicing the artificial elongation of the head, and doubtless in isolated districts the custom persisted for a longer period than in those regions which were more exposed to the influences of civilization (Bandelier 1910, p. 67).

Various works published in the early twentieth century mention the practice of cranial modeling in regions of Bolivia and Paraguay and relate the techniques used to accomplish this transformation to the importance of forming the heads into predetermined forms. These reasons approximate the arguments stated by native groups from colonial Mesoamerica further north. Dingwall recalls the work of *Nordenskiöld* who reports that:

> When a child was born at Ascención in north-east Bolivia, ... it had an unnaturally long head. The old woman who was assisting the mother during the confinement explained to her that this was of no consequence as the child's head could easily be made rounder, and among certain of the tribes in south-east Paraguay much attention is paid to the infant's head.

Details are provided on the exact measures taken.

> The child, when it is being washed, is thoroughly massaged with the finger tips, no part of the body being neglected. Each finger and toe is pulled and pressed and then the head is manipulated. Mayntzhusen reports that the majority of the children are born with very long heads, but he was not able to take any measurements in order to confirm this. This dolichocephaly is considered very ugly by the natives, and for this reason trouble is taken to make heads broader (Dingwall 1931, p. 203).

5.2.5 Guaycuru and Related Native Groups of Paraguay During the Nineteenth and Twentieth Centuries

Relatively late are the references to indigenous head-modeling practices among some of the Guaycuru or Guaykuru tribes of South America, a set of ethnic groups that at the time of the conquest spanned the present-day national territories of northern Argentina, east Bolivia, Paraguay, and south Brazil. Hutchinson refers to the Paraná:

> Again, further south some of the Guaycuru may have had the same custom at one time since Hutchinson describes some natives whom he thought to belong to this tribe as having an almost total absence of frontal development (Hutchinson 1868, p. 24).

Also G. Mazzoleni refers to the Guaycurus, stating that they have a "ferocious, imposing and terrifying look" (Mazzoleni 1876, p. 540; in Dingwall 1931, p. 202). And F. C. Mayntzhusen (1913, p. 409) adds:

> The form of forehead resulting from the treatment is said to give those who have submitted to it a pleasing and intelligent expression, although the shortened nose is not beautiful according to Western standards (Dingwall 1931, p. 203).

5.3 Final Considerations

Our purpose with this chapter was to make available to the broad academic community (and interested readership in general), the many and diverse references that appear in historic documental sources on cranial modifications among the different ethnic groups that did, and still do, populate a large part of the American continent. We are aware that many references and mentions of the ancestral head practices have long been forgotten, buried within the abundant colonial written legacy, which has given more attention to subjects deemed to be more relevant to the colonizers at the time. Some authors are more exhaustive in their descriptions of this cultural practice; others simply mention it more as an aside. It is clear that it all depends not just on the simple observation of a physical characteristic, but on being able to interpret the hidden or underlying significance at the time when the practice was common among different ethnic groups.

We have learned in the course of this compilation that almost all references in one way or another show an attitude that reproves the custom, including those who attempted to understand the reason for this modification of the heads of newborn babies (see also Chap. 10 and Tiesler and Aguirre 2011, 2013). It is clear that that condemnatory attitude prevented the authors' from understanding the cultural subtleties of why so many peoples attempted to change their physiognomy. Some of the simplistic interpretations that have been given by historic sources are: tribal distinction; a manifestation of social hierarchy; greater strength as a warrior; worldviews and ritual enactment; aspects of domination over the lowest strata of society to make them more obedient or being able to carry greater loads in carrying out their work; still others refer to prophylactic reasons. As stated in Chap. 10, although we cannot ascertain the underlying realities, it is clear that the head custom was deeply rooted among the different ethnic groups, as it continued despite all the prohibitions enacted for its eradication and even, as has been seen, to persist to recent years.

Acknowledgments I am grateful to my colleague, Vera Tiesler, for compiling part of the historical references and for kindly sharing with me the drawings and maps that illustrate this chapter.

References

Archivo General de Indias (A.G.I.) sixteenth to nineteenth century, Fondo Patronato 189, R. 40.

Acuña, C. de. 1641. Nuevo descubrimiento del Gran Rio de las Amazonias. Madrid.

Bandelier, Adolph F. 1910. The islands of Titicaca and Coati. New York: The Hispanic Society of America.

Calancha, Antonio de la. 1974 [1638]. Cronica moralizada del Orden de San Avgvstin en el Perv, con svcesos egemplares en esta Monarqvia, ed. Ignacio Prado Pastor. Lima: Universidad Nacional Mayor de San Marcos.

Casas, Bartolomé de las. 1967. Apologética historia sumaria, ed. Edmundo O'Gorman, 2 vols. Mexico: Universidad Autónoma de México.

Castillo, Cristóbal del. 2001. Historia de la venida de los mexicanos y de otros pueblos e historia de la Conquista. Mexico: Consejo Nacional para la Cultura y las Artes.

Church, George E. 1912. Aborigines of South America. London: Chapman and Hall.

Cieza de León, Pedro. 1984. La Crónica del Perú. Madrid: Historia 16.

Clavijero, Francisco Javier. 2009. Historia antigua de México. Mexico: Editorial Porrúa.

Cobo, Bernabé. 1893 [1653]. Historia del Nuevo Mundo, Vol.2. Sevilla: Sociedad de bibliófilos andaluces.

Comas, Juan. 1958. La deformación cefálica intencional en la región del Ucayali, Perú. In Miscelánea Paul Rivet 2, 109–119. México.

Cruz, Laureano de la. 1900 [1653]. Nuevo descubrimiento del rio de Marañón llamado de las Amazones. Jay I. Kislak Reference Collection.

D'Olwer, Luis Nicolau. 1963. Cronistas de las culturas precolombinas. Mexico: Fondo de Cultura Económica.

Diez de San Miguel, Garci. 1964 [1567]. Visita hecha a la provincia de Chucuito en el año de 1567. In Documentos regionales para la Etnología y Etnohistoria Andina, Vol. 1. Lima: Ediciones de la Casa de la Cultura.

Dingwall, Eric J. 1931. Artificial cranial deformation; a contribution to the study of ethnic mutilations. London: Bale & Sons & Danielsson.

Fernández de Oviedo, Gonzalo. 1945 [1557]. Historia general y natural de las indias. Paraguay: Editorial Guaranía.

Fournier García, Patricia. 1999. La arqueología del colonialismo en Iberoamérica: Balance y perspectivas. Boletín de Antropología Americana 34:75–87.

Guaman Poma de Ayala, Phelipe. 1944 [1615]. Primer nueva Crónica y buen gobierno (1580–1613). La Paz: Instituto Tiahuanacu de Antropología, Etnografía e Prehistoria.

Gillin, John. 1948. Tribes of the Guianas. In Handbook of South American Indians, Bulletin 143, vol. 3, 799–860. Bureau of American Ethnology.

Hernández, Francisco. 2001 [1576]. Antiguedades de la Nueva España. Madrid: Dastin.

Hutchinson, Thomas J. 1868. The Paraná; with incidents of the Paraguayan war and South American recollections, from 1861–1868. London: Edward Stanford, 6 and 7, Charing Cross, S.W.

Joyce, T.A. 1916. Central American and West Indian Archaeology. London: Philip Lee Warner.

Krickeberg, Walter. 1933. Los Totonaca. Mexico: Talleres Gráficos de Museo Nacional de Arqueología, de Historia y Etnografía and Museo Etnográfico de Berlín.

Landa, Diego de. 1982 [~1566]. Relación de las cosas de Yucatán (introd. A.M. Garibay). Mexico: Porrúa.
Leblond, Jean-Baptiste. 1813. Voyage aux Antilles. Paris: Arthus-Bertrand.
Lehmann, Walter. 1909. Reisebericht aus Managua. Zeitschrift fuer Ethnologie XLI:533–537.
López de Gómara, Francisco. 1987 [1552]. La Conquista de México. Madrid: Historia 16.
Lucena, Manuel. 2005. Atlas Histórico de Latinoamérica desde la Prehistoria hasta el siglo XXI. Madrid: Editorial Síntesis.
Maudslay, Anne C. and Alfred Percival Maudslay. 1899. A glimpse at Guatemala, and some notes on the ancient monuments of Central America. London: John Murray.
Mayntzhusen, F. C. 1913.Über Gebräuche bei der Geburt und die Namensgebung der Guayaki. (PICA, 18 Sess., 1912 [1913], 408–412)
Mazzoleni, G. 1876. Gli Indi Guaycurus. (BSG, 1876, Anno x, 2a Ser., i,[xiii], 539–543). pp. 540
Molina-Solís, Juan Francisco. 1943. Historia del descubrimiento y conquista de Yucatán. Vol. 3, Mexico: Ediciones Mensaje.
Montalvo, José R. 1884. Deformaciones artificiales del cráneo. La Habana: Soler & Alvarez.
Palka, Joel W. 2005. Unconquered lacandon maya. Ethnohistory and archaeology of indigenous culture change. Gainesville: University Press of Florida.
Paso y Troncoso, Francisco del. 1926. Códice de la Real Academia de la Historia Mexico: Museo Nacional de México fols. 46, 47 de Anticuarios Nova Hispania.
Pereira, Grégory. 1999. Potrero de Guadalupe: Anthropologie funéraire d'une communauté prétarasque du nord du Michoacán, Mexique. Oxford: International Series 816, British Archaeological Research.
Pim, Bedford, and Berthold Seemann. 1869. Dottings on the Roadside, in Panama, Nicaragua, and Mosquito. London: Chapman and Hall.
Purizaga Vega, Medardo. 1991. El rito del nacimiento entre los incas. Lima: Universidad de San Martín de Porres.
Raimondi, Antonio. 1862. Apuntes sobre la Provincia litoral de Loreto. Lima: Tipografía Nacional.
Reichlen, Paulette. 1961. La deformation de la tete au moyen dáppareils de type "vuita-nete" au Perou et en Malaisie. Travaux de l'Institut Francais d'Études Andines 8: 59–82.
Relación de Michoacán. 1977 [1541]. Morelia: Balsas Editores.
Riou, Édouard. 1867. Photo created for Le Tour du Monde, Paris. Web page: http://www.123rf.com/photo_15180189_cambeba-indian-old-engraved-portrait-brazil-s-amazon-valley–created-by-riou-published-on-le-tour-du.html
Roth, Walter Edmund. 1924. An introductory study of the arts, crafts, and customs of the Guiana Indians. Thirty-eighth Annual Report of the Bureau of American Ethnology to the Secretary of the Smithsonian Institution, 1916–1917. Washington: Government Printing Office.
Sahagún, Bernardino de. 2005. Historia general de las cosas de Nueva España, 4 vols. México: Editorial Porrúa.
Sala, Gabriel. 1897. Apuntes de viage del R. P. Fr. G. Sala, (exploración de los ríos Pichis, Pachitea y Alto Ucayali y de la región de El Gran Pajonal). Lima.
Santa Cruz Pachacuti, Juan de. 1995 [1613]. Relación de antigüedades de este Reino del Perú, ed. Carlos Araníbar. Lima: Fondo de Cultura Económica.
Shattuck, G.C. 1933. The Peninsula of Yucatan, medical, biological, metereological and sociological studies. Washington, D.C.: Carnegie Institution of Washington, no. 431.
Stedman, John G. 1729. The narrative of a five years' expedition against the revolted negroes of Surinam, 2 vols. London.
Tessmann, G. 1930. Die Indianer Nordost-Perus. Hamburg.
Tiesler, Vera, and Pilar Zabala Aguirre. 2011. El modelado artificial de la cabeza durante la Colonia. Una tradición maya en el espejo de las fuentes históricas. Estudios de Cultura Maya 38:75–96.
Tiesler, Vera, and Pilar Zabala Aguirre. 2013. Survival and abandonment of indigenous head shaping practices in Iberian America after the European contact. Submitted as a chapter contribution to the volume: Bioarchaeology of contact, colonial encounters and colonialism, ed. Melissa S. Murphy and Haagen D. Klaus. Gainesville: University Press of Florida.
Torquemada de, Fray Juan. 1969 [1615]. Monarquía Indiana, 2 vols. México: Editorial Porrúa.

Trias de Bes, Luis, Pedro Bosch Gimpera, Jose de C. Serra-Ráfols, D. José Batista y Roca, and D. Alberto del Castillo Yurrita. 1928. Las razas humanas. Su vida, sus costumbres, su arte, Vol. 2. Barcelona: Instituto Gallach.

Ulloa-Mogollón, J. de. 1885 [1586]. Relaciones geográficas de las Indias, Vol. 2. Madrid: Publícalas el Ministerio de Fomento.

Vega, Garcilaso de la, "El Ynca". 1982 [1609]. Comentarios reales I y II. Mexico: Secretaría de Educación Pública, Universidad Nacional Autónoma de México.

Vega, Manuel de la. 1826. Cristóbal Colón: Historia del descubrimiento de la América Septentrional, ed. Carlos María Bustamante. Mexico: Oficina de la Testamentaria de Ontiveros.

Veigl, Franz X. 1798. Gründliche Nachrichten über die Verfassung der Landschaft von Maynas. Nürnberg.

Yépez, Rosaura. 2001. El modelo cefálico intencional en los pobladores prehispánicos del barrio teotihuacano de la Ventilla. Exploración 1992–1994. Master diss., Universidad Nacional Autónoma de México.

Yépez, Rosaura, and Arturo Romano. 2008. Las malformaciones congénitas de la cabeza. Una diferencia con la modificación cefálica de tipo étnico-cultural. Anales de Antropología 42(1): 31–64.

Part II
Regional Practices and Their Cultural Meanings in Pre-Columbian Mesoamerica and Beyond

Unlike Part I, which discussed the manifold relevant aspects of New World cranial modifications from different angles, Part II of this volume targets the Mesoamerican cultural trajectory itself. I am therefore attempting to integrate the different sets of data, expressly, skeletal and contextual information, iconography, and pre-Hispanic and colonial ethno-historical accounts. Jointly, these allow a coherent appraisal of the nuances and changing roles of this body tradition within the Mesoamerican cultural sphere. The six chapters that comprise Part II provide a broad review of the evidence and the state of research, and draw a congruent interpretive synthesis for pre-Hispanic Mesoamerican head practices. I first discuss overarching Mesoamerican connotations and meanings for this body tradition. The following chronological retrospect of head practices, with comments, examines successive stages of Mesoamerican evolvement through the lens of head shapes and shaping. One chapter is specifically dedicated to the data-rich Maya area, from where a copious number of cranial series has been studied systematically. Beyond the data survey, here I seek to foster points of departure for further work, whether local or regional studies, on gender or social identity, on ideological themes, or simply Mesoamerican cultural history.

Chapter 6
Meanings of Head-Shaping Practices in Mesoamerica

As a tradition, cephalic modeling, even though possibly introduced to satisfy specific social needs according to circumstances, the locality, cultural area, or era, must—like all deeply rooted traditions in Mesoamerica—have responded over time to generic ideas that were impregnated with a unified and continuous religious world view. The practice's multifaceted and yet unified quality that is shared with all enduring traditions demands a nuanced approach that concedes a holistic understanding of it as an indissoluble element of the ideological system, relevant to other expressions of Mesoamerican thought (López-Austin 1989). Here I wish to tie cephalic practices to native body metaphors, to the intimate family sphere, and in general the "hard core" of Mesoamerican ideology (López-Austin 2001), supplying both a platform and a starting point from which the role of cranial formation and its deeply-rooted place in the cultural repertoire of indigenous precepts can be valued through the centuries.

After a brief historic overview of studies on Mesoamerican cephalic modification and its significance, this chapter explores some overarching guiding principles, meanings, and purposes of head transformation (and other directly related infant practices) from three different angles: namely, its "organo-plastic", procedural, and emblematic dimensions. Although this division might appear somewhat arbitrary, it is purely heuristic in nature, as it is intended to facilitate points of cultural intersection and conjunction; that is, the insertion of the practice within a unified ritual and ideological complex without which cephalic modeling would, doubtlessly, have been void of any cogent meaning for its practitioners and bearers.

6.1 Attempts to Document and Understand an Age-Old Mesoamerican Tradition

After colonization, a renewed interest in heretofore almost forgotten Mesoamerican cradleboard practices arose during the nineteenth century, now more naturalistic and archaeological in nature. The old and the new study object was no longer the living head-modeling practice itself, which had been abandoned in almost all parts of Mesoamerica long before then (see Chap. 5). Now it was the discovery of artificially shaped crania in abandoned ruins that attracted the curiosity of doctors, naturalists,

V. Tiesler, *The Bioarchaeology of Artificial Cranial Modifications*, Interdisciplinary Contributions to Archaeology 7, DOI 10.1007/978-1-4614-8760-9_6,

travelers, and collectors. The preceding academic community grasped the cultural and medical importance of ancient skull modifications early on, as is noted by a series of early case studies. Later, more extensive works were undertaken on craniological series: skeletal collections of the Aztecs and Teotihuacán, or of the "Old" and "New Maya Empire", as the different stages of Maya cultural development were being called. Samuel Morton's nineteenth century phrenological study of North and South American native skulls also includes Mexican pieces (from Otumba and Pames), the profiles of which display different artificial shapes (Morton 1839; see also Comas 1945, pp. 510–512). Early comparisons of Mexican head forms within North American Nachez skulls gave rise to speculation on pancontinental cultural diffusion and adoption of a cuneiform head shape (Comas 1945).

Unsurprisingly, those areas that once staged severe skull modifications, such as coastal Veracruz and Yucatán or the Usumacinta basin, dominated much of the attention of early scholarship. The literature highlights for example, the extreme bilobée and trilobée forms from Isla de Sacrificios in Veracruz, achieved by prolonged sagittal constriction (Gosse 1855; see also Comas and Marquer 1969, pp. 32–41). In the first volume of his acclaimed trip entitled *Incidents of Travel in Yucatan*, John L. Stephens elaborates on a "deformed" cranium that he had come across near Ticul in the 1840s (Stephens and Catherwood 1963/1963, pp 163–168). In a similar tone, Franz Boas (1890, pp. 350–357) reports severely elongated crania from the Yucatecan port of Progreso. Thereafter, Earnest Hooton (1940) calls attention to the enormous variety of artificial shapes among a series of skulls that Edward H. Thompson had unearthed from the Sacred Cenote of Chichén Itzá years before.

Early interpretations have insisted on the possible ethnic role played by this practice. At the turn of the twentieth century, the Norwegian explorer and ethnographer Carl Lumholtz already uses the criterion of cranial shape in his study on the burials from El Palacio, a site he excavated in Michoacán. From over a hundred recovered skulls he distinguishes (unshaped) Tarascan individuals from others with culturally produced shapes, and speculates on their cultural pertinence (Lumholtz and Hrlicka 1898; Pereira 1999, p. 163). By the mid-twentieth century, Peruvian anthropologist Pedro Weiss (1967) arrived in Mesoamerica from the Andes to recognize, with surprise, patterns that typified the ethnically differentiated ancient skulls in his own country. Arturo Romano (1974, p. 197) argues on the motives for Mesoamerican cranial modification from a more general perspective. He holds that this custom must have involved both ritual purposes and others, besides diverse aggregate meanings, given the manifold expression of skull shapes across the cultural territories that make up Mesoamerica.

Of special interest is that Frans Blom in the 1950s (1954, p. 131) had already alerts about a generic difference in artificial skull shapes in his analysis of the crania found in the caves of Chiapas. He distinguishes two specific types: type "a" modeling (frontal-lamboid, i.e., wide, low cranium) from type "b" (tabular oblique with a narrow, high cranium) and supposes that the latter denote earlier populations than erect shapes. A couple of years later, the 1960s, T. Dale Stewart (1974) made a noteworthy effort to systematize cephalic manipulations among Maya and coastal populations from Veracruz, and to understand the reason for the variety of forms and techniques used among ancient populations. His essay starts out by discerning the artificial modifications sustained by the skulls from the ancient city of Dzibilchaltún, nowadays in the Mexican state of Yucatán, and then compares the patterns with those described for other collections from Mexico, Guatemala, Belize, and Honduras. Stewart concludes his rigorous review by theorizing that the ancient practitioners must have used a combination of head-shaping techniques (correctly distinguishing the use of cradleboards and compression cribs), and confirms Blom's earlier observation that these preferences probably changed over time.

In the initial stages of my own coverage of pre-Hispanic Maya cranial modifications, I use the skeletal and contextual information of some 800 preserved and documented crania from different sites and times of circum-Maya cultural evolvement to approach the questions on social roles (Tiesler 1994, 1998, 1999). Some first results were presented as a master's thesis and highlight the adherence of specific head forms to cultural dynamics that are usually attested to by more conventional indications of the archaeological record. Such is the case for the introduction of narrow "Olmecoid" head shapes in different parts of the Maya territories by the Middle Preclassic, and their disappearance centuries later. Equally conspicuous is the change in head shapes and the loss of diversity concomitant to the drastic social changes during the so-called "Maya collapse". My doctoral dissertation on Maya biocultural attributes adds local facets to the prior regionally centered research. By examining contextualized evidence from different centrally located patio compounds from Classic Copán, Honduras, I examine the female craft of head shapes and their role in family succession and crafted group identity (Tiesler 1999, 2005; see Chap. 9 in this volume).

In general, the outcome of the above research on this body tradition and its social roles confirms the cultural weight of head practices in pre-Hispanic Maya society. Specific head shapes even revealed their use as chronological and geopolitical indicators. Unfortunately, the attempts at comparing the results of the study with other work performed in the area's skeletal register were problematic and reflect the difficulties noted decades before by T. Dale Stewart (1974). Equally limited was the understanding of the custom within a convincing ideological and ritual frames of reference in order to understand since the deeper values and beliefs that once motivated its daily enactment. Although the information that emerged from distribution patterns between the sexes, neighborhoods, settlements, regions, and eras of the Mayan world would have spoken to the importance of cephalic modeling as an expression of social reproduction, these parameters proved to be insufficient when attempting to trace deeper, *emic* motivations.

Apart from the research on Mesoamerican head shaping that is anchored in skeletal sets of data, recent Mesoamerican scholarship has increasingly turned its attention to the study of pre-Hispanic portraiture; some of this work combines reviews of ethno-historical accounts and native notions of materiality and semiology, of physical embodiment, and body theory (Bautista 2004; Bonavides 1992; Cyphers and Villamar 2006; Duncan and Hofling 2011; García and Tiesler 2011; Geller 2004, 2006; Houston et al. 2006; Martínez de León 2007; Pereira 1999; Romano 1987;

Sotelo and Valverde 1992; Winning 1968; see also Yépez 2006; Yépez and Arzápalo 2009). Most of these approximations consider the pre-Hispanic portraiture of artificial head looks either in its visible or its daily components. The latter address the role of head modeling in the context of child rearing (framed by transition ceremonies), while the visible outcome of head modeling is commonly examined in Mesoamerica's visual record: collections of small-scale sculpture, mural painting, and portraiture painted on ceramic ware, etc. Most of this work has focused on the Classic Maya and specifically its elite, arguing on the role of elongated head forms as signs of beauty and distinction, the emulation of the powerful jaguar or of the young Maize God.

6.2 The Head and its Parts Within the Mesoamerican Cosmological System

A viable approximation that grants insights into the possible individual and collective notions of native cranial modeling is the physical embodiment of native perception and of worldviews (López-Austin 1989). In a society such as Mesoamerica, imbued with deep religious feelings and whose daily life was governed by ritual practice, a study object, such as its almost omnipresent head-shaping tradition is suitably approached through the filter of native ideological systems and concepts of the human body. As with other beliefs concerning the body, its physical nature, and its meanings never provide static, immutable, or uniform truths that have been established by the society: Therefore their projections, as part of daily life or in the behavioral dimension (i.e., their performance), should not be conceived of as inherently dynamic and fluid (López-Austin 1989).

6.2.1 Head Practices in Mesoamerican Ideology

It is the expressed goal of this chapter to dissect some of the indigenous motives and meanings of the practice of cranial modification, and to understand those within a dynamic system of native worldviews and past Mesoamerican development in general. Naturally, it is arguably problematic to intend generalizations of these connotations and their potential social undercurrents across the enormous time and cultural space that spans Mesoamerica. Also the lens of modern scholarship, by definition, separates and detaches us from the study object by a great temporal, mental, and cultural divide. Linewise, the separation of the written, graphic, and ethnographic information comes to limit the possibilities of articulating and grasping an internal logic within a given ideological complex. To overcome some of the above constraints, I will intend cautiously to combine and discuss different sources of information, exemplified for Nahua and Maya ideology. The latter establishes a framework for grasping ancient ritual and ideological dimensions for head modeling that is, if not perfectly congruent, at least plausible and culturally akise. These native anatomical classifications provide meanings and connotations for the specific body segments involved in head shaping and assign them functions and value within a unifying system of reference of what is human.

To locate cephalic practice within the broadest social and ideological systems, let's turn to the framework set out by Alfredo López-Austin (1989; see also López-Austin 1998) in his seminal work *Cuerpo humano e ideología. Las concepciones entre los antiguos nahuas* (*The Human Body and Ideology. Concepts among the ancient Nahuas*). This work anchors around the concepts of the body among groups from the Mexican Central High Plains. Following López-Austin (1989, p. 23), we understand *worldview* as the "articulated set of ideological systems that are interrelated in a relatively congruent manner, through which an individual or social group at a moment in history attempts to understand the universe." It is precisely at the level of the *ideological system*, defined as an articulated set of ideological elements, i.e., individual and collective preferences, concepts, attitudes, and beliefs (López-Austin 1989, p. 23), where cephalic modeling finds its place. It is really here where the series of maneuvers with their meanings and visible outcomes would have operated. This accords with the individual or group *Weltanschaung*: the background that was to confer cultural relevance to the practice.

Importantly, the notion of worldview bestows significance upon dynamic change in ideological systems and upon conflicts that can occur between antagonistic communities within the social fabric (López-Austin 1989, p. 22). This connotation facilitates tenets regarding antagonistic dynamics among populations and their factions, such as imposition and domination, resistance, and subordination versus assimilation, which also are played out ideologically. Regarding Mesoamerican society itself, the groups that settled and defined it were in constant motion, just like other societies. Their beliefs and customs were in dynamic change over time and space and according to their social undercurrents. Changes in ideology and ritual tended to become more pronounced during times of social tension and crisis, as basic structural shifts were often the catalyst for their reconstitution and reorganization. Such phases of transformation and change identify for example the social and ideological impact of Teotihuacán in the Mayan Lowlands at the end of the fourth century B.C., the reconstitution of Highland society at the end of the Classic period (during the so called Epiclassic), and European Conquest—arguably the most serious collective catastrophe of all for Mesoamericans. Each of these episodes reveals contradictions and readjustments in belief systems and ideologies (López-Austin 1989).

These ideological readjustments are also patent in our chronological charts of Mesoamerica's cranial record, an aspect to be examined in detail in the following chapters. Shifts in or replacement of technical preferences and their visible outcomes are documented for example in the archaeologically retrieved cranial record from the Maya area, where the most dramatic changes in artificial head form surrounded the Maya collapse during the Terminal Classic and during the harsh aftermath of Iberian contact (Tiesler 2012; see Chaps. 5 and 10).

Before the Spanish conquest, head modifications in the form of cradleboarding and head splinting were practiced among the vast majority of Mesoamerican settlers and territories for millennia. Passed on from one generation to another, the custom participated in the collective and long-lasting construction of the group identity, constituting one of the deeply rooted Mesoamerican traditions following the definition assigned by López-Austin and López-Luján (1996, p. 62; see Chap. 2.3). While the techniques used in this cranial formation were subject to changes over time, the collective shifts in head looks decode underlying social changes, social transformation, and replacement over the centuries. Yet the infant practice itself, as a vehicle for ideological expression, shows a surprising permanence in the cultural repertoire, counting among the intimate elements, the "hard core" of the Mesoamerican ideological framework (López-Austin 2001).

6.2.2 Native Semantics of the Head

How to elucidate the ideological meaning of this "core" Mesoamerican tradition? In order to allow an integrated view of the reasons for culturally motivated cranial modifications, semantic connotations of the body and specifically the head with its spiritual components are pertinent sources of information. In the following paragraphs, I shall start out by summing-up a set of anatomic and specifically polysemic notions held among Maya with incursions into Nahua body notions and body semantics in Mesoamerica in general. I recognize that this enterprise is far from complete, given the many languages and language groups that span the Mesoamerican cultural landscape. But for now it is a stone of departure, which hopefully will facilitate further exploration of more specific regional and local connotations and metaphors.

The ethno-linguistic *modus operandi* that is used to interpret native notions human anatomy is a polysemic evaluation of the various meanings to anthropomorphic language that "embody" semantic categories according to a classification borrowed from the body and its functions (Brown 1976). This semantic interpretation of the human anatomy and its segments holds great potential for Mesoamerican language analyses given its broad use of the metaphor. This approach, at least in principle, allows conjecture on the attributes, values, and mutual relations between the organs and in general between the different parts of the human body, and using these as projections in thought and in spoken and written language. Cognitive approaches of this type in Mesoamerica have been successful in studies made of anthropomorphic concepts among both groups of today as well as the past (see for example, Bourdin 2007; Garza 1990; Houston et al. 2006; López-Austin 1989; MacLaury 1989), despite the limits and challenges inherent in dealing with lexicons that are extraordinarily complex, ambivalent, and in part nonexistent.

The concept of partonomy (the hierarchy of parts in relation to the whole) also plays an important role in interpreting the concept of the head and its components in the panorama of Mayan languages (Fig. 6.1). Mayan speakers frequently refer to body terminology to designate objects ("partitives") and to describe their location in

Fig. 6.1 Mother who carries her baby. The little one shows three isolated locks of hair on the head. (Adapted and drawn by M. Sánchez from Houston et al. 2006, p. 50, Fig. 1.53; Kerr Archive 7727)

space ("locatives") (Bourdin 2007, pp. 16–18), thereby signaling their body-centered thinking, in which the human body anchors all structure and interpretation of the world. General discourse normally grants not just topographical notions to the parts of the whole, but also establishes semantic correlations between their appearance and their function (Tversky cited in Bourdin 2007, p. 57). These semantic loans (or anthropomorphic projections) elevate the human topography to a model of reality that mirrors the native cosmos, and inversely facilitates the attribution of its cultural meanings.

Similar to the Tzeltal term *ghol*, in Yucatec Maya the head, *hool* or *pol* (in colonial Yucatec Mayan), *hó'ol* or *pòol* (in modern Yucatec Maya), is synonymous with "end point", "top", or "roof" (Barrera-Vásquez 1995; Bricker et al. 1998; Ruz 1992, p. 154). These commonly accepted meanings of "end point", "top", and "roof" are also found in other languages of the Mayan Lowlands such as Ch'orti', which uses the cognate *jor* for "head" (Alfonso Lacadena, personal 2009; Pérez et al. 1996; Wisdom 1950). This same word of *jo'l* ~ *jol* "head" (spelled as *JOL-la, JOL-lo, jo-lo*) is documented in hieroglyphic texts from the Maya Classic period. Then, the same semantic association was established between the "head" and the "roof", which recently led William Duncan (2009, pp 187–188; see also Duncan and Hofling

2011) to compare the artificial modeling and covering of the skull with the process of thatching a house (see also Tiesler 1999, p. 330).

Mayan writing from the Classic Period also used a second term to designate "head", as deciphered from the glyphic inscriptions. The term *Baah*, deriving from "head" or "forehead" (written as *B'AH-hi, b'a-hi, B'AH-ji*), uses the head as a personifier. It designates the human body as a whole, and beyond that, refers to the person (Houston et al. 2006, p. 6). One interesting note is that the Classic Period glyph that represents the term *baah* is different from the glyph that designates the human face, which appears as *bail* in Tzeltal (also used as the reflexive in Tzotzil and in other Mayan languages). This term is glyphically read as *ut*, *hut*, or *wut*, and the glyph is also used to identify the "face of a fruit" or "eye" (Houston et al. 2006, p. 28, p. 55, p. 60). This partonomy also plays a role in the Yucatec Maya lexicon with its double anatomical meaning of *ich* ("eye" and "face"). In this specific case, Bourdin (2007, pp. 202–203) proposes that the relationship, more than a contiguous association, denotes a "continent-content" relationship, arguing that the lexeme *ich* is used as the preposition "in". Also note that the jaw (or chin) is identified by at least three terms of the Yucatec lexicon, indicating the importance and frequent language designation of this anatomical region (Bourdin 2007, pp. 214–215).

The upper part of the face contains the forehead, identified in colonial Yucatec Maya by the lexemes *lec* and *thab* or *tab*. This area seems to harbor an ambivalent position between the face and the rear section of the head, establishing no relationship or belonging. Conversely, Tzotzil speakers use their own term, *quitba*, to refer to the forehead (Houston et al. 2006, p. 60). The Calepino de Motul (Arzápalo 1995) uses the term *thab,* which also refers to "bald" and "dandruff". Unlike Bourdin (2007), Houston et al. (2006, p. 60) assign a generic meaning to the "head" and the "face" that comes close in meaning and argue that it is the forehead and upper part of the head that permits the recognition of reflective thought and, ultimately, personal identity. This is also the locus where Classic period iconography places the exquisite diadems and headdresses on royal heads and where the personal name of the paramount is written who wears them (Houston et al. 2006, p. 68).

Apart from references to the top (*mollera*) or "roof" of the head and the face itself, the glyphic vocabulary is conspicuously silent regarding the back of the head or its occipital prominence, an aspect to be examined further down. As previously noted, Yucatec Maya does seem to establish a binomial lexicon between the face and the portion of the head that contains the brain. The latter, denominated in colonial Yucatec Mayan as *hool* or *pol*, is identified with the upper and posterior surface of the head; it refers both to the hair and to the top of the head without hair (Bourdin 2007, p. 214). The areas it involves are designated as *tzek'* (dry "skull") and *tz'oomel* ("brains", i.e., the wet part). Yucatec speakers of Hocabá refer to the bony cranium with the same term as a round gourd, or *leek*, identifying it with something round or spherical (Bricker et al. 1998, p. 167). Note that *hool* is combined with the term *kinam* (also "strength," "power") to describe a strong headache. There are also expressions that denote fright, fear, and other hazards prone to attack mental health. *Pol*, on the other hand, is used to refer to the qualities of ingenuity and understanding, although these facets of personality appear to be associated with the spirit, a distinct feeling that is found in the heart (Bourdin 2007; see also López-Austin 1989 for Nahua groups).

The semantic dichotomy between the face and the neurological sector of the head goes in tandem with Nahuatl ideology, which places the conscience and reason at the crown or top of the head (*cuaitl, atl*). This is not only the locus of joy but also for alteration, confusion, and anxiety (López-Austin 1989, pp. 210–211, p. 219). More anterior is the face or *ixtli* (both "eye" and "face"), which is recognized as the center of sentiment in the Nahua language and refers to the individual's state of mind. According to the same author, *cuaxicalli*, "the vessel of the head" is the "cranium", understood as the upper part of the human body, without the face. The related term *cuaxicalitzopyan* is translated as the "fontanel", "peak of the head", or "place of the puncture in the cranium". Similar to Maya lexica, the Nahuatl vocabulary is sparse when dealing with the rear or occipital section of the head in comparison with the breadth of the lexicon for the peak or "roof of the head" and for the face. Here, we find the term *cuexcochtla* ("occiput"), metaphorically called the "place for the skirt of sleep" (together with the note by López-Austin marking this as "very doubtful") (López-Austin 1989, vol. II, p. 153); and another word, *cuexcochtetl*, which refers to the formal properties of "acute", "prominent", and "protruding".

Returning to the Mayan lexicon, Mario Humberto Ruz (1992, pp. 154–163) translates "occiput" to Tzeltal as *pat gholol* or *pat gholal*. This term identifies the rear part of the head, an anatomical portion that is considered to be wrapped like a stalk or a shell. The *Diccionario Maya Cordemex* (Barrera-Vásquez 1995) translates the colonial Yucatec Mayan terms of *tzuk bak*, *pachka'*, *pachkaa'*, and *paachkab* alternatively as the "nape" or "back of the neck", "occiput", "brain," and "neck". All these terms identify the occiput therefore with the non-facial part of the head and, in particular, with the dry or "bony" segment of the head. Topographically, it is located in the rear section or "back" of the head *pach caa.* In addition to "*head*", *caa* also designates a round gourd or squash.

The Diccionario Maya de Motul, the first Mayan dictionary compiled after the Conquest, also shows the occiput as *tzuc bac* or *tzuc bacel*, where *tzuc* refers to a "hank of hair" and *bac/bacel* as "bone" (Victoria Bricker, personal communication 2009; see also Bourdin 2007, p. 316). The Calepino de Motul similarly offers the terms *tzotz* and *tzotzel* to designate hair of all sorts (Bourdin 2007, p. 216). The identification of the occiput as a combination of "hair" and "bone" may be relevant, knowing that at the time of contact it was common to shave the head except for the rear part. Don Juan Cueva Santillán notes in the Relación de Izamal that "[the locals] wear the hair on their head short, unless those of the neck which they allow to grow" (Barrera-Vásquez 1938, p. 191). Transcribed by Gaspar Pedro de Santillo, he similarly notes that: "perhaps for elegance, all the men grow their hair in the rear half of their heads very long like women, and the front hair is so short that it is cut almost to the root of the head, it is understood that they burned or singed them because this is the remedy for cutting hair for lack of scissors" (Barrera-Vásquez 1938, p. 199). An analogous custom occurred among Nahua groups, Totonac and Chichimecans during the times of European contact, as Bernadino de Sahagún (1989, p. 598, pp. 602–603) corroborates when referring to the rear strands of hair that youngsters let grow over the occiput, while shaving the remainder of their heads (Soustelle 1956; Stresser-Péan 2011, p. 143). The hairstyles among infants and children invite a number of further questions regarding the seats of vital energy in hair and the protective role of hair strands during infancy and beyond.

Another way of alluding to the back section of the head is *u yit u pol*: "the rear end of the head", although we do not know how this compares with the notion of *pach*, "back". The lexeme *it*, which is used in the expression *u yit u pol*, in today's Yucatec Mayan is translated as "rear end" or "bottom", and allude to both the toponomy and the attribute of having a convex shape. Yucatecan speakers use the same term when referring to soil sitting on a surface or liquid that rests in the bottom of a round vessel (Bourdin 2007, pp. 335–337). The Tzotzil folk of San Lorenzo Zinacantán identify the occiput in a similar fashion, referring to it as *pat jol(ol)*, the back of the head. The expression *paj-jol* describes a person who makes jokes or who has a long head. The syllable *pal* used in the expression *pat jol*(*ol*), means "to be molded" or "to mold with mud or something soft", perhaps alluding to the practice that centuries ago was used precisely to flatten the back part of the head (Laughlin 2007). Also note that in many of today's Mayan languages the term "occiput" refers to something that is "shameful" or "bad" (Otto Schumann, personal communication 1997).

After our brief polysemantic excursion through modern and historic speech and the ethno-historical terminology available on the occiput, we may ask: What partitive and locative qualities and indigenous spiritual qualities are attributable to it? Indigenous terminology appears to refer first to formal qualities and partitives without transmitting any spiritual value associated to this part of the head or its participation in any spiritually significant area, such as the crown or roof of the head, once—and now—considered an important spiritual entity for the being and the locus of conscience and reason. On a broader level, the connotations on the occiput seem to be related to the part covered by hair, the dry and bony part. Other accepted meanings have negative connotations or are frankly pejorative. We wonder if the back of the head could denote an anatomical sector that is given a lower spiritual value or level of vital forces or that is vulnerable, following the argument by López-Austin (1989, p. 217) on the potentially harmful role of head elongation among Nahua groups.

6.3 The Organoplastic Role of Cranial Modification

Our brief metaphorical entourage of Maya and Mesoamerican "cephalic phenomenology" underlines the important places of the body and, in particular, the head in indigenous world views, both past and present. Some anatomical segments discussed here, such as the crown of the head, the forehead, and the occiput, are worthy points of departure in attempting to assign meanings to past plastic head transformations. By providing semantic connotations of the occiput and its role in relation to the body and the Mesoamerican cosmos, ethno-phenomenological discourse gives us a potentially relevant clue to explain cephalic transformation in the *Mexica* lexicon, transcribed by Fray Alonso de Molina (1944). It includes a table with illnesses of the head including "cephalic elongation," an ailment described as the elongation of the head (long, narrow skull shape). In this transcription, cephalic elongation appears among such mental ills as insanity, fainting, and evil, similar to the illnesses previously mentioned here (fear, fright), things that affect the *hool* ("head") among the Maya of Yucatán.

Alfredo López-Austin (1989, p. 212) recognizes this categorization as a possible motive for cephalic modeling among Nahua groups and links it to an imbalance of the *tonalli*, a gap between the spiritual functions of the brain and the heart (the *cua* group). Without directly specifying which anatomical portion could be the possible *locus* of this imbalance (although it insinuates that it would be the dual allocation of thought in both above the heart and in the brain that was in imbalance), the author conceives that the peculiar "lengthening of the head" may have aroused concern, sufficient to engender "... cranial deformations which tended to cause the contrary effect, or shortening of the head". In that case we can suppose that the purpose of the practice was to increase the individual's mental functions, or at least to distance him or her from evil and madness" (López-Austin 1989, p. 212).

A couple of pages below, López-Austin (1989, pp. 224–225) describes the Nahua belief of loss of the infant spirit, according to which the infant *tonalli* is at risk of abandoning the body through the soft fontanels of the infant skull which had not yet ossified. The Nahua also believed that in addition to the fontanels, the occiput was a source of danger to the health and integrity of the child, and therefore recommended that haircuts at the rear part of the infant head should be avoided, hair being believed to protect the *tonalli* from leaving (Stresser-Péan 2011, pp. 129–130). Inversely, captives' and criminals' hair was cut by the victorious as a punitive measure or to obtain their life force. Prior to being sacrificed and flayed in honor of Xipe Totec, the Nahua earth deity, the captives were deprived of the hair on the top of their heads, believed to be the seat of the individual (Stresser-Péan 2011, p. 129). Also sources from Yucatán affirm that the hair covered mainly the occipital part of the head, while the front or upper part of the head was shaved or the hair cut in the form of a lock (Barrera-Vásquez 1938). This lock was considered to be filled with life force, and was retained, such as fingernails, until the death of a person had occurred, to be buried with the body (Guiteras 1986; Houston et al. 2006, pp. 25–26).

Returning to our discussion on the imbalance between the spiritual centers in the Nahua world vision and the pernicious role of the occiput; this notion appears also relevant in spiritual–physical beliefs of some Maya groups (Martínez 2007). According to the Northern Lacandon Maya, the heart or the animated spirit (*tukul*), is seated in three different parts of the body: in the forehead, under the ears, and directly over the *pixan,* which means "heart" as well as "spirit" (Boremanse 1998, p. 84). Mutual ties between these different spiritual sites allow the *tukul* to descend from the forehead to below the ears and from there to the part of the chest where the *pixan* is seated. Consequently, the *tukul* comes to rest over the *pixan*, achieving a healthy integration of feeling, thought, and animated spirit. The Lacandons believe that this balance of thought and spirit (*anima*) is acquired with maturity. It is held that infants do not have a *tukul* until they become older and more mature (Boremanse 1998, p. 81, p. 84). As with the Lacandon, the Tzotzil of San Pedro, Chiapas, affirm that the mind and the heart interact through the blood (Guiteras 1986). This interaction can occur in the form of cooperation or by struggle. Discord between the spiritual centers creates illness and emotional imbalance in need to be cured. It is therefore desirable that collaboration be established between the two sites, which occurs when "the soul is repeated in the heart", i.e., when the mind subordinates to the feelings of the heart. It is only then that true wisdom is achieved (Guiteras 1986, p. 235; see also López Austin 1989, pp. 211–212).

There are also specific situations that might threaten the integrity of life or its spiritual components. As with the Nahua, the highland Maya Tzeltal affirm that the spirit-heart (the *ch'ulel*) can escape the body through the mouth or the crown of the head (Pitarch 1996). Newborns are more susceptible to this, as they still have not accumulated sufficient vital heat to anchor the ch'ulel firmly in their bodies. Highland Maya Tzotzil thought is similar, maintaining that the infant *ch'ulel* has not been well anchored in the bodies of their babies; therefore, the spirit easily becomes volatile and separates from the body (Martínez 2007, p. 161; Page 2005, pp. 210–211). Teneek natives believe that a cold wind enters through the crown of the head as the spiritual energy exits the body (Martínez 2007, p. 161). The permanence of the *ch'ulel* is at risk, for example, when the "evil eye" (*mal de ojo*) puts the little one in peril or when fright or illness threatens its physical integrity and life (León 2005).

Maya communities know a series of remedies that they consider to be effective protection against the loss of their babies' *ch'ulel*. One such protection against the "evil eye" includes swaddling their babies and keeping them inside the house. A host of other, more specific measures are equally designed to retain and anchor the soul in the little body. Zinacantan Mayas, for example, tie their infants' arms and ankles with ropes in order to keep the body "closed" (León 2005, pp. 128–136) and thereby prevent the spirit from fleeing. The Tzotzil Mayas of Chenalhó fasten a wax disk to the crown of the head and then tie the babies' hands with cotton until they have completed their first month of life (Guiteras 1986, p. 102).

Putting the needs of natives to protect the infant head against spiritual loss into perspective with the negative characteristics that are attributed to the rear portion of the head, cephalic flattening appears in a new light; namely, that of containing and reducing the protrusion of the occiput or "crown of the head", a potentially vulnerable part of the human anatomy, especially in newborns. Francisco del Paso y Troncoso, translated from the original Latin, describes head modeling for colonial era Nahuas, who considered it beautiful to possess "small foreheads, rich with hair and a practically inexistent back of the head that was compressed by the midwifes by applying a weight from the time of birth, when the cranium is soft, and maintaining this form when the child is laid in the crib" (Paso y Troncoso 1926, p. 25). Still more explicit are testimonies regarding the Guatemalan groups who had adopted Nahua customs by the sixteenth century. Francisco López de Gómara affirms in his *Historia de la conquista de México*:

> The midwives manipulate the babies so they have no occiput, and mothers lay them in compression cribs so that they do not grow them because they consider them beautiful without them" (López de Gómara 1987/1987, p. 246).

This early colonial citation not only communicates native motives for cradleboarding, but also provides additional information on a potentially separate practice that was performed by the midwives although for the same purpose of eliminating the protruding back of the head. Unfortunately, the citation does not specify how the midwives actually managed to reduce the occipital protrusion.

Also Western Mesoamericans appear to voice their rejection of elongated heads. In the *Relación de Michoacán*, a native from the area of Lake Patzcuaro laments his being of small stature and feels ashamed for his rounded head, "which is not [the head shape] of men of braveness". Some lines further down, the account explains that among the important folk (*los Señores*), "the heads were compressed and flattened as a cake" (Anonymous 1977, p. 145; see also Pereira 1999, pp. 166–168). Again, the motive for head flattening appears to target the elimination of the occipital bun, considered a preventive of courage and valor.

On the other extreme of Mesoamerica's geography, early colonial eyewitnesses from Yucatán allude to the role of the occiput in the act of cranial modeling, although less emphatically (Contact Yucatecans specifically used the term *up' k'abtah* to describe the act of straightening out the head of the newborn and to mend or to adapt it (Barrera-Vásquez 1938, p. 901)). In the frequently cited reference from the *Relación de las Cosas de Yucatán*, Fray Diego de Landa (1982/~1566), p. 54, affirms that the occiput of the little ones was squeezed until the head was flat. Note that according to the *Relación de Izamal* (Barrera-Vásquez 1938, p. 191), the occiput is also the only part where Yucatecans let the hair grow during the time of European contact. This hair arrangement is similar to the aforementioned hairstyles from other parts of Mesoamerica, probably a complementary protective measure (López-Austin 1989, p. 225). Unfortunately, little historical source information is provided in ethnographic or ethno-historic literature on infant hairdos and haircuts to explore any potential role of hair in shielding the young ones from harm (Stresser-Péan 2011, p. 143).

Pre-Hispanic imagery depicts babies either wrapped, covered with cloth, or altogether without hair. Only a few representations show some isolated locks, among them a Classic Maya mother–child scene, which depicts an infant who sits astride the back of its mother. This scene is of interest because of the peculiar distribution of three distinct hair locks on the baby's head vault. One appears just over the forehead, another one covers the lambdic area of the head just behind *vertex*, and a third strand covers the flattened occiput behind the ear (Houston et al. 2006, pp. 49–50; Fig. 6.1). Anatomically, the two anterior locks sit on the locations expected for the anterior and posterior fontanel; both are soft and pulsating parts in the immature cranium (Furst 1995). The third strand of hair grows amidst the flattened occiput, on what would be the occipital prominence in a rounded skullcap. Recalling the historic references to the protective function of the hair and the vulnerability of the fontanels and the occiput in indigenous thought, this representation appears to suggest that this function could have found its expression in Maya infant hairstyles already back in the Classic period.

At this point, I bring to mind one particular cultural feature located in the center of the occiput, referred to in anthropological literature as the "suprainiac depression" or "suprainiac lesion" (see Sect. 4.4.4 for a detailed description). This depression, which is observed in most Mesoamerican skeletal series, shows different dimensions and contours. This depression is occasionally accompanied by signs of periosteal reaction, sometimes (although rarely) penetrating the occiput towards the endocranium, i.e., the internal surface of the skull. In most calottes, this mark is located directly over the occipital projection that identifies the *inial* point of craneometric literature.

Current literature links the presence of supra-inial depressions to the use of the cranial formation devices and argues that they might stem from the compression spots transmitted by knots, ties, or cushions. Other works have noted that this feature, although commonly linked to artificial cranial-vault modification, does appear also in unmodified craniums, sustaining the idea that artificial thinning of the occiput may also have been a practice independent of that of head modeling (Tiesler 1999, 2006; see also Lagunas 1970, 1972). Maybe this initial measure was designed to secure the compression vectors of the later daily adjustments of compression planes on the little one's head or to complement its flattening effects? Was it to neutralize what was thought of as a *locus* imbued with a pernicious or negative energy, a spot especially vulnerable to the entry of "bad air" or loss of vital energy?

Some notions of native head "phrenology" are also projected by native portraiture and come to add to our present discussion. Although these only confer preferences and trends whose immediate motives cannot be directly ascertained, they do trace individual or convention-driven artistic penchants for figurative rendering. These complement the above lines of argument on the native ideological significance given to the different portions of the head. Some of these trends go back to Preclassic times. Gulf-coast Olmec head sculptures, such as the known ceramic and jadeite figurines or the sculpted colossal head rocks, appear uniformly modeled with a high and narrow head vault (see for example Coe and Diehl 1980; Cyphers and Villamar 2006). These images present long, narrow foreheads, occasionally even buckled, while the posterior section of the head is rendered completely flat or even depressed. In profile, the back of the head is delineated as a straight or concave line immediately behind the auricular pavilion, thereby completely annulling the occipital protrusion (see Fig. 7.5). This rendering is not anatomically feasible, even in crania with extreme cultural transformations. The question arises whether the artists wished to exaggerate the visual effect of the occipital reduction caused by cephalic modeling, perhaps because this seemed culturally (and thereby aesthetically) pleasing to them.

Also later Classic and Postclassic portraiture follows this pattern in representing head profiles in Mesoamerica, quite independent of the artificial head form. Let us examine Central Lowland Maya ceramic painting, which stands out for its high degree of anatomical realism. This is coupled with artistic craftsmanship and elegance of stroke, facilitating an approximation of head countenance in the eyes of the artist of court scenes, as shown in the *Corpus of Photographs of Maya Vases*. Most head vaults are drawn with a reclined forehead and an artificially recessed capillary insertion line (Sánchez 2008). In profile, they are observed alternatively as conical or tubular, but invariably with no occipital volume or inion protrusion but a straight or reclined flat occiput instead. As in the early Olmec renderings, this part usually appears delineated immediately behind the auricular lobe. In some cases, the ear lobe itself delineates the line between the nape and the upper part of the head or *vertex*. Consequently, the neurocranial volume appears disproportionately reduced to that of the large face, which protrudes and dominates the features of the head. The hair line, pulled back to expose the high and receding forehead, highlights this impression, as do the different ornaments that are worn on the head such as circular wraps, diadems, ropes, and headdresses. These features likely emphasize desirable, beautiful attributes, like the out-jutting facial features and the receding forehead among Classic period Maya. By contract, less desirable attributes appear reduced, hidden, or eliminated in the artistic conventions, as we have argued for the round rear of the head in native Mesoamerican rendering.

6.4 Head Shaping in Child Rearing and Integration

Let us move on to the behavioral dimension of artificial vault modification and its performance in daily life. Again, eyewitness testimonies by Colonial writers of the New World, both from New Spain and Perú, offer valuable points of comparison that are fundamental in order to reevaluate the reasons for its enactment by Mesoamericans (see also Chaps. 4, 5 and 10). Witnesses in sixteenth century Perú, both Europeans and natives, view the native head-shaping practices not just as one of the many aesthetic recourses, but consider them of great sociocultural importance, such that they would visually confer an identity with social hierarchies or lineages and even entire ethnic groups (see for example Cieza de León 1984, p. 227; Cobo 1893, p. 175; Garcilaso de la Vega Weiss 1982/1982, p. 333). In fact, the apparent ethnic function of Andean cranial expansion would later become a resource and guide for developing the classifications that would be used in the archaeology and cultural osteology of South America (Weiss 1962). Centuries later, some early studies of Mesoamerican cranial-vault modifications assume *a priori* that the custom of head modeling in Mesoamerica must likewise have served as an ethnic emblem and a mark of privilege and exclusiveness (Krickeberg 1933, p. 47, 1961, p. 88, p. 269, p. 323, p. 339). However, not much information could be accrued to confirm direct or indirect conjectures on the role of artificial head modeling as visible identifiers of group identity as in the Andes. The paucity of colonial descriptions on native head practices within the Mesoamerican area is surprising, especially in light of the general eloquence of cultural narratives and the acceptance of cultural head modeling in many areas, as seen by the cranial collections from the Late Postclassic period (Dávalos 1951; Romano 1974).

The meager colonial references to the practice of cranial-vault modification and its motives do include a description made by Francisco del Paso y Troncoso (1926) on the Nahua, already referred to in the previous section, in which he speaks of the occipital reduction of the cranium (Chaps. 5 and 10). In the meantime, Fray Bernadino de Sahagún (1989, p. 606) mentions artificial head modification, as practiced in Veracruz, in a nonchalant statement that the Totonac "have a long face and flat heads and their lands are very hot". Fray Francisco de Bobadilla, repeats the idea that this modification deals with a docile nature and hardening of the head to support greater weight, basing these reasons on a dialogue construction, according to which

"When the children are born they have soft heads and they make them the way that you see that we have them with two planks of wood on the sides dividing it and a big gap remains in the middle of the head that divides both sides; because our gods said to our ancestors that in this way we are beautiful and good men, and the heads are tougher for the burdens that they bear" (Fray Francisco de Bobadilla, cited by D'Olwer 1963, p. 352). For the Yucatecan Maya, Fray Diego de Landa describes the techniques and inherent risks in this modeling process, and also introduces another idea related to the festivity that followed the conclusion of the process. Then the priest predicted the future of the child and the name was pronounced that the little one was to carry during infancy (Landa 1982/~1566, pp. 54–58; Chaps. 5 and 10).

Although succinct and ethno-centric, colonial testimonies on the Mesoamerican area offer valuable information on the behavioral roles of cranial modeling at the times of contact. In addition to occipital reduction, already discussed in the previous section of this chapter, many sources voice the discomforts suffered by the babies. Unlike the Andean testimonies with their detailed descriptions of the final forms, Mesoamerican scholars confer greater importance on the compression process itself, pointing to the cultural importance that this custom must have had as part of the child-rearing process, the devices and—Landa in particular—the festivities that culminated the modeling process.

From our incursions in colonial references from the Caribbean, South America, and Mesoamerica, it is patent that sixteenth century Mesoamerican practitioners must have conducted infant head modeling for purposes distinct from those of other geocultural spheres, specifically the Andean. Here, head compression seems to have been linked more to the process itself rather than to the final shape left in the head. This system, together with the specification on the first stage of infancy, invites reflections on the role of this practice in evolving personhood and the social integration of infants. These in turn tie in with numerous other aspects of ancient childhood experienced within the family unit, the community, and in society in general (see also Chap. 2).

The course of Mesoamerican infant life proceeded in successive stages whose transitions were often sanctioned by ritual, sometimes including ceremonies of great pomp and importance. The collectively sanctioned acts of transition and integration of the child legitimated its changing roles and conferred integration and identity to the little one living through them and growing up. Many pre-Hispanic infant ceremonies are still recognized today, such as the ceremonies surrounding birth, the first haircut, Yucatecan, and Lacandon *hetzmek* ceremonies or the ritual of the *caputzuhil*, for example (Bonavides 1992). Let us analyze the possible roles of artificial head modifications within this system of transitions and transformations, starting with birth. Much has been written on pregnancy and conception among pre-Hispanic and post-contact Mesoamericans, their taboos and prohibitions, preparations and care, many of which are still observed in communities today.

As in other cultural spheres of Mesoamerica, among Maya groups the pertinent care was ideally supervised by family members with the assistance of two people from outside the nuclear family that are repeatedly mentioned in literature (Nájera 2000). One was the midwife who was responsible for the wellbeing of the mother and her child before, during, and immediately after birth (Fig. 6.2; see also Fig. 4.11).

Fig. 6.2 Elderly Maya woman fondles the forehead of a baby that is strapped into a crib. (The Museo Popol Vuh, Universidad Francisco Marroquín, Guatemala catalogue no. 0296; photo by V. Tiesler)

The other person who was granted importance later on in infancy was the priest, who discerned the life path of the baby boy or girl (see Bunzel 1952; Landa 1982/~1566; Villa Rojas 1978). In the time after delivery, it was the midwife who played a predominant role, especially during the puerperium of the mother, a threshold phase that marked the birth of a new being in the nuclear family (Cosminsky 2001; Nájera 2000). As in other Mesoamerican cultural sectors, the puerperium was and still is considered a time of imbalance, of vulnerability and risk to the *ch'ulel* (the spiritual "sacred heart") among the Chiapanec highland Maya, especially for the mother and the child (Vogt 1965, p. 29). This condition made the reclusion of the new mother essential, while the midwife cared for her and for the physical and spiritual integrity of the mother and newborn. At the end of this period, which normally lasted for days or even weeks, the woman was reintegrated with the community and reincorporated into her daily activities, now in her new role as mother. This return could be accompanied by acts of a physical or a spiritual cleansing performed with the motive of reestablishing the mother's vital heat. It may also have been the object of prayers or celebrations during which the midwives were compensated and thanked for their services, as is still performed among some Quich'e and Cakchiquel communities in Guatemala (see Nájera 2000, pp. 237–241).

Putting the beginning of a baby's life and its care in perspective with the custom of artificial vault modification, we recognize that this was a transgenerational custom. Those who cared for the head of the newborn were second or third generation women. If we believe the ethno-historical accounts by Diego de Landa, it should have been specifically the midwife who first initiated the manipulation of the soft infant head, along with other measures designed to protect the child from evil forces, to prevent him from losing his warmth and vital energy, and to guarantee his physical and spiritual integrity (Fig. 6.2). As related by different sources of information, these measures may have consisted of repeated massages, bandaging or wrapping, including compresses made of hard materials, and scraping as described earlier in the section on "supra-inial lesions" (Sect. 4.4.4). In all probability also the midwife instructed the mother-to-be in the general care of herself and her baby, such as daily cleaning and feeding. In pre-Hispanic and colonial times these instructions must have also included the responsible use of the compression wedges, maybe remarking on possible signs of risk and considerations in handling the device, together with issues involved with cleaning and nursing the newborn.

According to ethno-historic sources, cephalic compression itself was initiated shortly after the birth of the child and was soon left directly in the hands of the mother or other members of the nuclear family. Colonial accounts in Mesoamerica are silent on potential acts or festivities that may have socially sanctioned the first placement inside the crib or of head-splint collocation. Colonial sources from South America are more eloquent. The Inca celebrated the "presentation of the crib to divinity" (*huahua, quirau*) (Latcham 1929, p. 542, p. 1937; Purizaga 1991, pp. 43–45; see Chap. 5). The maternal uncle or older brother of the mother were responsible for fabricating the child's compression crib while invoking the Huaca or "totem" of the family to shelter the child and protect it against harm. There are references of celebrations held throughout the Inca Empire around the seminal act of placing the child in the prepared crib. Each group appears to have adapted the cradleboard in a predetermined way in order to give the head the shape that would visually identify the child's relevant group. Unfortunately, we can only speculate on the possible enactments of this practice in the Mesoamerican sphere or on the potential rituals during the crafting of the crib or the child's first placement therein.

As a daily operation that lasted months or years, compression of the infant head falls between ceremonies held at the time of birth and others that were performed months or years later. Landa explains that "once born, the children are bathed and then, after they were removed from the torment of flattening their foreheads and heads they went with them to the priests so that they could see its luck and pronounce the prayers . . . and provide the name that they would bear during the time of their childhood, . . ." (Anonymous 1900, p. 350). Given the timing of the measures during the early life course it follows that the practice of artificial head modification must have prepared the baby boy or girl for the childhood ceremonies to follow (Bonavides 1992; Nájera 2000; Tiesler 1998). Childhood festivities generally consecrated the identity and the place that the child later would hold in the community and in the cosmos, anchoring the *ch'ulel* and the "entering of understanding" (Cervera 2007). In this context, the manipulation of the infant head must have been seen as a measure to actively strengthen the child's development (promote the warmth of the spirit, fix the spiritual essences) and seal the body, as well as to prevent the child from possible harm from external ferces sources. These were believed to take the form of cold or pernicious winds that would blow away the still volatile spiritual essences of the child, thereby endangering his health and even putting his life at risk, or of internal origin (lengthening of the occiput as a dangerous spiritual locus), as discussed for the organoplastic motives of the head shaping.

The custom of artificial head molding therefore promoted the infant's development and prepared it for the liminal ceremonies that consecrated its social and spiritual integration. Marion (1994, pp. 31–32) reports that for the Lacandon folk of today's community of Lacanjá, the rite of *mek'bir*, described as the equivalent of the Yucatek *hetz mek* ceremony, was performed during the same period when the child received his name, although the two occasions do not necessarily coincide. It is noteworthy that name assignment for the Lacandon child "means recognizing a social identity by the other members of the community and also revealing his gender, as until that time nothing is known about the newborn" (Marion 1994, p. 31).

The very process of wrapping and covering the head, critical components in the native ritual liturgy (Guernsey and Reilly 2006), must have acquired importance in infant upbringing. William Duncan's (2009, pp. 187–188) recent polysemic interpretation of the significance of Maya modeling is pertinent here. By comparing the head with the roof of a house, the author suggests that the compression and covering of the infant's head vault (similar to the roof and its consecration) must have "ensouled" the new member of the community. Like a roof, over which pernicious winds pass, the head of the child required protection against the loss of its spiritual essences; namely, the *alter ego* or the *tonal*. According to common thought the *tonal* or *wayjel*, an animal that accompanied the individual from the time of his birth, also shared his fate until his death (Bonavides 1992; Garza 1990; Guiteras 1986; Nájera 2000).

Also the naming of the child designates an occasion that was, and still is, given much importance in traditional communities, as the name would define his or her identity and symbolic place in the Mayan cosmos as well as destination in life (Roys 1940). Names, like surnames and ritual identifications, complied with spiritual functions and also acted as cohesive elements in ancient society. The name assigned could follow family traditions for patrilinear transmission, such as those reported by ethnographic histories of the Tzotziles and Lacandones (Boremanse 1998, p. 87; Vogt 1965, p. 29). The Maya may have held various names throughout their lives. Postclassic period Yucatecan Maya are reported to have possessed several names: a family name, (*cha kaba*), transmitted by the paternal line; the maternal name (*naal*), inherited from the maternal line; their childhood name (*paal kaba);* along with a nickname (*coco kaba*) (Roys 1940). Sometimes, a title would be added to these, designating social range, profession, or position held by the person.

As with Maya naming ceremonies, *hetz mek* initiation ceremonies ("placing the child for the first time astride the hip") marked an important step in the Yucatec Maya's infant's life. Like other measures and precautions that were observed in the first years of life, the purpose of the ceremony was to fix the *ch'ulel* in the body of the child and protect him from situations that could cause the vital energy to leave him or her (Boremanse 1998, pp. 80–85; Guiteras 1986, pp. 229–234; Vogt 1965,

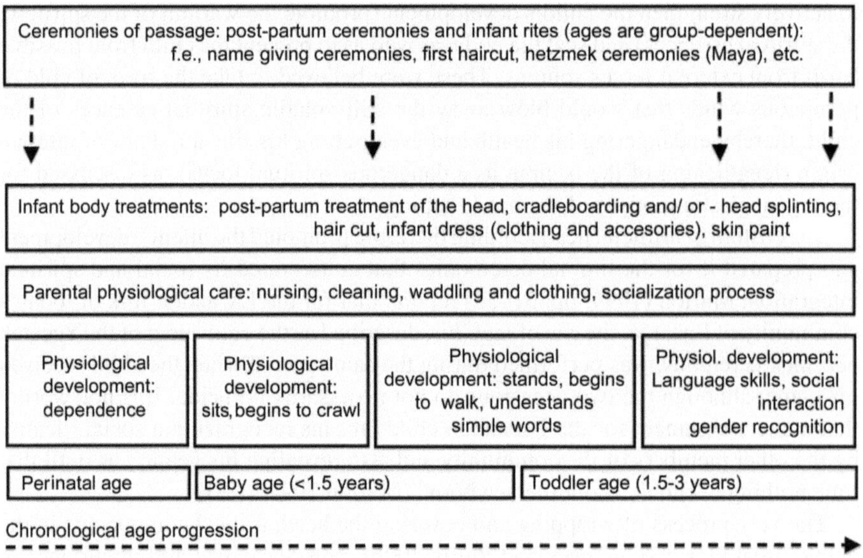

Fig. 6.3 Schematic diagram of chronological, developmental and social age progression of babies and toddlers, along with corresponding measures of care and socialization (adapted from Lewis 2007, p. 6; drawing by V. Tiesler)

p. 29). The culmination of the *hetz mek* consisted of sitting the child on the hip for the first time. This rite, which is still observed in the communities of Yucatán and Chiapas (Boremanse 1998, pp. 80–86; Villa Rojas 1978, pp. 412–415), was commonly performed at 3–4 months after the birth of the child, although it could be delayed until the child was 7 or 8 years of age in the case of the Lacandons. Some ethnographics raise the meaning of the ceremony to that of rebirth, a new the beginning of life as a person. This sort of rebirth culminates a long process of preparation and spiritual instruction for the little ones, with a religious tutor or *meek'ul* specifically appointed for this task (Boremanse 1998, pp. 84–86). On the day of the *meek'chähäl*, the parents give thanks, pray, burn copal, and sing, all in thanks to the gods for their help in bringing their child into the world and keeping him alive and converted into his own person.

The *hetz mek* ceremonies that once accompanied or culminated head compression routines also had—and still preserve—an engendering role. The boy or the girl are presented with the respective powers to carry out the work that they will perform as adults and are given their future work tools for the first time (Bonavides 1992; Marion 1994; Redfield and Villa Rojas 1967; Villa Rojas 1978). This division expresses the traditional belief that females and males complement each other. Male labor was dedicated to producing raw materials, while women's work was to transform these into articles for use and consumption. Today, traditional Lacandons still express this complementary nature in their prayers during the *meek'chähäl* (Boremanse 1998, pp. 84–86).

In summary, the daily dimension of head shaping acquires importance as an explanation of its role within child rearing and social integration (Fig. 6.3). As I have delineated for different Maya groups, this process designates becoming a person. Framed between the birth ceremonies and the rites that will define social identity and the symbolic place that he or she will occupy in the indigenous cosmos, the practice of cephalic modeling acquired a role as a daily preventive and protective measure that prepared minors for the spiritual as well as professional integration as part of the community, sanctioned by means of acts such as the naming ceremony or the *hetz mek* among Yucatecan communities. This latter act especially establishes the identity in the native cosmos even as it marks the differences between the trajectories that will be followed by the boys and girls, as we noted a few lines earlier. In terms of the subject discussed here, the three notions—spiritual, professional, and gender—constitute starting points for the formulation of a series of questions on head modeling and its function in the reproduction of social structure and organization.

6.5 Emblematic Head Shapes, Identities, and Ideology

Up to this point we have explored ideologically ascribed anatomical motifs for head modeling in the Mesoamerican cultural sphere and their native daily enactment for infant protection and integration. Both ideological notions deal with cultural dynamics that in principle must have remained relatively unchanged, considering the information of the pan-Mesoamerican skeletal record, which speaks of continuity of head practices. However, neither of the two cultural dimensions previously discussed provide any convincing answers on the considerable morphological variety observed in the pre-Hispanic cranial record, especially during the first millennium of our era. Earnest Hooton commented, amazed by the cranial collection gathered by Edward H. Thompson from the Sacred Cenote of Chichén Itzá, that "the varieties of cranial deformation are so numerous that they are bewildering" (Hooton 1940, p. 273). This response has resounded again and again in more recent cranial studies in the area (Pereira 1999; Stewart 1974, p. 222; Tiesler 1998, 1999; Tiesler and Romano 2008).

Given the diversity of form, there should have been additional, more ephimeral purposes for head shaping, prompted by their highly visible result in the head. Head form should have identified its carrier with group identity or ethnicity, as we have argued for Classic Period Usumacinta Maya (Tiesler 2012). It is no coincidence that artificial pear-shaped head forms, reminiscent of Olmec head rendering, are ascribed to the Preclassic era in what has been documented for Mesoamerican skull shapes (Tiesler 2010). As noted in previous phases of my own investigation on the ancient Maya, head forms seem to identify family or community traditions, at least during the Classic Period. When examined regionally, these variations in family practices translate into disparate head morphologies at the site level, and into different preferences in terms of technique and form when their modalities are compared among sites and areas (see Tiesler 1998, 1999, 2005). It is also apparent that the frequency of each artificial modification was destined to change according to the socio-political geography and over time, delineating broad ethnic divisions in the head choice, at least for the Maya. On the other hand, it is surprising that the practice of cephalic manipulation was apparently uniform between the social sectors, as there is no palpable evidence for status or gender divisions, indicating that cephalic devices were not used to designate exclusiveness, social position, status, or gender (Tiesler 1999). The female practitioners from the Maya territories, for example, seem to have modeled the heads of their baby girls the same way they did with their infant boys, an aspect to be elaborated on in Chap. 9.

Beyond notions of ethnicity and unspecific cultural identification or aesthetic choice, the visible outcome of head shaping must have been imbued with deeper aggregated emblematic meanings in those areas that staged diversity, such as in most parts of Oaxaca, Veracruz, and the Maya area. As regards the Maya, different venerated patron deities apparently established charters for human head looks (García and Tiesler 2011). Young deities, such as the Maize God (God E), are consistently represented with a slanting oblique head in its Classic anthropomorphic rendering. In all probablitity, this profile conformed to the aesthetic ideals of beauty and nobility among Classic Lowland Maya and was appropriated by the courtiers (Houston et al. 2006, p. 18). Unlike the Maize Deity, the artistic conventions of showing the Old Gods, such as Chac (the god of rain and lightning) mostly designated by high and broad, even natural head profiles, as is the case for God A (García and Tiesler 2011).

Pre-Hispanic Maya portraiture generally renders the aristocracy without an occiput, with slanting foreheads and receding hairlines, an elongated almost tubular vault and a prominent, protruding facial profile. This iconographic convention, enhanced—with visual recourses—the morphological effect of the oblique shapes that the Maya population displayed and thereby established a powerful symbolical means to aggrandize royal self-assigned attributes of beauty and the sacred. Note that reclined, tubular head forms were reproduced especially in the iconography of the Usumacinta basin and some connected coastal stretces towards the north. In real life, these extreme forms were produced by combining head splints with tight head wraps over a prolonged period (see Chap. 4).

Also God L, the patron deity of Maya merchants (García and Tiesler 2011; Tiesler et al. 2010) is represented distinctively by native artists, assigning him a pronounced superior flattening of the head. This rendering comes close to the head forms described for the Classic period Mixtequilla populations from El Zapotal in Veracruz, providing a strong motif of emulation of this sacred patron deity. This form was also to become popular among the merchant folk that settled along the traditional trading routes of the Río Grijalva valleys, where it was documented in series of crania dated from the Middle Classic period onwards. Also, the trader communities on Yucatán's coastal fringes from the centuries before the Maya collapse and well into the Postclassic era, also reproduced this head form in their little ones (Tiesler 2012; Tiesler and Cucina 2010; Tiesler et al. 2010), an aspect to be explored further in Chap. 10.

Later in time, during the Postclassic era, the visible result of native head-shaping practices was to become increasingly uniform in the territories that comprised Mesoamerica, at least technically speaking (Tiesler and Zabala 2011). By the time of

Fig. 6.4 Postclassic codex rendering of the Maya Maize God, who is displayed no longer with a reclined forehead, but with an up-right head form. (Madrid Codex, page 68a; from Rosny 1883; courtesy photograph by G. Vail)

European contact, most head-splinting practices had vanished, giving way to standardized broad, shortened looks, produced by cradleboarding in most parts of the Mesoamerican sphere (except for suscribed areas, as described in Chap. 10). It is unsurprising therefore that the focus of the colonial chroniclers' report of this native tradition was not the artificial shapes seen in the natives' heads, but the daily head practice itself along with occipital reduction. From the above, it follows that at the time of contact, in the sixteenth century, it was the longstanding ritual and "organoplastic" motifs, both "hard core" elements of Mesoamerican ideology and ritual, and no longer the diverse visible outcome of cranial modification, which should have motivated the practice. This is patent for example in the Maya area, where the Maize God, formerly shown with a characteristically elongated, reclined head profile (see Fig. 9.6), is portrayed with an erect, shortened head, like all other deities (Fig. 6.4).

A related question, which has not been addressed so far but is important as a potentially visible signifier, denotes the categorical distinction that must have persisted through time between the predominant portion of individuals with culturally modified versus those few with unflattened heads. Unfortunately, there is no systematic, let alone unified or standardized, inventory of global Mesoamerican modification frequencies (see also Chaps. 7–11 for more specific information on each epoch and area). From the published coverage, we presume that the great majority of

Postclassic period Maya—perhaps excluding Belizean communities—still practiced cradleboarding. When the groups of people with artificial head looks are compared to those who were not shaped, there are no indications of discrimination in favor or against any group (Chap. 9). Similarly high frequencies are also reported from Veracruz during the second millennium (Montiel 2012; Romano 1965) and for most Central Mexican Highland populations. On those, there is no indication regarding possible emblematic connotations of ethnicity or identity of either flattened or natural, round heads.

More towards the vast northern and northwestern fringes of the broader Mesoamerican sphere, cradleboarding with its visible results was more circumscribed and heterogeneous in its distribution than further east and south. Note that Chichimec populations are generally seen with rounded, unshaped heads unlike most local folk from Central and West Mexico. This distinction should have translated into visible attributes when contemplated in the context of the multiethnic landscapes that emerged during and after the massive migrations in northern Mesoamerica. This conclusion is drawn by Grégory Pereira (1999, pp. 167–168) in his interpretation of archaeological and skeletal data from Northern Michoacán, which is upheld by ethno-historical accounts from the area; namely, by references to desirable head attributes mentioned in the *Relación de Michoacán* (Anonymous 1977/1977). This early colonial transcription recounts the conflicts and population movements around Lake Patzcuaro (state of Michoacán). At that time, this area participated in the corridor of broader population movements between Chichimecans, who did not practice head modifications, and local Western Mesoamerican Purepecha folk (Chap. 5). In the Relación, the local aristocracy ("los Señores"), holds artificially flattened heads to be the desirable attributes of bravery and gallantry and look down upon the recently established Chichimecan Uacusecha, whose rounded heads do not provide their human carriers with the credentials of courage (Pereira 1999, pp. 167–168). This notion denotes a common ethno-centric theme in anthropology, that of disagreement with outsiders and with otherness. It is this interethnic difference that surfaces in the blunt assertions of rounded versus flattened head shapes in the *Relación* and not any internal social validation of distinctiveness. The latter connotations would be a rare endorsement of mainstream Mesoamerican ideology, especially during the Postclassic era, as we have argued earlier.

In synthesis, this rough chart of head shapes from the pre-Hispanic iconographic and skeletal registry reflect some of the long-standing motives for head modeling during the first two cultural eras (i.e., Preclassic and Classic periods) and should have been different from those staged during the Postclassic. We conclude that the emblematic connotations must have reached their most differentiated expression—a la par with recognition—during the second part of the first millennium. As we have seen, there are distinct parallelisms when comparing formal preferences in artificial vault modification with those presented in the portraiture of human and mythical personages. Even more relevant than the artistic conventions are the parallelisms established between changing cephalic long-term trends and the concurrent artistic practices.

However, in contrast with the Andean world where cranial-vault modifications were used to confer individual distinction and status in many parts (Blom 2005; Torres-Rouff 2002; Yépez 2006), it seems that Preclassic and Classic Mesoamerica (such as the Maya and Totonac areas and perhaps parts of Western Mesoamerica (Pereira 1999)) used this recourse in a more inclusive fashion. In this quality, it physically embodied recourses to show family and clan membership through the emulation of supernatural forces, as we have argued for the old versus young Maya deities during the Classic period. Some of these visible identifications could have designated professions and vocations in society, such as ethnicity or merchant vocations, but not so much the notion of exclusivity. The recognition of the forms that were frequently reproduced in the heads of venerated gods establishes a powerful causal link and at the same time, a window to knowledge of the religious identifications of the families and their daily reproduction in the heads of their offspring. The fact that it was the women who were entrusted with crafting the desired forms into the heads of their infants casts new light on collective religious life from an angle that has been relatively unexplored until now: the female perspective, and their participation in ideologically conferred ethno-genesis, in cultural reproduction, and perhaps social change.

We close this outline of shifting connotations of visible head shapes with the colonial era, by which time the "emblematic" attribute of this physical practice paradoxically had taken on a new but contrary relevance. If in pre-Hispanic times it indicated group membership of a family, a community, group, or just vaguely, being Mesoamerican, now it indicated marginalization, a shameful "otherness" in the reproving eyes of the recently installed Hispanic society, especially in the crowded urban spaces shared by Spaniards and natives. These and other transcultural dynamics were surely a factor behind the post-contact shifts in its use, as we have discussed extensively for Hispanic America in Chap. 5 and again in Chap. 10.

References

Anonymous. 1900. Relaciones de Yucatán. Madrid: Impresos de la Real Casa.

Anonymous. 1977 [1541]. Relación de Michoacán (trans: J. Tudela.). Morelia: Balsas Editores.

Arzápalo, Ramón (ed.). 1995. Calepino de Motul. Mexico: Instituto de Investigaciones Antropológicas, Universidad Nacional Autónoma de México.

Barrera-Vásquez, Alfredo. 1938. Introducción. In Relación de las cosas de Yucatán. Mérida: E. G. Triay e hijos.

Barrera-Vásquez, Alfredo. 1995. Diccionario Maya Cordemex. Mexico: Porrúa.

Bautista, Josefina. 2004. Evidencias de deformación cefálica intencional en figurillas prehispánicas de México, Ph.D. diss., Universidad Nacional Autónoma de México.

Blom, Deborah E. 2005. Embodying borders: Human body modification and diversity in Tiwanaku society. Journal of Anthropological Archaeology 24:1–24.

Blom, Frans. 1954. Ossuaries, cremation and secondary burials among the Maya of Chiapas, Mexico. Journal de la Société des Américanistes 43:123–135.

Boas, Franz. 1890. Cranium from Progreso, Yucatan. Proceedings of the American Antiquarian Society 6:350–357.

Bonavides, Enrique. 1992. Ritos de pasaje entre los mayas antiguos. Estudios de Cultura Maya 19:397–425.
Boremanse, Didier. 1998. Hach winik: The Lacandon Maya of Chiapas, Southern Mexico. Institute for Mesoamerican Studies 11. Albany: University of Albany.
Bourdin, Gabriel Luis. 2007. El cuerpo humano entre los mayas. Mérida: Universidad Autónoma de Yucatán.
Bricker, Victoria, Eleuterio Po'ot Yah, and Ofelia Dzul de Po'ot. 1998. A dictionary of the Maya language as spoken in Hocabá, Yucatán. Salt Lake City: University of Utah.
Brown, Cecil. 1976. General principles of human anatomical partonomy and speculations on the growth of partonomic nomenclature. Journal of the American Ethnological Society 3:400–424.
Bunzel, Ruth. 1952. Chichicastenango. A Guatemalan village. Locust Valley: J. J. Augustin Press.
Cervera, María Dolores. 2007. Etnoteorías parentales, alma y enfermedades infantiles entre los mayas de Yucatán. Estudios de Antropología Biológica 14(2):731–750.
Cieza de León, Pedro. 1984. La crónica del Perú. Madrid: Historia 16, ed. Manuel Ballesteros.
Cobo, Bernabé. 1893. Historia del Nuevo Mundo, Sevilla, ed. Marco Jiménez de la Espada, 4 vols, Sociedad de Bibliógrafos Andaluces.
Coe, Michael D., and Richard A. Diehl. 1980. In the land of the Olmec. 1. The archaeology of San Lorenzo Tenochtitlan. Austin: University of Texas.
Comas, Juan. 1945. Osteometría olmeca. Informe Preliminar sobre los restos hallados en el Cerro de las Mesas, estado de Veracruz. Anales del Instituto Etnográfico Amerericano 6:169-20.
Comas, Juan, and Paulette Marquer. 1969. Crâ nes déformés de l'île de Sacrificios (État de Veracruz, Mexique). Bulletins et Mémoires de la Société d'anthropologie de Paris 4(3):209–257.
Cosminsky, Sheila. 2001. Maya midwives of Southern Mexico and Guatemala. In Mesoamerican healers, ed. Brad R. Huber and Alan R. Sandstrom, 179–210. Austin: University of Texas.
Cyphers, Ann, and Enrique Villamar. 2006. La representación del modelado cefálico en las figurillas de San Lorenzo. In Presencia zoque: una aproximación multidisciplinaria, ed. Dolores Aramoni, Thomas A. Lee and Miguel Lisbona, 257–272. Tuxtla Gutiérrez: Universidad de Ciencias y Artes de Chiapas, Consejo de Ciencia y Tecnología del Estado de Chiapas, Universidad Autónoma de Chiapas, Universidad Nacional Autónoma de México.
D'Olwer, Luis Nicolau. 1963. Cronistas de las culturas precolombinas. Mexico: Fondo de Cultura Económica.
Dávalos Hurtado, Eusebio. 1951. La deformación craneana entre los tlatelolca. B.A. and Master diss., Escuela Nacional de Antropología e Historia, Instituto Nacional de Antropología e Historia.
Duncan, William N. 2009. Cranial modification among the Maya: Absence of evidence or evidence of absence?. In Bioarchaeology and Identity in the Americas, ed. Kelly Knudson and Christopher M. Stojanowski, 177–193. Gainesville: University Press of Florida.
Duncan, William N., and Charles A. Hofling. 2011. Why the head? Cranial modification as protection and ensoulment among the Maya. Ancient Mesoamerica 22:199–210.
Furst, Jill L. 1995. The natural history of the soul in ancient Mexico. New Haven: Yale University.
García Barrios, Ana, and Vera Tiesler. 2011. El aspecto físico de los dioses mayas. Arqueología Mexicana 112:59–63.
Garza, Mercedes de la. 1990. El hombre en el pensamiento religioso náhuatl y maya. México: Centro de Estudios Mayas, Universidad Nacional Autónoma de México.
Geller, Pamela L. 2004. Transforming bodies, transforming identities. Ph.D. diss., University of Pennsylvania.
Geller, Pamela L. 2006. Altering identities: body modifications and the pre-Columbian Maya. In Social Archaeology of Funerary Remains, ed. Rebecca Gowland and Christopher Knüsel, 279–291. Oxford: Oxbow.
Gosse. Louis. 1855. Déformations artificielles du crâne. Bailliere: Paris.
Guiteras, Calixta. 1986. Los peligros del alma: Visión del mundo de un tzotzil. Mexico: Fondo de Cultura Económica.

Guernsey, Julia, and Frank Kent Reilly III. 2006. Introduction. In Sacred bundles. ritual acts of wrapping and binding in Mesoamerica, ed. Julia Guernsey and Frank Kent Reilly III, 5–16. Barnardsville: Boundary End Archaeology Research Center.
Hooton, Earnest A. 1940. Skeletons from the Cenote of Sacrifice at Chichen Itza. In The Maya and their neighbors: Essays on Middle American anthropology and archaeology, ed. Clarence L. Hay, R. L. Linton, S. K. Lothrop, H. Shapiro, and G. C. Vaillant, 272–280. New York: Appleton-Century.
Houston, Stephen, David Stuart, and Karl Andreas Taube. 2006. The memory of bones: Body, being and experience among the Classic Maya. Austin: University of Texas.
Krickeberg, Walter. 1933. Los Totonaca. Mexico: Secretaría de Educación Pública.
Krickeberg, Walter. 1961. Las antiguas culturas mexicanas. Mexico: Fondo de Cultura Económica.
Lagunas, Zaid. 1970. Nota sobre el hallazgo de cráneos con lesión supra-iniana en Cholula. Boletín del Instituto Nacional de Antropología e Historia 39:1–3.
Lagunas, Zaid. 1972. La trepanación suprainiana en cráneos de Cholula, Puebla. Comunicaciones Proyecto Puebla Tlaxcala 8:47–50.
Landa, Diego de. 1982 [~1566]. Relación de las cosas de Yucatán. Mexico: Porrúa.
Latcham, Ricardo. 1929. Creencias religiosas de los antiguos peruanos. Santiago de Chile: Bacells.
Laughlin, Robert M. 2007. El gran diccionario tzotzil de San Lorenzo Zinacantán. Mexico: Centro de Investigaciones y Estudios Superiores en Antropología Social, Consejo Nacional para la Cultura y las Artes.
Lee, Thomas A. 1985. Los códices mayas. Edición Conmemorativa X Aniversario. Tuxtla Gutiérrez: Universidad Autónoma de Chiapas.
León, María de Lourdes de. 2005. La llegada del alma: Lenguaje, infancia y socialización entre los mayas de Zinacantán. México: Instituto Nacional de Antropología e Historia.
López-Austin, Alfredo. 1989. Cuerpo humano e ideología: Las concepciones de los antiguos nahuas. México: Instituto de Investigaciones Antropológicas, Universidad Nacional Autónoma de México.
López-Austin, Alfredo. 1998. Los ritos. Un juego de definiciones. Arqueología Mexicana 34:4–17.
López-Austin, Alfredo. 2001. El núcleo duro, la cosmovisión y la tradición mesoamericana. In Cosmovisión, ritual e identidad de los pueblos indígenas de México, ed. Johanna Broda and Félix Báez-Jorge, 47–65. Mexico: Consejo Nacional para la Cultura y las Artes, Fondo de Cultura Económica.
López-Austin, Alfredo, and Leonardo López-Luján. 1996. El pasado indígena. Mexico: El Colegio de México, Fideicomiso Historia de la Américas, Fondo de Cultura Económica.
López de Gómara, Francisco. 1987 [1552]. La conquista de México. Madrid: Historia 16.
Lumholtz, Carl, and Ales Hrdlicka. 1898. Marked human bones from a Prehistoric Tarasco indian burial place in the state of Michoacan. Bulletin of the American Museum of Natural History 10: 61-79.
MacLaury, Robert. 1989. Zapotec body part locatives: Prototypes and metaphoric extensions. International Journal of Amercan Linguistics 55(2):119–154.
Marion, Marie-Odile. 1994. Identidad y ritualidad entre los mayas. México: Instituto Nacional Indigenista.
Martínez de León, Blanca. 2007. Estudio de la deformación cefálica intencional tipo Zapotal. B.A. diss., Escuela Nacional de Antropología e Historia, Instituto Nacional de Antropología e Historia.
Molina, Alonso de. 1944. Vocabulario en lengua castellana y mexicana. Madrid: Cultura Hispánica.
Montiel, Mireya. 2012. Modificaciones corporales en la huaxteca prehispánica. Master diss., Universidad Nacional Autónoma de México.
Morton, Samuel George. 1839. Crania Americana. Filadelfia: John Pennington.
Nájera, Martha Ilia. 2000. El umbral hacia la vida. El nacimiento entre los mayas contemporáneos. Mexico: Universidad Nacional Autónoma de México.
Page, Jaime Tomás. 2005. El mandato de los dioses. Etnomedicina entre los tzotziles de Chamula y Chenaló, Chiapas. Mexico: Universidad Nacional Autónoma de México.

Paso y Troncoso, Francisco del (ed.). 1926. Códice de la Real Academia de la Historia. Mexico: Museo Nacional de México.

Pereira, Grégory. 1999. Potrero de Guadalupe: Anthropologie funéraire d'une communauté pré-tarasque du nord du Michoacán, Mexique. Oxford: British Archaeological Research International Series 816.

Pérez, Vitalino, Federico García, Felipe Martínez, and Jeremías López. 1996. Diccionario ch'orti', Jocotán, Chiquimula. La Antigua Guatemala: Proyecto Linguístico Francisco Marroquín.

Pitarch, Pedro. 1996. Ch'ulel: Una etnografía de las almas tzeltales. Mexico: Fondo de Cultura Económica.

Purizaga, Medardo. 1991. El rito del nacimiento entre los Inca. Lima: Universidad de San Martín de Porres.

Redfield, Robert, and Alfonso Villa Rojas. 1967. Chan kom a Maya village. A classic study of the basic for culture in a village in Eastern Yucatan. Chicago: University of Chicago.

Romano, Arturo. 1965. Estudio morfológico de la deformación craneana en Tamuín, S.L.P., y en la Isla del Ídolo, Veracruz. Serie de Investigaciones 10. Mexico: Instituto Nacional de Antropología e Historia.

Romano, Arturo. 1974. Deformación cefálica intencional. In Antropología física, época prehispánica, ed. Juan Comas, 197–227. Mexico: Instituto Nacional de Antropología e Historia.

Romano, Arturo. 1987. Iconografía cefálica maya. In Memorias del Primer Coloquio Internacional de Mayistas, 1413–1474. Mexico: Universidad Nacional Autónoma de México.

Rosny, Leon de. 1883. Codex Cortesianus. Manuscrit hiératique des anciens indiens de l'Amérique Centrale, conservé au Musée Archéologique de Madrid. Libraires de la Société d'Ethnographie: Paris.

Roys, Ralph L. 1940. Personal names of the Maya of Yucatan. Contributions to American Anthropology and History 31:35–48.

Ruz Sosa, Mario Humberto. 1992. Copanaguastla en un espejo. Un pueblo tzeltal en el virreinato. Mexico: Instituto Nacional Indigenista, Consejo Nacional para la Cultura y las Artes.

Sahagún, Bernadino de. 1989. Historia general de las cosas de Nueva Espaa. Mexico: Porrúa.

Sánchez, Mirna Beatriz. 2008. Emblema visual de identidad y distinción. Reconstrucción de rostros mayas (fisionomía y rasgos bioculturales). Master diss., Universidad Autónoma de Yucatán.

Sotelo, Laura, and Carmen Valverde. 1992. Los señores de Yaxchilán: Un ejemplo de felinización de los gobernantes mayas. Estudios de Cultura Maya 19:187–214.

Soustelle, Jacques. 1956. La vida cotidiana de los Aztecas en vísperas de la Conquista. México: Fondo de Cultura Económica.

Stephens, John, and Frederick Catherwood. 1963 [1843]. Incidents of Travel in Yucatan. New York: Dover.

Stewart, Thomas D. 1974. Human skeletal remains from Dzibilchaltun, Yucatán, México, with a review of cranial deformity types in the Maya region. Archaeological Investigations on the Yucatan Peninsula, Middle American Research 31:199–225.

Stresser-Péan, Claude. 2011. De la vestimenta y los hombres. Una perspectiva histórica de la indumentaria indígena en México. Mexico: Fondo de Cultura Económica, Centro de Estudios Mexicanos y Centroamericanos, Fundación Alfredo Harp Helú, Museo Textil de Oaxaca.

Tiesler, Vera. 1994. La deformación cefálica intencional entre los mayas prehispánicos: Aspectos morfológicos y culturales. Master diss., Escuela Nacional de Antropología e Historia, Instituto Nacional de Antropología e Historia.

Tiesler, Vera. 1998. La costumbre de la deformación cefálica entre los antiguos mayas: Aspectos morfológicos y culturales. Mexico: Instituto Nacional de Antropología e Historia.

Tiesler, Vera. 1999. Rasgos bioculturales entre los antiguos mayas: Aspectos arqueológicos y sociales. Ph.D. diss., Universidad Nacional Autónoma de México.

Tiesler, Vera. 2005. Prácticas bioculturales y organización social en los sitios de Copán, Honduras, y Xcambó, Yucatán. Estudios de Antropología Biológica 12(1):103–117.

Tiesler, Vera. 2006. Bases conceptuales para la evaluación de restos humanos en arqueología. Mérida: Universidad Autónoma de Yucatán.

Tiesler, Vera 2010. "Olmec" head shapes among the Preclassic period Maya and cultural meanings. Latin American Antiquity 21:290–311.

Tiesler, Vera. 2012. Transformarse en maya. El modelado cefálico entre los mayas prehispánicos y coloniales. Mexico: Universidad Nacional Autónoma de México, Universidad Autónoma de Yucatán.

Tiesler, Vera, and Andrea Cucina. 2010. La deformación craneana como emblema de identidad, etnicidad y reproducción cultural entre los mayas del Clásico. In Identidades y cultura material en la región maya, ed. Héctor Hernández and Marcos Pool Cab, 111–134. Mérida: Universidad Autónoma de Yucatán.

Tiesler, Vera, and Arturo Romano. 2008. El modelado del cráneo en Mesoamérica. Emblemática costumbre milenaria. Arqueología Mexicana 94:18–25.

Tiesler, Vera, Arturo Romano, and Carlos Pallán. 2010. Las formas cefálicas en las vísperas del periodo Clásico: Implicaciones para el cambio social en el área maya. Los Investigadores de la Cultura Maya 18:83–96.

Tiesler, Vera, and Pilar Zabala. 2011. El modelado artificial de la cabeza durante la Colonia. Una tradición maya en el espejo de las fuentes históricas. Estudios de Cultura Maya 38:75–96.

Torres-Rouff, Christina. 2002. Cranial vault modification and ethnicity in Middle Horizon San Pedro de Atacama, North Chile. Current Anthropology 43(1):163–171.

Vega, Garcilaso de la, el Inca. 1982 [1609]. Comentarios Reales, 2 vols. Mexico: Secretaría de Educación Pública, Universidad Nacional Autónoma de México.

Villa Rojas, Alfonso. 1978. Los elegidos de Dios: Etnografía de los mayas de Quintana Roo. Mexico: Instituto Nacional Indigenista.

Vogt, Evon Z. 1965. Zinacanteco souls. Man 29:33–35.

Weiss, Pedro. 1962. Tipología de las deformaciones cefálicas de los antiguos peruanos, según la osteología cultural. Revista del Museo Nacional 31:15–42.

Weiss, Pedro. 1967. Ensayo de osteología cultural en Guatemala. Antropología e Historia de Guatemala 19:14–26.

Winning, Hasso von. 1968. Process of head deformation shown by Mesoamerican figurines. The Masterkey 42(2):53–58.

Wisdom, Charles. 1950. Materials on the Chorti language. Microfilm collection of manuscript materials on middle american cultural anthropology 28. Chicago: University of Chicago.

Yépez, Rosaura. 2006. La práctica cultural de modelar la cabeza en dos culturas andinas del Antiguo Perú: Paracas y Chancay. Un estudio de los procesos de significación de la cabeza modelada intencionalmente. Ph.D. diss., Universidad Nacional Autónoma de México.

Yépez, Rosaura, and Ramón Arzápalo. 2009. La práctica cultural de modificar el cuerpo como un texto de información e interpretación social para la antropología física. Una perspectiva semiótica. Papeles de trabajo 15:75–108.

Chapter 7
Emulating Olmec Gods Through Head Form. Origins and the Preclassic Period

This chapter explores the early evidence of infant head modeling in the broader Mesoamerican sphere, where the skeletal record testifies to cultural skull modifications for almost 10,000 years. This information is complemented by drawing insights from figurative head representations. The remote origins of cradleboard, wrapping, and head-splinting practices in Preceramic and Preclassic times are reviewed briefly for sites with documented cranial series that exemplify the preferences in the different cultural territories of Mesoamerica. The skulls from the Preclassic period of Monte Negro and Monte Albán (Mexican state of Oaxaca), Tlatilco, Cuicuilco y Tlaltenco (Mexico City), in particular, make the increasing cultural embedding of the practice apparent, along with its growing medley of regional and local preferences and, with it, a growing diversity in forms. Additional skeletal information comes from the Isthmus of Tehuantepec and the Maya area. Although Gulf Olmec head traditions have not preserved well in the material record of the harsh tropical environs, its rich iconographic patrimony clearly illustrates the crafting of specific pear-shaped head forms to emulate the sacred forces of Olman. High and narrow silhouettes were also adopted outside the Olmec coastal plains and practiced in tandem with a host of increasingly diversified modeling forms. The last part of this chapter discusses the evolving social and possibly religious roles of Preclassic body modifications in Mesoamerica, arguing that the appearance of Olmecoid pear-shaped head styles most likely reflected syncretic ideological adoption among those populations that dwelled outside or past Gulf Coast Olman occupation.

7.1 Origins of Head-Compression Practices in Mesoamerica and Beyond

If we believe the archaeological record, infant head modeling has been known in the broader Mesoamerican territories for almost 10,000 years now (Fig. 7.1). Six early cases of clear tabular erect modification have been dated as 8,800–7,000 years B.P. (Anderson 1965, p. 496, 1967; Lagunas 1989, p. 33; Romano 1974a, p. 210). The findings were made during an archaeological exploration in the Central Highland

V. Tiesler, *The Bioarchaeology of Artificial Cranial Modifications*,
Interdisciplinary Contributions to Archaeology 7, DOI 10.1007/978-1-4614-8760-9_7,

Fig. 7.1 Map of Mexico showing sites of Archaic and Formative skull findings described in the text. (Map by B. Ceballos and V. Tiesler)

Basin of Tehuacan (Mexico). This area was at that time foraged by hunters-and-gatherers with strong signs of dental attrition due to their abrasive diets. Anderson reports a slightly asymmetric occipital flattening of one skull, likely to have stemmed from cradleboard use.

Another pre-ceramic specimen (Burial 1/65-1) with cranial modification, presumed to be cultural, stems from the Texcal Cave in the Valley of Valsequillo, Puebla (Mexico). Burial 1 was attributed an antiquity of around 5,000 years B.P. by the on-site archaeologists Noemí Castillo and Roberto García Moll by its stratigraphic association (Romano 1972a, 1974a). The individual is described by Arturo Romano as a relatively robust and probably female adult. Her skull cap displays almost symmetrical lambdic flattening, a distinct bulging of the lower occipital portion with a horizontal sulcus towards *inion*. According to Romano, the craniometric polygon measurements confirm the cultural origin of the morphological changes by manifesting shifts in the natural balance of various cranial measurements. From here, the author suggests that a cradleboard was used as the principal compression device in the early times and infers from the sulcus below the occipital prominence that the head had been tied to the apparatus below the nuchal crest. Apart from the above specimens, there are no convincing documented cases of cultural change in cranial vaults dated before the onset of agriculture. The well-preserved Mexican Tepexpan skull, the crania from Tlapacoya and Santa María Astahuacan, the skull of the so-called "Woman of El Peón" and from San Nicolás, the findings from San Vicene Chicoloapan or the Cave of the Tecolote are not noted for any artificial modeling according to Romano (1970, 1974a, b, 1978).

Further south, in Ecuador, Juan Munizaga (1973) screens a large series of crania, which include eight restored pieces belonging to the Vegas Complex, dated around 6000 B.C. None of these bore any signs of artificial flattoning. It is only in the following phase (Chorrera), dated to the first millennium B.C. that cultural changes are evident in the heads of early Americans. In another study (Munizaga 1965), he documents the earliest confirmed cases of flattening for Ecuador in the site of Machalilla. The cranial vaults are dated around 3,400 B.P. and presumed to belong to formative agricultural folk. It is noteworthy that the compression apparatus of these pristine cranial modification was not a cradleboard but a head-compression device. Its application did not flatten lambda but instead led to the complete elimination of the occipital bun below lambda, generating a sulcus in the middle of the compression plane. Comparatively less expressed is the forehead, which almost retains its convex surface. Munizaga coins this type of cranial transformation "cuneiform type".

Still further south, along the Chilean coast, Juan Munizaga (1974) and Patricia Soto-Heim (1987) explore the onset of cranial modifications for northern Chile. Apparently, preceramic head modeling in South America was, at best, scarce; the cranial vaults of five specimens described for the site of Pisagua Vieja, dated to some 5,000 years before present, do not appear to be transformed culturally in any of their segments. Also, Soto-Heim (1987, p. 135) estimates the earliest case in the skeleton of a child somewhat late, at around 3,000 years ago. It was recovered at the site of Camarones 15 (Chinchorro complex) among other remains that had been radiocarbon dated around 1,100 years B.C. This subadult vault showed both frontal and occipital flattening. Much earlier is the evidence reported by Cardich and Bórmida for the Peruvian Highlands for the so-called "Man of Lauricocha", dated over 4,500 years B.P. Of eleven recovered skeletons, Burial 6 displayed clear signs of tabular erect cranial shortening (Yépez 2009, p. 527). Barrientos (Barrientos and L'Heureux 2009), Pérez and colleagues (2009) and Béguelin et al. (2006; see also Menéndez 2010), have complemented the panorama of early head practices for northern Patagonia. They find the first traces of head flattening by constriction for the Arroyo Seco 2 Phase of the Middle-Late Holocene, dated around 7,000 years B.P.

From the above review, we can cautiously presume that head modifications appeared in the cultural territory that later became Mesoamerica roughly around the same time as in South America. The seminal cases, documented from Texcal and Tehuacan (Mexico) also suggest that cradleboards were the first implements used in this process and that head shaping was still isolated, given the many unmodified preceramic crania recovered from the area of Mesoamerica. But what was the role of cradleboards back then? What practical advantages did the tight baby carriages have for their mothers? How were they used during foraging? These and other questions still await scrutiny by the academic community. Unfortunately, the lack of suitable skeletal evidence and the absence of detailed head portraiture (or really any figurative imagery) before the onset of the Preclassic period, translates into persisting gaps in our understanding of the pristine uses of cradleboards and the roles of cranial flattening in the preagricultural societies that occupied the Mesoamerican territories before becoming sedentary.

7.2 Cranial Modification Turned Tradition: The Preclassic Period

With the advent of the Early Preclassic era (1400–1000 B.C.), probably many native populations already practiced cradleboarding as an established daily routine in Mesoamerica. I presume that, by then, the predominant technique must have been related to body devices, perhaps effected by wraps or bandages, applied to the child's head while it rested inside the crib. These procedures should have produced a variety of slight lambdic and frontal flattenings, most of them leading to asymmetric results (see Sect. 3.3). The following centuries comprise the Mesoamerican Preclassic Period. It saw population growth and an increasing cultural regionalization in the cultural areas that encompass its territory; some of its territories experienced a rise in social complexity and hierarchy.

Although the following sections are not intended to provide a complete review of the many studies or mentions of local Preclassic cranial modification, I will provide rough trends by surveying published information on large skeletal series from different sites and areas. Unfortunately, most publications do not provide systematic listings of attributes that are relevant for making comparisons, such as presence vs. absence of cultural change, of variety, secondary bands and degree of morphological modification. The combined results from each site testifies to a diversification of artificial head shapes, especially during the first millennium B.C. The proportion between erect and oblique modification types and the evidence of horizontal circular and sagittal wraps on skulls from some territories, but not from others, appear to reflect upon locally diverse and also some regionally specific preferences. It is noteworthy that, unlike some of the territories north and south of Mesoamerica, the overwhelming majority of Mesoamerican skulls do not exhibit true annular forms during the Preclassic era. When used, compression bands or wraps were attached to tablets or boards.

7.2.1 The Mexican Highlands

The first millenium B.C witnesses a diversification of artificial head shapes in the Central Highlands (Tlatilco; Ecatepec) and further south (Bautista 2004, 2005; Peña and López-Wario 1989; Romano 1972b, 1973, 1974a, 1979). Arturo Romano (1972b; see also Faulhaber 1965) analyzed a large skeletal series from the Tlatilco site, in the Valley of Mexico, after four field seasons of burial recovery. This important collection incompasses some 500 individuals, dated between 1200 and 200 B.C.

and mostly derived from Middle Preclassic contexts. Of the 232 preserved cranial vaults, some 95.6 % showed tabular erect shaping, with some additional few described as tabular oblique and mimetic. Importantly, Romano notes that some of the rigid anteroposterior flattening was complemented by annular wraps. These bipolar constrictions elongated visibly the few tabular oblique vaults that have been documented and heightened and also narrowed many of the common tabular erect shapes. No mention is provided as to whether the latter looked like the pear-shaped heads described for the Olmec Gulf Coast plains; a question that awaits scholarly attention in view of the ideological relevance this peculiar variety should have held in eastern Mesoamerica (Sect. 7.3). Faulhaber (1965, pp. 87–89) adds information on Tlatilco's skeletal series in a separate study by specifying that the forms and degrees of artificial head modeling appear equivalent in men and in women. She also notes that only some 15 % of the series (43 of $N = 232$) did not show any culturally induced morphological change, thereby testifying to its almost generalized practice already during the first Preclassic phase.

Slightly more recent is the chronological sequence of Cuicuilco (700 B.C to 150 A.D.), another important Preclassic site, nestled in the southern part of the Mexican capital. The skeletal collection from Cuicuilco was examined by Patricia Sánchez (1971) under the supervision of Arturo Romano. Among other analyses, she scored and measured some 68 of the 150 individuals for cranial shape. All of these specimens from Cuicuilco displayed cultural modification. Again, tabular erect shapes predominate (90 %), albeit to a slightly lesser degree than recorded in Tlatilco (Fig. 7.2). Again, different oblique variants complement the range of erect head looks among Cuicuilco's settlers who also combined circular wraps with rigid compression devices (Fig. 7.2).

Close to Cuicuilco, in the southern part of the Central Mexican Basin lies the Late Preclassic site of Tlaltenco, reported by Mari Carmen Serra, Magali Civera, and Arturo Romano (Serra et al. 1982; Fig. 7.3). Here, both tabular oblique and tabular erect shapes occur. Case studies of the two measured crania from Tlaltenco highlight the severity of the morphological change and conclude for both skulls that different constriction and compression devices were combined to transform the shape of the cranial vaults, noticeably a deep horizontal sulcus that visibly divides the inferior and superior portion of the skull cap. Equally, Late Preclassic and possibly Early Classic is two other cranial series from the site of T-358, Tlaxcala, and Jalapasco, Puebla. Here, Rosa María Peña (1982, p. 352) documents seven additional artificially modified cranial vaults from that period, two being tabular oblique, five others erect. Comas et al. (1981) report 19 skulls which evince a variety of different degrees and varieties of tabular erect and oblique shapes, confirming, once again for the Central Highlands the trend toward diversity and inclination of head form towards the Classic period.

In Preclassic western and northwestern Mexico, oblique cranial-vault shapes from head splinting appear in similar—possibly even higher—proportions than in the aforementioned Central Highland series. Arturo Oliveros (1971) documents this type in the Early Preclassic community of El Opeo, in the present-day Mexican state of Michoacán. Grégory Pereira (1999, pp. 165–169) describes Late Formative period skulls for Guadalupe (also in Michoacán), which show equally erect and

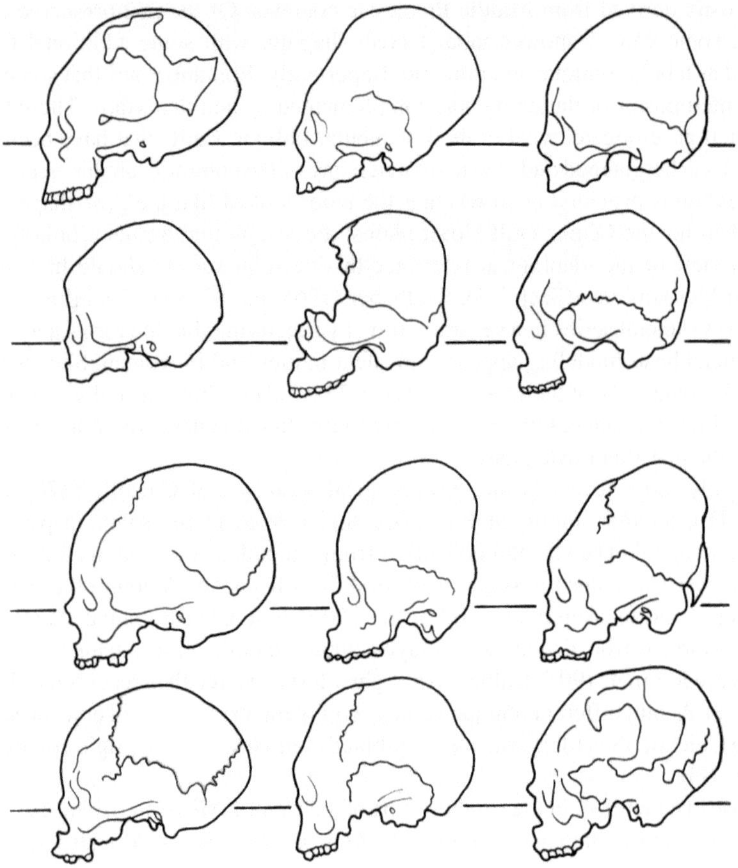

Fig. 7.2 Set of representative cranial profiles from the site of Cuicuilco, Mexico City. (Adapted by B. Ceballos from Sánchez 1971, Figs. 6 and 7)

oblique head forms. The author argues that for the six tabular oblique specimens, a rigid head compressor must have been employed and infers in analogy to figurine representations that the boards had been fastened over the top of the head with a strap or buckle, possibly leaving a sagittal groove.

North of Michoacán lies the important Late Formative site of Chupicuaro, Guanajuato, presumed to be an early Chichimeca center that evolved on the shores of the Lerma River, on the northern fringes of Mesoamerica. Several well-preserved skulls have been documented from its monumental area, which was excavated during the 1940s. The precise dates of these burials are uncertain but in all probability accord with the peale in occupation that Chupicuaro experienced between 500 B.C. and 300 A.D. (Darras and Faugère 2010). Regarding the preserved cranial assemblage from Chupicuaro, lambdic flattenings and also some inclined, elongated pieces make up this collection (Dávalos 1951). A recent reexamination of this series by Cybéle David (2000, p. 58) specifies some 23 individuals of which 87 % bear the effects of cultural flattening, of which sixteen display an erect form, and four more an oblique form.

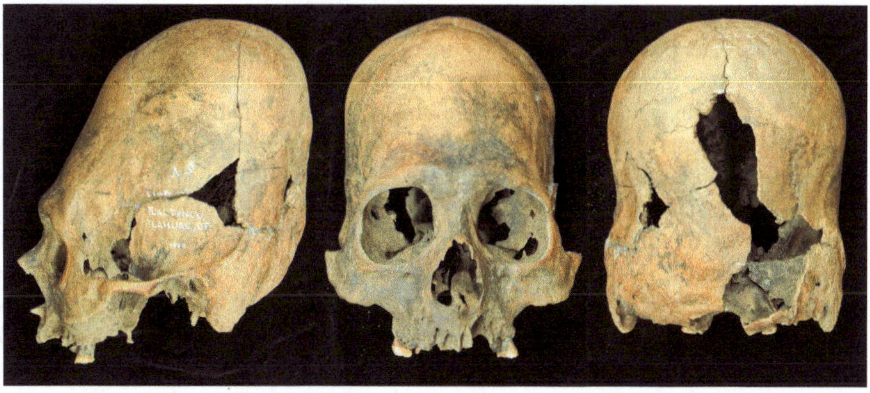

Fig. 7.3 Different views of cranium of Late Preclassic Burial 1A, Tlaltenco, Mexico City. (DAF/INAH; photo courtesy of A. Romano and J. Gómez-Valdés). The calotte displays a straight frontal plane and marked lambdic and occipital flattening. The posterior view reveals clear signs of annular constriction that left a horizontal sulcus, which divides the skull in a superior and an inferior lobe. (See also Serra et al. 1982)

On the other side of the Mexican Central Plateau lies today's state of Oaxaca, an area that had already become one of Mesoamerica's main cultural hubs (Winter 1989) by the Early and Middle Preclassic period. San Mateo Etlatongo in Oaxaca's Nochixtlan District counts among the early local developments to be archaeologically explored. From here, Romano (1987) reports two unmodified crania and another three that display different degrees of erect shaping from cradleboard use. As with the settlers of the Valley of Mexico, Etlatongo's practitioners used to combine the effect of annular constriction frequently with rigid cradleboard compression. Its constricting effects on the skull recall the peculiar, narrow Olmec head profiles excavated towards the Gulf coast. Javier Romero (1950, 1951, 1983) provides additional information for those cranial series that were recovered from Preclassic Monte Albán contexts and the Late Preclassic site of Monte Negro. In both skeletal assemblages, the author confirms abundance and diversity in artificial cranial shapes, albeit tabular erect varieties appear to prevail. In the Zapotec capital of Monte Albán, which, by the Late Preclassic period had turned into the preeminent Zapotec sociopolitical and economic center, head splinting appears to make its debut at the onset of the Early Classic period, during Monte Albán IIIA phase (A.D. 200–500). From then on, the material record testifies both oblique and erect head silhouettes in burial remains (Romero 1983, p. 102).

In synthesis, head modification practices among Mesoamerican Highland settlers became almost generalized in the course of the two millennia that breach the Preclassic era. At the beginning, its women probably used standard cradleboards that led to erect fronto-occipital and lambdic flattenings in their infants. The great majority of Preclassic period crania from the Mesoamerican Highlands still bears these erect forms; at least, this is the tendency in the combined record of the Preclassic period series curated by the National Museum of Anthropology in Mexico City, which make up some 95 % of 221 individuals surveyed (Romano 1974a, p. 213). Chronological comparisons among Highland collections also indicate subtle shifts in formal preferences and in the use of compression kits over the two Preclassic millennia. Towards the Late Preclassic period, distinctive forms of head splints and wraps were increasingly incorporated in some areas, possibly functioning as part of the cradleboard kit. This diversity is upheld by the finely modeled small-scale figurines from Tlatilco, Chupicuaro, and other pre-Hispanic Highland sites, which display babies predominantly in cribs or carriages. The top of the head usually appears either covered by a little slab or a cushion or may appear as wrapped in cloth. Other foreheads seem to be actively manipulated by a female caretaker (see Chap. 4). Less frequent among the ceramic imagery than the crib scenes are representations of women carrying a baby in their arms. In many of these pair scenes, the adults' head style is curiously mirrored by that of the little one; when covered, semirigid rings or broad circular bands circle their heads. Naturally, we cannot be sure if the infants' headwear was put on purely for adornment or was used as a constricting tool as well.

Among the insipient signs of diversification and regionalization, the combination of tight wraps and cradleboard compression stands out. Romano documents its outcome in a number of individuals from Tlatilco and Cuicuilco in the present-day urban area of Mexico City, and from Tlaltenco and Etlantongo in Oaxaca. The appearance of these high and narrow vaults comes close to Olmecoid head portraiture and technically qualifies as the tabular erect pseudocircular variety (Sect. 4.3), although Romano is careful to clarify that the horizontal constrictions do not appear as expressed as in the Middle Preclassic skull from Pampa el Pajón, in the present-day state of Chiapas. The latter, which appears extremely narrow and rounded, with a bulging, towering forehead and a noticeable horizontal separation between upper and lower neurocrania, will be referred to extensively in the following paragraphs and Sect. 7.3 of this chapter. For the sake of flow, we will also discuss other series that surround the Mesoamerican Isthmus area in the context of the Formative Maya period, although, strictly speaking, these belong to the Mexican Highlands that are under review here.

7.2.2 The Maya Area

East of Oaxaca, the terrain descends and opens towards the Isthmus of Tehuantepec, a transition zone between the Gulf of Mexico and the Pacific. This natural strait instituted major Mesoamerican merchant routes between the Gulf of Mexico and the Pacific Coast and specifically towards the Soconusco area, wedged between the Sierra Madre mountain ranges of Chiapas and the ocean. This was formerly a prime spot of Mesoamerican long-distance exchange and trade. From remote times onward, the Isthmus was also an important transit area between the Central and Maya Highlands. Further east and towards the Yucatecan Peninsula lie the vast territories that during the Preclassic period saw the rise of complex hierarchical political units that encompassed the Maya Lowlands and Highlands.

During the last part of the second millennium B.C. and the first millennium B.C., the agricultural settlers of these diverse territories shared some elements of their cultural repertoire with the neighboring Olmec hegemonies of the Gulf Coast, although there are controversies on the form and extension of the exchange and influence (Clark 1990; Pool 2007). According to recent interpretations, Middle Preclassic Maya society already looked back to a culture and sociopolitical tradition of its own that was relatively independent of the Gulf Coast hegemonies to its west. Architectural monuments in early sites of northern Yucatán, such as Poxilá, Xocnaceh, and Xtobó, with their *sacbeob* (white roads), acropolis, ball courts, triadic, and type E groups, already then denote the presence of an elite that had the capacity to mobilize a collective labor for public works (Diehl 2004; Pool 2007). In any event, expansive networks from the lands of Olman, regional and long-distance exchange, and redistribution must have also catalyzed a mutual understanding and adoption of ideological elements throughout the Preclassic period, probably in a similar fashion as happened with the Mesoamerican Highland development west of Olman.

In these evolving cultural landscapes of the Mesoamerican east, the earliest evidence for modeling dates at least to the Middle Preclassic period, but should have appeared still earlier in the settlers' cultural repertoires. Unfortunately, Preclassic period skeletal remains are very reduced in numbers and suffer from poor preservation due to the erosive soil and the harsh environmental conditions that prevail in a large part of the Maya sphere. Those early cases of cranial modification that are positively identified correspond in part to artificially elongated, inclined head profiles and intermediate tabular erect forms, along with the "Olmecoid" pseudocircular erect shapes (Romano 1977; Saul 1972; Saul and Saul 1991, 1997; Tiesler 2010, 2012). Frank and Julie Saul (1997; Saul and Saul 1997) record the earliest Maya evidence of cranial modification in a Phase Xe individual from the site of Altar de Sacrificios, along the shores of the Río Usumacinta, and for Swasey/Bladen phase crania from Cuello, Belize. The latter date to the Early Middle Preclassic period and display both elongated and erect head shapes. Further east, another pristine example of Soconusco head modifications is published by Arturo Romano at Pampa el Pajón, in the present-day state of Chiapas (1977, 1980; see also Salas 1980). We will take up his case study again in the next section of this chapter. A second early culturally modified specimen (in the form of pseudocircular tabular erect flattening) integrates the material record of Caucel, a major Preclassic settlement from the northern Yucatecan Peninsula, recently excavated by Fernando Robles (INAH) near the city of Mérida (Tiesler 2009; Tiesler and Rodríguez 2009; Robles and Ligorred 2008). This individual was recovered from Structure 52 of the site and was accompanied by the Early Nabanché phase (800/700–400/300 B.C.) ceramics.

It is noteworthy that the above described pristine cranial modifications appeared at roughly the same time as dental modifications in the Maya archaeological record. Saul and Saul (1997, pp. 45–46) document some of these primeval dental alterations for the Early Bladen Phase (900–800 B.C.) of Cuello, in Belize. The early tooth decorations consist of deep grooves that had been filed into the maxilla dentition of a female adult. Also, Javier Romero (1958) describes early material evidence of Maya dental practices such as inlays in the drilled teeth of an individual who lived at the central Lowland site of Uaxactún (Petén, Guatemala) during the Mamon Phase (600–350 B.C.)

Upon integrating the information documented by different authors for Preclassic skeletal assemblages, we can cautiously intend some regional generalizations on

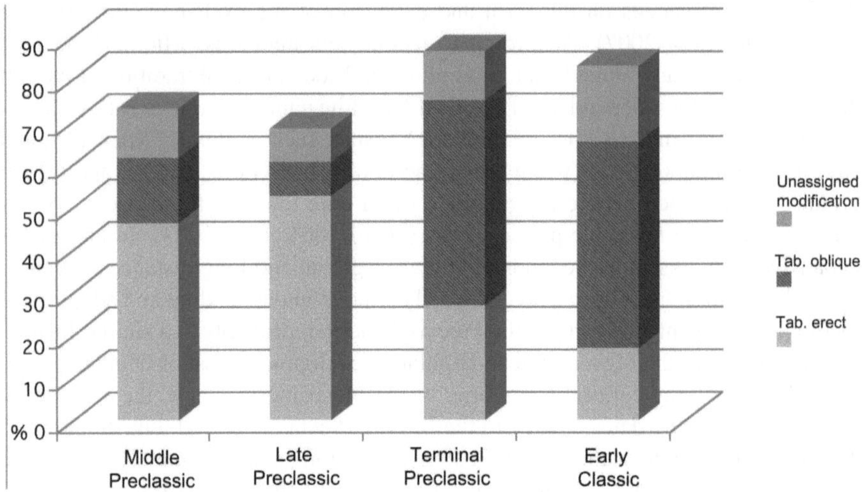

Fig. 7.4 Preferences (erect, oblique, unspecified) and overall frequencies of head morphologies during the Middle and Late Preclassic, the Protoclassic, and Early Classic in the Maya area. (Histogram by Vera Tiesler; see also Tiesler 2010, 2012)

early Maya head practices and their preferences. Unfortunately, these trends are heavily biased towards the information that comes from those archaeological sites with preserved cranial vestiges. Most of these places line the Caribbean coastline or come from the northern part of the Yucatán Peninsula, the Maya Highlands, and the central and southern Maya Lowlands. At least here, after making its appearance during or before the Maya Middle Preclassic period, the modeling was carried out by employing both compression cribs and head devices. As in the Highlands, back then, compression splints still appear to have been used relatively less among Maya practitioners than later, during the Classic period. On the regional level, the increase in frequency and diversity appears to be accompanied by a gradual change in techniques. In those Preclassic Maya samples of skulls that were systematically scored by the author ($N = 49$), the custom still appears to be less varied and less severe than later, Early Classic cranial assemblages express. During the Preclassic era, head shaping not only seems less diversified but also less common, with a proportion of less than 70 % of the crania shaped during the Middle Preclassic ($N = 17$) and the Late Preclassic periods ($N = 22$), compared to over 80–90 % during the Classic period (Fig. 7.4).

The full range of artificial cephalic shapes that would later characterize Classic Maya society apparently was still unknown during the first millennium B.C. This situation is evident for the Maya Highlands, for example, as an assemblage from the large Formative site of Kaminaljuyú, nowadays part of Guatemala City, illustrates. Two individuals that date back to its early occupational phases during the Middle Preclassic period exhibit an isolated frontal flattening. The flat planes probably stem from tumpline use during early age, as its position is untypically high for being the product of cradleboarding. There is also no evidence for complementary flattening on the back (Gervais 2001; Tiesler 1999). The third cranium, dated to the same period as the first two, shows no sign of cultural change at all. Further west, none of the four individuals from the Middle Preclassic period of San Mateo in the Chiapanecan Angostura basin exhibited artificial manipulation, indicating in principle that the custom was not yet used in this area.

Beyond these broad trends, Preclassic Maya site assemblages are less informative than comparisons on a regional scale, principally for lack of sufficient sample sizes. This is because the early dated crania tend to represent only small proportions of the overall skeletal series from important Maya sites, such as Copán, Tikal, or Calakmul, where the majority of individuals are invariably dated to the Late Classic period. Possibly a welcome exception for the purposes of reviewing local head-shaping practices is the archaeological urban site of Chiapa de Corzo on the eastern fringes of the Maya region with its relatively large Preclassic series (Agrinier 1964; Lowe and Agrinier 1960). Here, the five earliest of 23 examined crania date to the French Phase (450–250 B.C.). Two of these had been artificially reclined by use of headsplints, combined in each case with compression bandages, probably to accentuate inclination and elongation. Later, during the Protoclassic period (Guanacaste and Ismo Phases, 300 B.C.–200 A.D.), some 15 of 18 evaluable skullcaps show signs of artificial transformation, indicating its widespread use already in those centuries. Both head splints and cradleboards were common then and both were often reinforced with constricting bandages, including the "Olmecoid" shapes, to be discussed in the following section of this chapter.

The first two centuries A.D. closed the Preclassic period in the Maya area, and among the settlers of Chiapa de Corzo, the advent of the Classic era was accompanied by a visible change in head form that followed the elimination of narrow "Olmecoid" shapes and that of head inclination and elongation. At the end of this process, the only implement that seems to have remained in use among practitioners was the compression crib. Classic era inhabitants of Chiapa de Corzo showed uniformly erect and broad head forms. This style is in inverse proportion to the trends observed in the remainder of the Maya area, where the compression crib seems to have been gradually replaced by head presses adjusted directly over the calotte of the infant and thereby leading to the collective look of back-bent heads, portrayed in Classic imagery and evident in the cranial record. This was especially true in the Usumacinta Basin which borders the Chiapanec highlands. In sites such as Yaxchilán or Palenque, over 90 % of the population displayed mostly severe oblique and narrow head models. The discrepancy in head form of Chiapa de Corzo's settlers in the east invited questions regarding the cultural and possibly ethnic borders of this area, which is presumed to have been settled not by Western Maya but by Mixe-Zoque-speaking folk, who maintained close cultural ties with the Isthmus and the coastal plains (Lee 1969, 1993).

In synthesis and viewed from a pan-regional perspective of the Maya culture, it seems that the cultural head modeling became a generalized body tradition in most parts of Mesoamerica during the Late Preclassic period. As for the Maya, it is then that the formal diversity and average degrees of modification (around or exceeding '2', Sect. 4.3.) would have reached high levels of collective acceptance by the population, with a frequency of above 80 % of the population showing artificial modeling. Similarly high frequencies would be retained in the cultural repertoire during the next millennium, at least speaking for the Maya (Tiesler 2012). Practically all facets of the head-shaping practices of the later Classic period were already present here during the Protoclassic, with a single exception: the tabular erect pseudocircular or "Olmecoid" form, which appears to emulate the cephalic imagery of the Olmecs. If we believe the skeletal record, this modality was abandoned altogether in the Maya sphere during the first two centuries A.D.

7.3 Olmec and "Olmecoid" Head Forms

The Olmec characterize the earliest large hierarchically administered society in Mesoamerica. Known for its rich artistic legacy and permanent city–temple complexes with ball courts, such as those at San Lorenzo Tenochtitlán, La Venta, and Tres Zapotes, Olmec culture flourished from around 1400–400 B.C. In that time, it dominated much of the Early and Middle Formative periods in Mesoamerica and lay the precedent for much of its subsequent cultural development. Although, strictly speaking, the archaeological remnants left by the Gulf Coast Olmec folk of Oman is geographically limited to the Gulf Coast territories of lower Veracruz and Tabasco, the Isthmus of Tehuantepec and maybe the Soconusco area, Olmec objects and Olmec-style artifacts and symbols came to circulate through most of Mesoamerica during the Middle Preclassic period (900–400 B.C.; Coe and Diehl 1980; Diehl 2004). The form and time sequence of pan-regional interaction and the driving forces behind the assimilation of Olmec cultural elements have been the subject of ongoing scholarly debate, commonly identified with the "sister cultures" vs. "mother culture" controversy (see, for example, Stark 2000, pp. 35–43; Pool 2007, pp. 12–17; Cheetham et al. 2009). By all accounts, exchange between the Olmec (or their intermediaries) with populations east and west of the Soconusco must have been intense by the Middle Preclassic period and was most likely accompanied by a rich material and intellectual exchange; a dynamic cultural tapestry of direct or syncretic religious and life-style adoptions (Diehl 2004, pp. 135–136).

7.3.1 Distribution of Pseudocircular Erect Head Shapes

The pervasive nature of this interaction is illustrated by a highly visible tradition that physically embodies Olmec heritage: infant head modeling. Richard Diehl (2004) describes the anthropomorphic Olmec head sculptures found on the Gulf Coast as distinctly pear-shaped, narrow, and high (Fig. 7.5). Considering this shape, the predominant technique should have been cradleboard use in combination with tight circular head wraps, apt to produce a variety of narrow and heightened head profiles. Unfortunately, no well-preserved skeletal record allows us to witness this custom directly from the highly erosive coastal stretches once occupied by Olman's people (Villamar 2009).

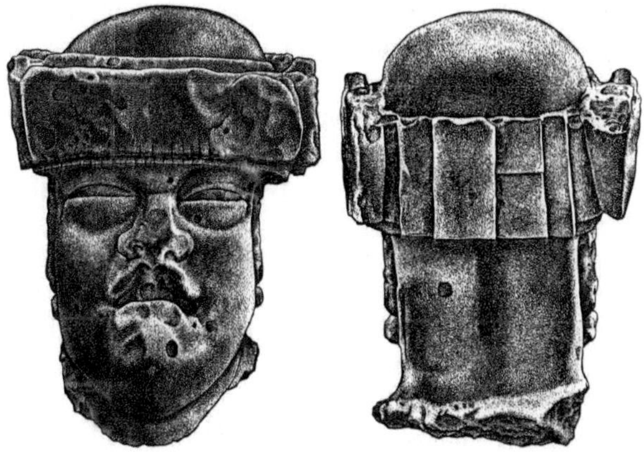

Fig. 7.5 Colossal Olmec head, Monument 6, San Lorenzo. (Adapted from Coe and Diehl, 1980, Fig. 429, 311, and drawn by M. Sánchez)

In fact, the first association between the past living custom of artificially changing head form and the cephalic morphology of the Olman heartlands outside the Gulf area were noted by Arturo Romano in his seminal case study of Burial 1 from Pampa el Pajón, dated to the Middle Preclassic Encanto Phase (750/700–600 B.C.; Paillés 1978; Romano 1977, 1980; Fig. 7.6). At that time, Pampa el Pajón functioned as a thriving estuarine center on the Soconusco coastal plain of Chiapas with direct ties to La Venta (Paillés 1978, p. 93). Romano describes the skull of a young male with an extreme tabular erect modification. Its morphology displays the constrictive action of the band, which increases the upward projection of the compressed head, while effectively limiting its bilateral expansion. This form characterizes pseudocircular fronto-occipital modifications according to Imbelloni's scheme, achieved by a combination of rigid and bland compression materials. Romano specifically notes a tight horizontal headband that constricted the child's head while resting in a compression cradle device, constraining any backward inclination of the infant skull. In the case

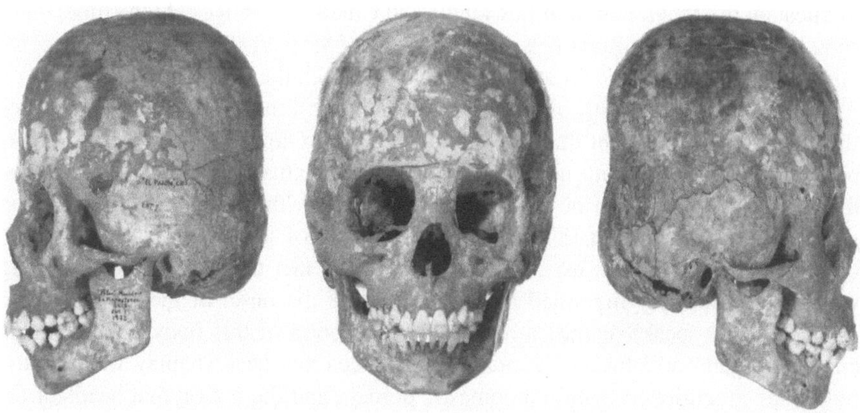

Fig. 7.6 Skull from Pampa el Pajón, Chiapas, Mexico, in different views. (Photo by A. Romano)

of Burial 1 from Pampa el Pajón, the deep horizontal groove left by the band visually separates the cranium into an inferior and a superior lobe, quite similar to the formation of the heads of the anthropomorphic sculptures in the heart of the Olmec lands which encourage Romano to speak of an "Olmec" cranial modification in his work.

Other modifications of this type have been described over the years in the series from the coastal mountain ranges of Veracruz (Carlos Serrano, personal communication, 2001), and for the Mexican Highland series, some of which have been reviewed in this chapter. Although no collection has been systematically scored for pseudo-circular erect shapes, many are characterized by the same attributes described for the skull from Pampa el Pajón. Specifically, Romano notes the artificially reduced width of many tabular erect skulls from Tlatilco, the horizontal grooves in the two detailed case studies on the Late Formative Tlaltenco skulls, and in more cases from the Preclassic period Oaxaca.

An additional nine specimens with this peculiar cranial-vault shape come from Chiapa de Corzo in Chiapas and the Maya Lowlands; namely, from Caucel and Dzibilchaltún (Mexican state of Yucatán), Seibal, and Altar de Sacrificios (Río de la Pasión, Guatemala; Tiesler 2010, 2012; Fig. 7.7). In all cases, these are associated with the Preclassic period. The pseudocircular tabular erect modifications of these skull vaults were produced in all cases by combining a cradling device with a horizontal wrap. It is noteworthy, however, that none of the cranial pieces described for the broader Maya area come close to the extreme degree of morphological transformation that the youngster from Pampa el Pajón displays; all of these appear higher than simple tabular erect forms and are narrow, not broad. All show horizontal circular constriction in the form of lateral grooving and many display additional bulging towards the occipital foramen (pseudo-xifobasia). Chronologically, the nine cases span almost a millennium (from the Middle Preclassic to the Protoclassic periods), before being eliminated from the cultural repertoires of ancient Maya practitioners at the onset of the Classic period (Fig. 7.4). Geographically, the pseudocircular erect shapes come from different sectors of the Maya region; namely, the northern stretches of Yucatán, the Maya Southern Lowlands, and Chiapanec fringes. At the time, these sites functioned as political or exchange centers. Most of them show evidence of interaction with Highland Guatemala and contact with the Olmec region.

Considered jointly with other information on the Preclassic Maya repositories for the dead, it is evident that Maya enactment of "Olmecoid" head shaping was performed equally on both boys and girls. The contextual indications also suggest that it was one shape out of many different shapes with no distinct consideration given to individuals with this head form in terms of grave structure, offerings, or corpse treatment. This aspect, together with the fact that cranial-vault shaping was already a deeply ingrained family tradition at the onset of the Maya Middle Preclassic era, speaks against a restrictive application of this body practice. This conjecture gains still further support from the idea that head shaping was enacted by female practitioners within a domestic domain and the notion that heads of female and male babies were equally the objects of Olmecoid shaping. Considering the above, I advocate rather—at least for the territories east and south of Oman—that the

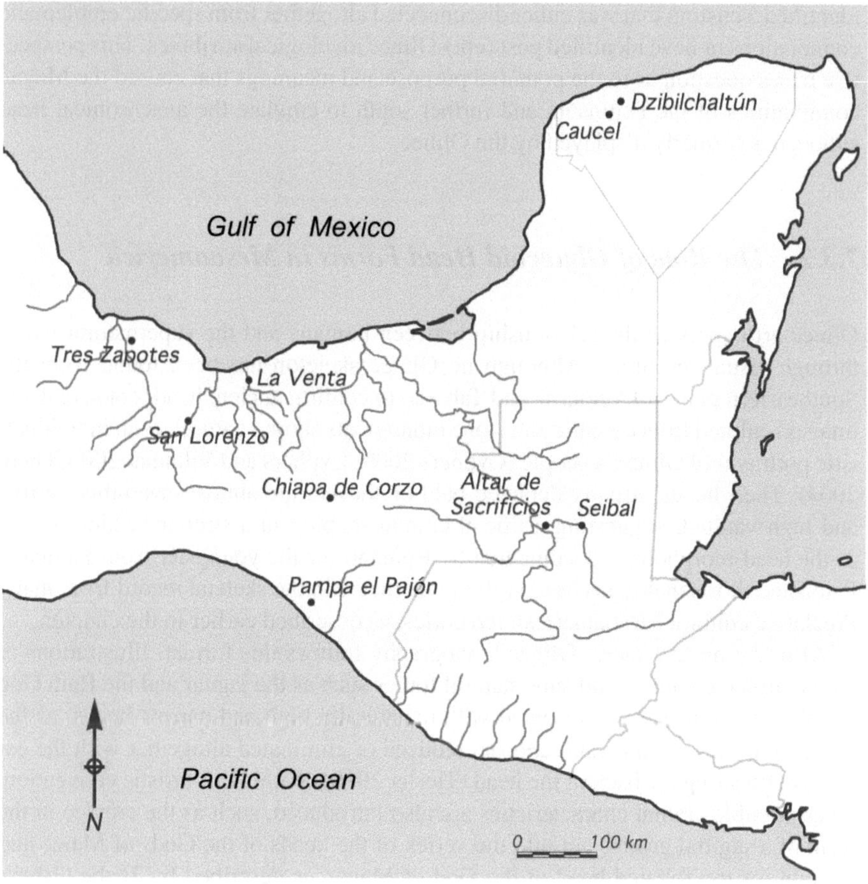

Fig. 7.7 Regional distribution map of Olmecoid (pseudoannular tabular erect) head shapes in the Maya area. (Map by V. Tiesler)

emblematic meaning(s) ascribed to artificially molded heads must have been in any case inclusive of all segments of society. A similar view is expressed by Cyphers (2009, pp. 12–16; see also Cyphers and Villamar 2006), who examines the pseudocircular erect modification in figurines from Mesoamerican territories outside the Gulf Coast (specifically the Basin of Mexico, Chalcatzingo, and Oaxaca), as social identifiers with affinities to the Isthmian cultures. She also sees the formal heterogeneity as an expression of political decentralization in view of social renovation and adoption.

Regarding the chronological association of the Olmecoid crania from the Maya territories and their distribution, it is noteworthy that the earliest cases of this sort of cranial modification occurred not during but after the flourishing hub of Olmec society, especially of the earlier San Lorenzo and La Venta hegemonies. In fact, most of the specimens with this shape postdate the fall of the Olmec by centuries. I therefore believe that the peculiar head forms in the Maya territories must have identified a custom that was either disconnected altogether from specific emblematic connotations or have identified post (epi)-Olmec ideological attributes. This perspective raises question as to the potential purpose and meanings that caused the Mayan communities of the Peninsula and further south to emulate the anachronical head silhouettes formerly displayed by the Olmec.

7.3.2 The Role of Olmecoid Head Forms in Mesoamerica

Olmec art expresses the relationship between humans and the supernatural world through visual resources. Although no Olmec skeleton has been found from the Southeastern plains of Veracruz and Tabasco to confirm ancient head looks, realistic images sculpted from ceramic and stone monuments show a natural, even individualistic portrayal of Olman's people (Cyphers 2009; Cyphers and Villamar 2006; Diehl 2004). Their heads, usually depicted bald or shaved, are almost invariably narrow and high vaulted, suggesting artificial cranial shaping in a similar fashion to that of the head morphology documented by Romano for the youngster from Pampa el Pajón and those analogous head styles perpetuated in the skeletal record from many Preclassic cultural Mesoamerican territories, as described earlier in the chapter.

Also, the more esoteric Olmec iconography follows this format. Illustrations of metamorphic creatures and supernatural forces such as the jaguar and the Rain God are almost exclusively represented with unnaturally high and narrow heads, as the posterior portion of the vault appears reduced or eliminated altogether with the ear lobes delineating the back of the head (Tiesler 2010). Within this artistic convention, other possible cranial characteristics are also introduced, such as the crevice in the form of a sagittal groove cut into the vertex of the heads of the Gods of Maize and of Rain, or the foliated head of the God of Maize, as described by Taube (1996). Prominently placed bands commonly tie the heads of both anthropomorphic images of the supernatural and venerated rulers, leading some authors to grant importance to the headband as a symbol of politically and religiously conferred power (Reilly 2006). Some investigators even affirm that artificially transformed heads are an indication of status (Clark 1993; Diehl 2004; Hansen 2005). Although there are no clear arguments to confirm or deny this affirmation for the Olmecs because of the lack of evaluable material, at least for the areas outside the Olmec heartlands, any connotation of prestige in the pear-shaped Olmec heads should be considered doubtful, given their lack of relevance in the burial record. I conclude that the artificial form in itself, which was probably the common denominator in modification techniques within the core territories of the Olmecs of the Gulf Coast, would not have held any prestigious connotation.

Still another notion to be considered concerns those cultural processes, which led to the apparent embodiment of Olmec ideology in the cultural repertoires of greater Formative Mesoamerica, as documented for crania from the Maya area, the Tehuantepec straight and the Soconusco, as well as for some of the Central Mexican series. The sociocultural dynamics and processes that led to the adoption and practice of head constriction in cradleboard devices across the Mesoamenerican cultural landscape was surely as complex as it is multifold. In terms of the local adoption of the artificial Olmec forms, these would have added most probably to already embedded family traditions with their preexistent, more autochthonous meanings (Tiesler 2010). This situation speaks more of syncretism and reinterpretation of Olmec heartland ideology than direct imposition.

One aspect that strengthens this reading of the skeletal registry is the fact that Olmecoid head forms continued to be used in Central Highland sites, as documented at the Late Formative site of Terremote-Tlaltenco or the Mayan territories long after any direct Olmec influence had languished. The visible reproduction of these models by many generations of women over a span of at least a millennium provides a unique view on the many ways in which Olmec and Epi-Olmec cultural elements, just like any other ideological traits, were adopted and incorporated into daily life. This dynamic also highlights the active role of the female practitioners in conferring family and group identity through this visible practice, an aspect that we will return to when discussing Classic period dynamics in Chaps. 8 and 9.

One last aspect concerns the possible emblematic meanings of Olmec head morphology, especially its potential role as emblematic personifiers of the supernatural. To what extent did the Preclassic Olmec heads project any clear connotation such as signifiers related to venerated deities (Taube 2000)? We will discuss this point briefly for the Mayan world here, where more nuanced statements are possible thanks to its naturalistic iconographic corpus (see Sect. 6.5; Tiesler 2010). As we have argued already in Sect. 6.5, beyond quotidian and organoplastic motives, the artificial head forms could have also borne visible testimony to the cosmovision of their practitioner groups. This association becomes more palpable in those cultural areas that fostered realistic imagery. Karl Taube (1996, pp. 54–62) has made a case, for example, for the evolving veneration of the Olmec Maize God in southeast Mesoamerica, arguing that many Mesoamerican maize deities would have originated in Olmec Maize God "prototypes". In his argument, the Classic period profiles of the Isthmian Maize God still represent this Maize God with an erect cranial vault and a vertical forehead contour, similar to earlier Preclassic representations from the area (Taube 1996, Fig. 19a–d). These conventions still appear to imitate anachronously the Olmec prototypes described earlier.

Centuries later, the Maya Maize God would be represented not as erect but with a reclined, elongated head profile by the generations of Classic Period artists (Taube 1992, Figs. 20 and 21). The tonsured version of this young deity makes this new convention still more obvious, considering the transverse postcoronary sulcus that Classic period artistic rendering places on the top of the god's shaved head (see Fig. 9.6). From an anatomical perspective, this groove appears to be the physiological side effect of severe and prolonged head splinting, which suggests that the Maize God's artificial head rendering was inspired by the observation of human infant head modeling (see Chap. 3). It is noteworthy in this regard that the inclined tubular head forms that characterize most Maya Maize God representations during Classic period do not really appear until the onset of this era. I conclude this line of thought by assuming that the shifting display of head looks during the final stages of the

Preclassic, must have accompanied and promoted ethnogenesis and forged Maya kin group identities during times of social change (Tiesler 2010).

References

Agrinier, Pierre. 1964. The archaeological burials at Chiapa de Corzo and their furniture. Papers of the New World Archaeological Foundation, Vol. 16. Provo: Brigham Young University.

Anderson, James E. 1965. Human skeletons of Tehuacán. Science 148(3669):496–497.

Anderson, James E. 1967. The Human Skeletons. In The Prehistory of the Tehuacan Valley, Vol. 1, ed. Douglas S. Byers, 91–113. Austin: University of Texas.

Barrientos, G., and L. L'Heureux. 2009. Una aproximación bioarqueológica al estudio del poblamiento prehispánico tardío del sudeste de la Región Pampeana. Revista Intersecciones en Antropología 2:3–18. Buenos Aires: UNCPBA.

Bautista, Josefina. 2004. Evidencias de deformación cefálica intencional en figurillas prehispánicas de México. Ph.D. diss., Universidad Nacional Autónoma de México.

Bautista, Josefina. 2005. La deformación cefálica intencional en el México prehispánico. Estudios de Antropología Biológica 12:795–809.

Béguelin, Marien, Valeria Bernal, Mariano Del Papa, Paula Novellino y Gustavo Barrientos. 2006. El poblamiento humano tardío del sur de Mendoza y su relación con el norte de Patagonia: una discusión bioarqueológica. Anales de Arqueología y Etnología 61:5–25.

Cheetham, David, Susana E. Gonzáles, Richard J. Behl, Michael D. Coe, Richard A. Diehl, and Hector Neff. 2009. Petrographic analyses of Early Formative Olmec carved pottery. Mexicon 31:69–73.

Clark, John. 1990. Olmecas, olmequismo y olmequización en Mesoamérica. Arqueología 3:49–56.

Clark, John. 1993. ¿Quienes fueron los olmecas?. In Segundo y Tercer Foro de Arqueología de Chiapas. Serie Memorias, 45–55. Tuxtla Gutiérrez: Gobierno del Estado de Chiapas.

Coe, Michael D., and Richard A. Diehl. 1980. In the land of the Olmec: The people of the river, Vol. 2. Austin: University of Texas.

Comas, Juan, Carlos Serrano, and María Teresa Jaén. 1981. Craneología de Jalapasco, Puebla. Anales de Antropología 18(1):229–249.

Cyphers, Ann. 2009. Face to face, head to head, exploring the significance of Olmec figurines. Paper presented at the Braunstein Figurine Symposium, Las Vegas, January 18, 2009.

Cyphers, Ann, and Enrique Villamar. 2006. La representación del modelado cefálico en las figurillas de San Lorenzo. In Presencia Zoque, Una aproximación multidisciplinaria, ed. Dolores Aramoni Calderón, Thomas A. Lee Whiting, and Miguel Lisbona Gullén, 257–272. Tuxtla Gutiérrez: Universidad de Ciencias y Artes de Chiapas, Consejo de Ciencia y Tecnología del Estado de Chiapas, Universidad Autónoma de Chiapas and Universidad Nacional Autónoma de México.

Darras, Véronique, and Brigitte Faugère. 2010. Chupicuaro and the Preclassic shaft tomb tradition. Mexicon 32:22–30.

Dávalos, Eusebio. 1951. La deformación craneana entre los tlatelolca. BA and Master diss., Escuela Nacional de Antropología e Historia/Instituto Nacional de Antropología e Historia.

David, Cybéle. 2000. Balance de los conocimientos sobre las prácticas funerarias en Chupícuaro, Guanajuato. Master diss. In pre-Columbian Archaeology, Sorbonne University.

Diehl, Richard A. 2004. The Olmecs. America's First Civilization. London: Thames & Hudson.

Faulhaber, Johanna. 1965. La población de Tlatilco, México, caracterizada por sus entierros. In Homenaje a Juan Comas, Vol. 2, 83–121. Mexico: Instituto Indigenista Interamericano.

Gervais, Veronique. 2001. El uso del mecapal en Guatemala y sus consecuencias sobre la columna cervical: primeras observaciones. In Memorias del XIV Simposio de Investigaciones Arqueológicas en Guatemala 2000, Vol. 2, 1089–1091. Guatemala: Museo Nacional de Arqueología y Etnología, Ministerio de Cultura y Deportes, Instituto de Antropología e Historia, and Asociación Tikal.

Hansen, Richard D. 2005. Perspectives on Olmec-maya interaction in the Middle Formative period. In New perspectives on Formative Mesoamerican cultures, ed. Terry G. Powis, 51–72. Oxford: British Archaeological Research International Series 1377.
Lagunas, Zaid. 1989. Los antiguos habitantes de Cholula: Prácticas osteoculturales. Notas Americanas 11:8–50.
Lee, Thomas A. 1969. The Artifacts of Chiapa de Corzo, Chiapas, Mexico. Papers of the New World Archaeological Foundation, Vol. 26. Provo: Brigham Young University.
Lee, Thomas A. 1993. El protoclásico: un problema de definición. In Segundo y Tercer Foro de Arqueología de Chiapas, Serie Memorias, 119–125. Tuxtla Gutiérrez: Gobierno del Estado de Chiapas.
Lowe, Gareth W., and Pierre Agrinier. 1960. Excavations at Chiapa de Corzo, Chiapas, Mexico. Papers of the New World Archaeological Foundation, Vol. 8–11. Provo: Brigham Young University.
Menéndez, Lumina Paula. 2010. Patologías bucales en cráneos humanos del noreste de Patagonia: Tendencias temporales en la dieta y estado de salud de las poblaciones del holoceno-tardío del valle inferior del Río Negro. Magallana 38:115–126.
Munizaga, Juan R. 1965. Skeletal remains from sites of Valdivia and Machalilla Phases. Smithsonian Contributions to Anthropology 1:219–234.
Munizaga, Juan R. 1973. Intentional cranial deformation in the precolumbian population of Ecuador. American Journal of Physical Anthropology 45:687–694.
Munizaga, Juan R. 1974. Deformación craneal y momificación en Chile. Anales de Antropología 11:329–336.
Oliveros, Arturo. 1971. Excavaciones de las tumbas en el Opeo, Michoacán. B.A. diss, Escuela Nacional de Antropología e Historia.
Paillés, Maricruz. 1978. The process of transformation at Pajón: A Preclassic society located in an estuary in Chiapas, Mexico. In Prehistoric coastal adaptations ed. Barbara L. Stark and Barbara Voorhies, 81–96. New York: Academic Press.
Peña, Rosa María. 1982. Restos óseos humanos en una tumba de la fase Tenanyécac en Tlaxcala. Estudios de Antropología Biológica 1:335–362.
Peña, Rosa María, and Luis Alberto López-Wario. 1989. Un cráneo deformado del Preclásico de Ecatepec, Estado de México. Estudios de Antropología Biológica 4:609–616.
Pereira, Grégory. 1999. Potrero de Guadalupe: anthropologie funéraire d'une communauté prétarasque du nord du Michoacán, Mexique. Oxford: British Archaeological Research International Series 816.
Pérez, Sergio Iván, Claudia Della, Paula Novellino, Paula N. González, Valeria Bernal, Estela Cuneo, and Adam Hajduk. 2009. Deformaciones artificiales del cráneo en cazadores-recolectores del Holoceno medio-tardío del noroeste de Patagonia. Magallana 37:77–90.
Pool, Christopher A. 2007. Olmec archaeology and early Mesoamerica. Cambridge: Cambridge University Press.
Reilly, F. Kent. 2006. Middle Formative origins of the Mesoamerican ritual act of binding. In Sacred bundles. ritual acts of wrapping and binding in Mesoamerica, ed. Julia Guernsey and F. Kent Reilly III, 1–21. Barnardsville: Boundary End Archaeology Research Center.
Robles, Fernando, and Josep Ligorred. 2008. Informe del proyecto salvamento arqueológico en áreas de crecimiento urbano de la ciudad de Mérida, Yucatán. Etapa ciudad Caucel (2004–2006). Submitted to Estéban de Vicente Chab, Mérida, Mexico.
Romano, Arturo. 1970. Preceramic human remains. In Physical anthropology. Handbook of Middle American Indians, Vol. 9, 22–34, ed. Robert Wauchope and T. Dale Stewart. Austin: University of Texas.
Romano, Arturo. 1972a. Un cráneo precerámico con deformación intencional. Boletín del INAH 1(2nd series):35–36.
Romano, Arturo. 1972b. Deformación craneana en Tlatilco, México. In Religión en Mesoamérica. Memoirs of the XII Mesa Redonda de la Sociedad Mexicana de Antropología, 415–419.
Romano, Arturo. 1973. Deformación cefálica intencional en la población prehispánica de Cholula, Puebla. Comunicaciones Proyecto Puebla Tlaxcala 8:49–51.

Romano, Arturo. 1974a. Deformación cefálica intencional. In Antropología física, época prehispánica ed. Juan Comas, 197–227. México: Instituto Nacional de Antropología e Historia.
Romano, Arturo. 1974b. Restos óseos humanos precerámicos de México. In Antropología Física, época prehispánica, Vol. 3, ed. Javier Romero, 29–81. México: Colección México: Panorama Histórico y Cultural, Instituto Nacional de Antropología e Historia.
Romano, Arturo. 1977. Cráneo del Pajón, Chis. Revista Mexicana de Estudios Antropológicos 23:365–394.
Romano, Arturo. 1978. La población prehispánica de Querétaro. In Simposio Problema del Desarrollo Histórico de Querétaro, 64–68. México: Museo Regional de Querétaro, Dirección del Patrimonio Cultural del Gobierno del Estado de Querétaro.
Romano, Arturo. 1979. El material osteológico humano de Toniná, Chiapas. Estudio morfológico, descriptivo y comparativo. In Toniná, une cité Maya du Chiapas (Mexique). Études Mesoaméricaines ed. Pierre Becquelin and Claude Baudez, 179–192. México: Centro de Estudios Mexicanos y Centroamericanos.
Romano, Arturo. 1980. Appendix 3. The Skull from El Pajón, Chiapas. In Pampa El Pajón, an early estuarine site, Chiapas, Mexico, ed. Maricruz Paillés, 95–114. Papers of the New World Archaeological Foundation. Provo: Brigham Young University.
Romano, Arturo. 1987. Análisis osteológico (Apéndice A). In Excavaciones de un sitio preclásico en San Mateo Etlatongo, Nochixtlán, Oaxaca, México ed. Roberto Zárate. Oxford: British Archaeological Series Internacional Series 322.
Romero, Javier. 1950. El material osteológico de Monte Albán, Mexico. Boletín Bibliográfico de Antropología Americana 12(1):166–168.
Romero, Javier. 1951. Monte Negro, centro de interés antropológico. In Homenaje a Alfonso Caso, 317–329. México: Instituto Nacional de Antropología e Historia.
Romero, Javier. 1958. Mutilaciones dentarias prehispánicas de México y América en general. Mexico: Instituto Nacional de Antropología e Historia.
Romero, Javier. 1983. Las tumbas y los entierros prehispánicos de Oaxaca. Anales de Antropología 20(1):91–113.
Salas, María Elena. 1980. Appendix 2. Burial 1 from El Pajón, Chiapas. In Pampa El Pajón, an early estuarine site, Chiapas, Mexico, ed. Maricruz Paillés, 95–114. Papers of the New World Archaeological Foundation. Provo: Brigham Young University.
Sánchez, Patricia. 1971. Cuicuilco: estudio osteológico de la población prehispánica. BA diss., Escuela Nacional de Antropología e Historia/Instituto Nacional de Antropología e Historia.
Saul, Frank P. 1972. The Human Skeletal Remains of Altar de Sacrificios. Papers of the Peabody Museum of Archaeology and Ethnology, Vol. 63, Num. 2. Cambridge: Harvard University.
Saul, Frank, and Julie Mather Saul. 1991. The Preclassic population of Cuello. In Cuello, an Early Maya Community in Belize, ed. Norman Hammond, 134–158. Cambridge: Cambridge University.
Saul, Frank, and Julie Mather Saul. 1997. The Preclassic skeletons from Cuello. In Bones of the Maya: Studies of ancient skeletons, ed. Stephen L. Whittington and David M. Reed, 15–50. Washington: Smithsonian Institution.
Serra Puche, Mari Carmen, Magalí Civera, and Arturo Romano. 1982. Entierros en un sitio formativo del sur de la Cuenca de México: Terremote-Tlaltenco, D.F. Anales de Antropología 19(1):55–92.
Soto-Heim, Patricia. 1987. Evolución de deformaciones intencionales, tocados y prácticas funerarias en la prehistoria de Arica, Chile. Chungara 19:10–213.
Stark, Barbara L. 2000. Framing the Gulf Olmecs. In Olmec art and archaeology in Mesoamerica, ed. Mary E. Pye and John E. Clark, 31–53. Washington: National Art Gallery.
Taube, Karl A. 1992. The major gods of ancient Yucatan. Washington: Dumbarton Oaks Research Library and Collection.
Taube, Karl A. 1996. The Olmec maize god: The face of corn in Formative Mesoamerica. In Res 29/30 Spring/Autumn 1996. The Pre-Columbian, ed. Francesco Pellizzi, 39–81. Philadelphia: University of Pennsylvania Museum Publications.

Taube, Karl A. 2000. Lighning celts and corn fetishes: The Formative Olmec and the development of maize symbolism in Mesoamerica and the American Southwest. In Olmec art and archaeology in Mesoamerica, ed. Mary E. Pye and John E. Clark, 297–37. Washington: National Art Gallery.

Tiesler, Vera. 1998. La costumbre de la deformación cefálica entre los antiguos mayas: aspectos morfológicos y culturales. México: Instituto Nacional de Antropología e Historia.

Tiesler, Vera. 1999. Rasgos bioculturales entre los antiguos mayas: Aspectos arqueológicos y sociales. Ph.D. diss., Universidad Nacional Autónoma de México.

Tiesler, Vera. 2009. Reporte del análisis de deformación cefálica en el sitio preclásico de Caucel, Yucatán. Submitted to the Proyecto Arqueológico de Caucal, INAH; on file, Mérida: Instituto Nacional de Antropología e Historia (Center Yucatán).

Tiesler, Vera 2010. “Olmec” head shapes among the Preclassic period maya and cultural meanings. Latin American Antiquity 21:290–311.

Tiesler, Vera. 2012. Transformarse en maya. El modelado cefálico entre los mayas prehispánicos y coloniales. México: Universidad Autónoma de Yucatán, Universidad Nacional Autónoma de México.

Tiesler, Vera, and Mónica Rodríguez. 2009. “Olmec” head shapes among the Preclassic maya and cultural identity. Paper presented at the 74th Annual Meeting of the Society for American Archaeology, Atlanta.

Villamar, Enrique. 2009. Estudios osteológicos y tafonómicos de entierros olmecas del Período Preclásico de San Lorenzo, Veracruz. B.A. diss., Escuela Nacional de Antropología e Historia.

Winter, Marcus. 1989. El Preclásico en la región oaxaquea. In El Preclásico o Formativo, avances y perspectivas, seminario de arqueología “Dr. Román Piña Chan”, ed. Martha Carmona Macías, 461–479. Mexico: Instituto Nacional de Antropología e Historia.

Yépez, Rosaura. 2009. El simbolismo de la modificación cultural de la cabeza en la cultura andina de Paracas del antiguo Perú. Estudios de Antropología Biológica 14:526–543.

Chapter 8
Head Shapes in Classic Period Mesoamerica

This chapter explores the broad geographic and chronological trends in Mesoamerican head-shaping practices during the Classic period (Tiesler 2012a, 2012b). Back then, the eastern and western parts of Classic period Mesoamerica (Michoacán, Veracruz, Oaxaca, and the Maya Lowlands) harbored the strongholds of cranial modeling in terms of popularity, visibility, and diversity of the culturally produced head morphologies. Central highlanders used predominantly cradleboards, often only evident in weak lamboid flattening. Only in Teotihuacán (north of Mexico City), is the use of body devices notably diverse. Regional preferences for head forms were surveyed systematically across the Maya area. Here, extremely narrow and slanted heads were preferred in the western sphere of the Maya Lowlands; namely, in the Lower and Middle Usumacinta region (Palenque, Chiapas) and along the Gulf coast fringes (Jaina, Campeche). Conversely, tabular erect forms were in vogue among those who lived along the eastern shores, and in the Guatemaltecan and Chiapanecan Highlands. The Mexican state of Veracruz is recognized for a preference for pronounced superior (obelionic) flattening, as witnessed at the site of El Zapotal, which was settled during the Late Classic period. Apart from superior compression, the cranial record of El Zapotal and other sites in Veracruz display other extreme oblique and erect skull modifications, among which the literature highlights the *bilobée* and *trilobée* forms.

8.1 The Mesoamerican Highlands

8.1.1 Central Highlands

During the Early and Middle Classic periods, Mesoamerica's highland plateau was dominated by far-reaching road networks that emanated from Teotihuacán. Considering the importance of this and other Classic period highland centers and the amount of research that has been conducted on Highland Mesoamerica, the lack of coverage of Classic head-shaping practices comes as a surprise. Also, the urban and suburban neighborhoods of Teotihuacán themselves await a systematic survey of cranial forms.

V. Tiesler, *The Bioarchaeology of Artificial Cranial Modifications*, Interdisciplinary Contributions to Archaeology 7, DOI 10.1007/978-1-4614-8760-9_8,

Important work in this direction has been promoted or conducted recently by Kanjou and Andrade (2003), Manzanilla and Serrano (2003), Yépez (2001) and González-Miranda (2009), among others, all of whom have been assembling findings on this subject.

Taken together, these descriptions appear to communicate both a general preference in looks towards broad, shortened heads from cradleboarding, albeit diversity in head looks is apparent up to the end of Teotihuacán's functioning, to be followed by uniform cradle practices after its fall. For instance, González-Miranda mentions 15 tabular erect head morphologies in his catalogue of the Proyecto Arqueológico Teotihuacán, spanning the phases of the Classic and Postclassic periods. Exclusively erect head shapes were also apparently sported by the residents of the Atetelco palaces, a compound which was resettled centuries later, during the Postclassic period.

By contrast, the expressions of infant head flattening are multifarious in some of Teotihuacán's central *barrios*. Serrano and colleagues (2003; Serrano and Lagunas 1999) report a set of differently formed crania from Teotihuacán's La Ventilla neighborhood (see also Yépez 2001). This series includes tabular erect models with annular constriction although the bipolar reduction is light. There is also an isolated case of superior flattening described by Yépez (2001) from this *barrio* compound, which is dated back to the Middle Classic period (Late Xolalpa Phase). Also, the residents of Teohihuacán's merchant district displayed a variety in artificial head morphology. Here, Rattray and Civera (1999) note erect and oblique shapes in selected mentions of the practice. The recent work by Alvarado (2013) is based on a systematic exploration of a barrio center of Teopancasco, a group of artisan families residing in the metropolis. Laudable for including explicit mentions on the many naturally rounded crania, Alvarado distinguishes five tabular erect specimens and two severely oblique cranial morphologies. The latter show a good fit with the broad-brimmed head splints that are so prominently displayed in small-scale figurine representations from the area (Fig. 4.10).

Still scarcer than at Teotihuacán is the coverage of Monte Albán, an important Zapotec center in the present state of Oaxaca, which outlived Teotihuacán (Martínez et al. 1996; Winter et al. 1996). Tabular erect head shapes stem mostly from here. Isolated additional forms are described in a set of crania with signs of trephining from Monte Albán's residential neighborhoods. These display elongated head forms (Tiesler 2006). A more systematic survey of the custom's expression was incorporated into a study on social identity among households of Monte Albán by González Licón (2003), although no specific information is itemized for reference.

Also, settlement reports on other Central Highland series such as the one from Cholula (Lagunas 1989; Romano 1973), point towards dominance in erect head looks over artificially elongated morphologies. In the large skeletal series of Cholula, Puebla, Romano notes that 119 of 121 cranial vaults exhibit the signs of having been culturally flattened and specifies that only two out of nine modeled Classic period specimens display head elongation rather than shortening (Romano 1973, p. 49). This is also the trend conveyed by the survey on Classic period crania, most of which are curated at the National Museum of Anthropology. Romano (1973, p. 206) identifies most of these as erect, but recognizes at the same time that oblique and mimetic forms show up in the record. Generally speaking, the popularity of head compression and its visible expression in cranial vault morphology appears reduced among Highlanders when compared to the folk that lived east and west. Naturally, given the lack of systematic regional coverage, these observations are premature, awaiting confirmation and interpretation by future research.

8.1.2 *Western and Northern Highlands*

Cranial data on the Western and Northern fringes of Mesoamerica are similarly isolated as on the territories that breach the Central Highlands. For the Classic period contexts of the Zacapu Basin of Michoacán, Gervais (1989; supplemented by Pereira 1999, p. 165) notes mostly reclined, oblique modifications of the cranial vault, most of which are expressed slightly or moderately. At the sites of Loma Alta and Guadalupe, also in Michoacán, this head model exists together with a host of other forms, many of them tabular erect variants. During the Late Classic era, superior head flattening also appears in the area's cultural repertoire, just as in other Mesoamerican areas we have covered. Also, here, top-flattened heads continue to be seen during the Postclassic period, while other forms are no longer reproduced, giving way to an almost uniform look of frontal-occipital flattening during the later stages of occupation (Pereira 1999, p. 166; see Chap. 10). Gabriela Uruñuela (Acosta and Uruñuela 1997) notes a change in artificially produced cranial forms at the end of the first millennium in the area. This is evident when she compares the shapes between the Classic and the Postclassic periods (concretely between the Sayula and Amacueca phases) in the Valley of Sayula, Michoacán. Uruñuela argues that the replacement of head models accompanies deeper shifts in population and ethnic shift, identified later with Tarascan folk.

On the other side of the northern Highland Plateau lie the Classic Huaxtec communities of Cerro de Silva and Río Verde in the present Mexican state of San Luis Potosí. The Huaxtecan series from San Luis Potosí are part of a larger cultural area on the northern borderlands of Mesoamerica. Carlos Serrano and Rosa María Ramos (1984; see also Montiel 2013, p. 79) note that 80 % of the skulls of these series bear the signs of artificial modeling and among these, they document both tabular erect and oblique morphologies. The reclined skull vaults are noted as especially prominent in Vista Hermosa and surpass in numbers other tabular erect vaults. In pre-Hispanic times, the Huaxtec territories extended towards the coast of the Gulf of Mexico and will be noted again in Chap. 10 for its Postclassic period evolvement.

8.1.3 *Maya Highlands*

The eastern territories of the Mesoamerican Highlands were home to Mayan speakers. Those human remains available for our systematic regional survey are relatively reduced in number and hold only scarce chronological information. They come from the Guatemaltecan settlements of Acul, Kaminaljuyú, Zaculeu, Zacualpa, Cobán, Nebaj, Los Cimientos, Los Cerritos, and Chagüites. On the Mexican side, we scored the Maya Highland sites from Chiapa de Corzo and the valleys of Chicoasen and Angostura, along with several dated cave sites. Given the reduced number of dated burials, it is speculative, at best, to generalize on any shifts in head modeling preferences in the Highlands (see Tables 8.1 and 8.2), beyond noting a frank predominance of tabular erect forms.

This predilection for erect cephalic shapes seems to have been shared on both sides of today's border and was to become the common denominator of head modification during the Early Postclassic period. This trend is evident for example in the Classic era population of Nebaj, presently located in the Guatemalan Department of El Quiché. Some 64 % ($N = 14$) show artificial modifications in our survey, all of which display tabular erect forms in their intermediate and lambdic variants (see also Gervais 1989).

Just a few kilometers from Nebaj lies Acul, a large settlement was excavated by the French Mission (CEMCA) in the valley known by the same name. This research produced 14 skulls (six dated to the Late and Terminal Classic periods (Gervais 1989)). Similar to Nebaj, all cases of artificial modification displayed tabular erect forms in its intermediate and lambdic variants. In a similar fashion, T. Dale Stewart (1953, pp. 296, 297) notes the strong preference in the Maya mountain area for erect forms among skulls from Zaculeu, dated to the Atzan and Chinaq phases (second half of the Classic period). Of those, some 68.75 % (34 crania, $N = 48$) show cultural modification. Apart from another six skulls described by Stewart as pseudocircular and that may have corresponded to oblique variants, there is a frank partiality here for erect forms, noting that the average degree of modification (Qankyak and Xinabahul phases) still increases towards the end of Zaculeu's occupation. By then, erect models—referred to as "erect frontal-occipital" by Stewart—monopolize Zaluleu's head looks.

On the Mexican side of the Maya Highlands, the Chiapanec foothills and mountain ranges communicate west with the Central Depression of Chiapas and the isthmus towards Oaxaca, still further west. Here, the Classic period residents of Chiapa de Corzo all display erect modification in our series. Similarly, a recent study of over 100 skulls recovered from the Classic period cave sanctuary of Las Banquetas, near the Tzeltal community of La Trinidaria, reveals a strong predominance of tabular erect crania (Romano et al. 2011; Rodríguez and Sonora 1984).

Also, peculiar for the Maya Highlands, is an elevated number of sagittal constriction bands, shown by 58.1 % of the surveyed regional population. Some practitioners, all from the valley of La Angostura in Chiapas, actually achieve dramatic lateral bulging of the parietal lobes of their infants above *lambda*. In general, the use of sagittal bands seems to have been more popular among Highlander Mayas than among their Lowland Maya neighbors. There, only 23.6 % of the crania of the same period show sagittal grooves, and none actually achieves a visible bipolar separation seen in some of the highland crania. Again, it would appear that—beginning in the second half of the Classic period—mountain populations used similarly strongly constrictive sagittal bands that characterized the practices of some populations further

Table 8.1 Frequency of artificial cranial-vault modifications among the Classic period Maya

	Clásico Temprano (%, *N*)	Clásico Tardio (%, *N*)	Clásico Terminal (%, *N*)	Clásico total (%, *N*)
Maya Gulf Coast and Caribbean				
Frequency (%)	91.4, 58	79.1, 369	Not identified, 3	81.7, 498
Frequency of circular wrap	63.0, 27	26.2, 130	Not identified, 3	33.1, 181
Frequency of sagittal groove	18.5, 27	42.0, 143	Not identified, 2	36.4, 195
Tabular oblique/erect	90.63, 32	69.2, 185	Not identified, 3	74.0, 262
Average degree of modification (0–4)[a]	2.17, 13	1.62, 89	Not identified, 2	1.69, 114
Yucatec peninsula				
Frequency (%)	Not identified, 8	84.2, 19	Not identified, 4	93.8, 80
Frequency of circular wrap	Not identified, 2	Not identified, 8	Not identified, 4	34.8, 23
Frequency of sagittal groove	Not identified, 2	Not identified, 8	Not identified, 3	33.3, 24
Tabular oblique/erect	Not identified, 5	23.1, 13	Not identified, 4	58.8, 51
Average degree of modification (0–4)[a]	Not identified, 4	N.A., 7	Not identified, 1	1.95, 29
Central Lowlands				
Frequency (%)	80.8, 26	78.3, 138	85.2, 94	82.9, 281
Frequency of circular wrap	Not identified, 9	43.4, 53	50.0, 38	42.7, 103
Frequency of sagittal groove	20, 10	26.9, 52	23.1, 39	23.6, 106
Tabular oblique/erect	45.5, 11	68.8, 77	64.5, 62	65.4, 162
Average degree of modification (0–4)[a]	Not identified, 7	1.97, 56	2.03, 39	1.99, 188
Western Maya periphery				
Frequency (%)	Not identified, 0	88.6, 70	100, 11	90.7, 118
Frequency of circular wrap	Not identified, 0	52.8, 36	36.4, 11	40.85, 71
Frequency of sagittal groove	Not identified, 0	13.3, 30	54.5, 11	14.02, 107
Tabular oblique/erect	Not identified, 0	68, 50	45.5, 12	5.3, 85
Average degree of modification (0–4)[a]	Not identified, 0	2.45, 39	2.50, 11	2.24, 68
Maya Highlands				
Frequency (%)	76.7, 30	Not identified, 9	90.9, 11	81.9, 74
Frequency of circular wrap	Not identified, 9	Not identified, 7	Not identified, 5	27.6, 29
Frequency of sagittal groove	63.4, 11	Not identified, 5	Not identified, 7	58.1, 31
Tabular oblique/erect	11.8, 17	Not identified, 7	18.2, 11	24.5, 49

Table 8.1 (continued)

	Clásico Temprano (%, *N*)	Clásico Tardio (%, *N*)	Clásico Terminal (%, *N*)	Clásico total (%, *N*)
Average degree of modification (0–4)[a]	Not identified, 9	Not identified, 5	Not identified, 7	2.04, 34
Southeastern Maya periphery				
Frequency (%)	Not identified, 4	67.7, 65	Not identified, 3	77.5, 155
Frequency of circular wrap	Not identified, 4	29.0, 31	Not identified, 0	21.3, 61
Frequency of sagittal groove	Not identified, 4	16.1, 31	Not identified, 0	34.3, 70
Tabular oblique/erect	Not identified, 4	71.1, 38	Not identified, 0	73.8, 80
Average degree of modification (0–4)[a]	Not identified, 3	1.74, 30	Not identified, 0	1.86, 61

N number of cases. All results below $N = 10$ were scored as "not identified", *N.A.* not available
[a]Only adult crania were scored

northwest, on the coast of Veracruz. Here, extreme oblique and erect skull modifications with furrows and strong lateral bulging are reported, among which the literature highlights the *bilobée* and *trilobée* forms (Comas and Marquer 1969; Gosse 1855; Martínez 2007, 2009).

Aside from the preferred use of compression cribs with sagittal constriction among Classic Maya Highlanders, these also set themselves apart from coeval inland Lowlanders by staging particular cultural models. By the middle Classic period, some heads display superior flattening. Some of these come close to the extreme degrees of flattening observed at the Mixtequilla site of El Zapotal in Veracruz, to be described in more detail in the following section.

8.2 The Gulf Coast

Classic period Mesoamerican cultures along the Gulf Coast are generally distinguished along its north–south axis. The northern portions were home to Huastec-speaking folk already during the Classic period. However, no skeletal series dated to the first millennium A.D. has been documented from here (but see also the previous section for coverage of Highland Huaxtec cultures (Martínez 2007, p. 12; Montiel 2013, p. 84). South to the Huaxtecan territories lies El Tajín, a major center of Mesoamerica, which flourished during the second half of the first millennium. Unfortunately, also this major archaeological site coverage of head-shaping practices from its portraits and skeletal record, which has been examined recently by Yamile Lira and Jaime Ortega (2004) and by Mair Sittón (2010, p. 111).

Table 8.2 Frequency of artificial cranial-vault modifications in different areas of the Classic period Central Lowlands

	Clásico Temprano (%, *N*)	Clásico Tardío (%, *N*)	Clásico Terminal (%, *N*)	Clásico total (%, *N*)
Middle and lower Usumacinta Basin				
Frequency (%)	Not identified, 1	93.0, 43	Not identified, 0	95.1, 74
Frequency of circular wrap	Not identified, 0	70.4, 27	Not identified, 0	67.9, 28
Frequency of sagittal groove	Not identified, 0	13.6, 22	Not identified, 0	13.0, 23
Tabular oblique/erect	Not identified, 0	84.8, 33	Not identified, 0	85.3, 262
Average degree of modification (0–4)[a]	Not identified, 0	2.43, 26	Not identified, 0	2.37, 27
Northern Lowlands				
Frequency (%)	Not identified, 8	89.5, 38	100, 26	94.5, 91
Frequency of circular wrap	Not identified, 2	28.6, 14	41.7, 12	30.6, 36
Frequency of sagittal groove	Not identified, 2	31.3, 16	27.3, 11	25.6, 39
Tabular oblique/erect	Not identified, 5	73.9, 23	45, 20	61.3, 62
Average degree of modification (0–4)[a]	Not identified, 5	1.9, 18	1.67, 13	1.85, 41
Southern Lowlands				
Frequency (%)	Not identified, 4	75.9, 29	89.3, 28	82.0, 61
Frequency of circular wrap	Not identified, 3	N.A., 9	56.3, 16	37.0, 27
Frequency of sagittal groove	Not identified, 3	16.7, 12	21.4, 14	17.9, 28
Tabular oblique/erect	Not identified, 2	70.6, 17	66.7, 21	65.0, 40
Average degree of modification (0–4)[a]	Not identified, 2	2.35, 12	2.00, 15	2.11, 29
Eastern Lowlands				
Frequency (%)	Not identified, 5	63.3, 49	86.1, 36	71.1, 97
Frequency of circular wrap	Not identified, 1	37.5, 16	N.A., 6	40.9, 22
Frequency of sagittal groove	Not identified, 1	7.7, 13	20.0, 10	12.5, 24
Tabular oblique/erect	Not identified, 0	63.6, 23	82.4, 17	70.7, 41
Average degree of modification (0–4)[a]	Not identified, 0	1.84, 15	N.A., 8	1.98, 24

All results below $N = 10$ were scored as "no data"
N.A. not available
[a]Only adult crania were scored

Towards the Mixtequilla region, in what is now the south-central part of Veracruz, the settlements of Higueras, El Zapotal, and Cerro de las Mesas once held a concentration of religious, political, and economic power. There are a number of Classic

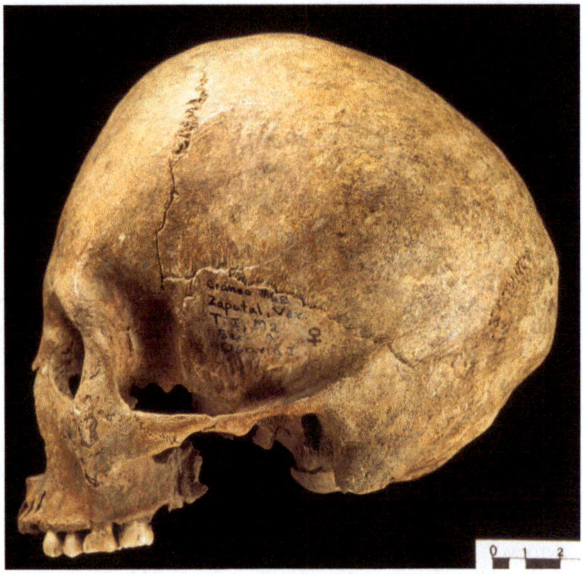

Fig. 8.1 Lateral view of cranium from El Zapotal, Veracruzo, displaying strong superior flattening (DAF/INAH; photo by A. Romano)

period skeletal site series that represent this area; namely, those from Tlalixcoyan and El Zapotal (Martínez 2007; Romano 1973; Tiesler 2012c). The latter flourished during the Late to Terminal Classic period, between A.D. 600 and A.D. 900. During the 1970s, Mexican archaeologist Manuel Torres (2009) and his team discovered an extraordinary earthen shrine at Mound 2 of El Zapotal, dedicated to an early version of the Aztec God of Death, Mictlantecuhtli. Immediately to the north of the shrine, although believed to date later in time (Tiesler et al. 2013), a large cylindrical ossuary, known as Ossuary I, stacked abundant amounts of disarticulated or partly articulated human bones with mainly crania and long bones presumed to stem from victims of sacrifice, a number of which had been flayed.

It was Arturo Romano who first documented the crania of El Zapotal and termed the peculiar top-flattened skull forms among them as "Zapotal-type head shapes" (Romano 1974, 1977; see also Martínez 2007; Tiesler et al. 2013; Fig. 8.1). A recent reexamination of the human remains from El Zapotal at the INAH Department of Physical Anthropology confirms and complements earlier observations by Romano and is described in detail in other work (Tiesler et al. 2013). Each of the examined crania of this series exhibits the effects of artificial modeling (100 %; $N = 50$), most of them severe. Tabular erect shapes predominate over tabular oblique forms. The latter appear in extreme expressions and seem to be deposited later in time, because they were recovered almost exclusively in the shallow parts of Ossuary I (Fig. 8.2). The receding head profile, produced by head splints, sometimes still appears emphasized by a visible division of the two parietal lobes.

More frequent overall, and present at all levels of Ossuary I, are tabular erect shapes from cradleboarding. In the series under study, this practice led to mostly severe changes in form, by shortening the anteroposterior length and by increasing

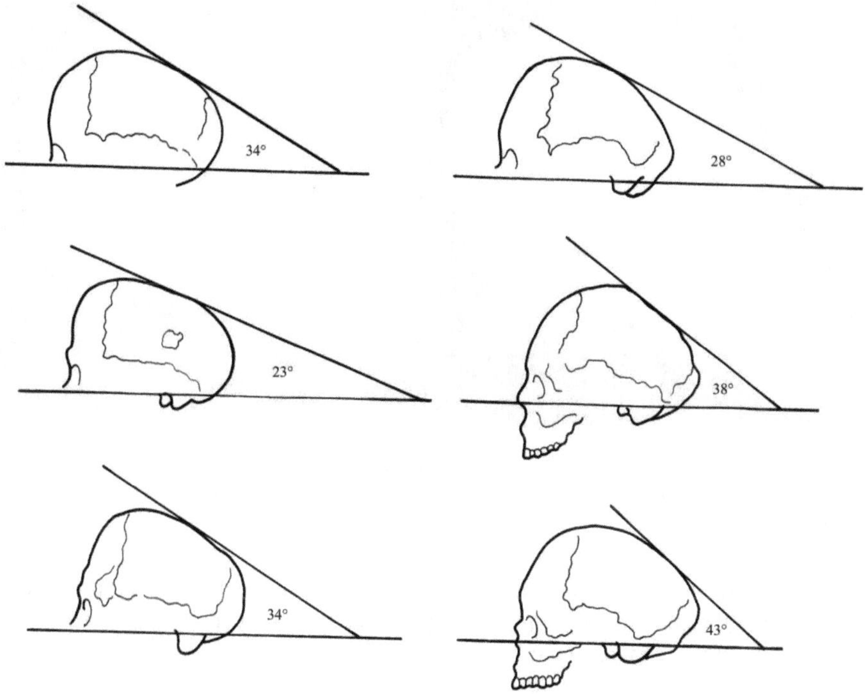

Fig. 8.2 Technical drawings of top-flattened crania from El Zapotal, Veracruz. (Adapted and redrawn from courtesy material of A. Romano by B. Ceballos)

biparietal width (Fig. 8.3). Among them are numerous skulls that display strong obelionic or superior flattening. The latter make up roughly half of the overall series from El Zapotal (47.92 %, $N = 48$) and an impressive two-thirds of those of its lower depositional portion. Note that the superior flattening characterizes both men and women from the lower and therefore earlier portion of the ossuary. Many of these top-flattened crania display marks of having been flayed, materializing their presumed sacrificial origin (Tiesler et al. 2013).

Regarding the compression devices used to flatten the top of the head, there is controversy in the literature as to the device responsible (Martínez 2009; Nelson and Mandimenos 2010; Pereira 1999; Tiesler 2012c). As argued in Chap. 4, both head splinting and cradles can be used for flattening the upper portion of the cranial vault, leading either to an inclined or broadened head form (see also Pereira 1999). At least at El Zapotal itself, I assume that the extreme forms of parietal compression were produced with conventional cradleboards. Here, maybe the child's head was tied to the backboard in an overextended position, maybe an additional tablet was fixed on top. This scenario gains strength when considering the asymmetry observed in many top-flattened crania at El Zapotal, a characteristic by-product of crib-transmitted compressions. Another argument in this same direction can be made from the orientation of the anterior and posterior compression planes that are readily visible on the

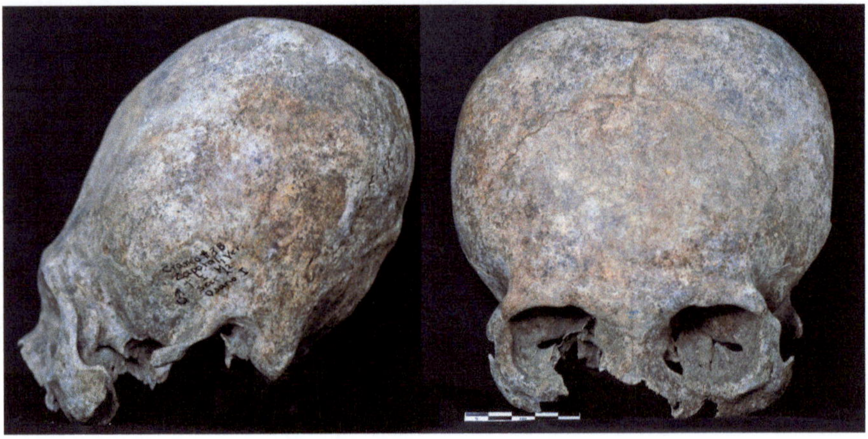

Fig. 8.3 Lateral and frontal views of the cranial vault of a male from El Zapotal, Veracruz, showing severe reclination and flattening of the calotte with sagittal grooving (DAF/INAH; photo by V. Tiesler)

skull vaults from El Zapotal. While the frontal plane is not inclined towards the back but straight, the occipital plane is typically located in the upper occipital area, above inion, sometimes leading to the bulging of the inferior portion (Tiesler et al. 2013). Again, this would speak of a cradle device, not a head splint. Surely, iconography would provide a cleared image of the compression kits used for superior flattening.

Regarding the cultural identity of the male and female crania assembled in the lower portions of Ossuary I, we hold that these still showed the local body insignias of El Zapotal and its surrounding area in the form of their top-flattened artificial shape, which reflects the earlier, local preference in the Mixtequilla area. Heads with this particular profile appear in the skeletal record or are displayed in the mural portraiture from the Río Blanco and Río Hondo valleys; namely, Tlalixcoyan, Nopiloa, Apachital, and Dicha Tuerta (Martínez 2007, p. 101; 2009). Similarly, the murals from Classic period Las Higueras and the anthropomorphic clay statues from Paso de Ovejas express the strong superior cephalic flattening that was so popular in the region at that time (Franco 1993). Note that in the later stage of use of Ossuary I, only remains of females were added on top of the earlier remains. None of these crania bore the artificial superior flattening that appears to represent the local pattern, which could imply either a nonlocal and culturally distinct origin or a shift in local preferences late in the occupation at El Zapotal.

These lands were home to Mixtequilla folk, who were known for their growing trade network along the coast and towards the Highlands since the Early Classic period, but more so after the fall of Teotihuacán (Wyllie 2002). They exhibited highly visible head looks as murals and sculpture appear to communicate. This idea opens new perspectives towards the interpretation of top-flattened head morphologies within broader social and economic contexts during the second half of the Classic period. Ideological elements that were introduced, like flows of materials such as obsidian and the presence of foreign personages, similarly express the intensity of cultural contact to and from the areas lining the Gulf coast of Veracruz and Tabasco (Grube et al. 2009; Pallán 2009; Wyllie 2002). Given the expansive propagation of top-flattened head looks during the later stages of the Classic period, and specifically the "Putún" dominated Maya coast, we might wonder if that these artificial cephalic forms could have visibly expressed the new economic and affiliated military networks that expanded their power and influence during the last centuries of the first millennium. We know that the long-distance dynamic was strongly mediated by Gulf Coast groups during the second half of the first millennium, an aspect to be taken up in the following paragraphs.

Further east, the Gulf Coast plain opens towards the Grijalba and Usumacinta basins. The Grijalba River delineates part of the Isthmus of Tehuantepec, which represents the shortest distance between the Gulf of Mexico and the Pacific Ocean, a natural passage that has always served as a convenient route for intercoastal trade. Unfortunately, there are few published skeletal collections from these lands to grant glimpses on locals' way of infant head modeling during the Classic era. Among the few documented series counts the Del Lazo cave context of subadults, explored by the Río La Venta Project in the El Ocote Forest, on the western slopes of the Tehuantepec Isthmus (Domenici and Lee 2009). All the children, whose bodies had been deposited during the Late Classic Period, had sustained head elongation after birth, leading to different expressions of tabular oblique types.

8.3 Maya Lowlands

East to the Grijalba River, the circumpeninsular coastline, densely settled by hierarchically organized Mayan-speaking groups, organized into networks of regionally organized hegemonies watched over by aristocratic ruling elites (Figs. 8.4 and 8.5; Sharer and Traxler 2006).

8.3.1 Yucatán Peninsula and Coastal Populations

Our regional survey of the Maya cranial record points towards high frequencies of artificially shaped crania both among the settlers of the coastal and inland territories of Yucatán (above 80 % of all artificially modified crania).

8.3.1.1 Inland Populations

Regarding the Maya settlements in their karstic interiors, it is problematic to assume any shifts in preferences given the reduced number of individuals dated securely to the Early Classic period. Consequently, I will attempt a global survey for Classic period inland forms ($N = 80$). Generally speaking, the preferences in head form appear to be analogous to those of the Central and Northern Petén, further south.

Fig. 8.4 Location of sites covered or mentioned in the regional survey of head-shaping practices among the ancient Maya. (Map by V. Tiesler)

Site populations from Dzibilchaltún, Oxkintok, Chac, Yaxuná, and Nohbec, in the present state of Yucatán, all exhibit both erect and reclined head looks. Intermediate, pseudocircular, and slight posterior or frontal flattening are popular. Conversely, no cases of superior flattening were scored.

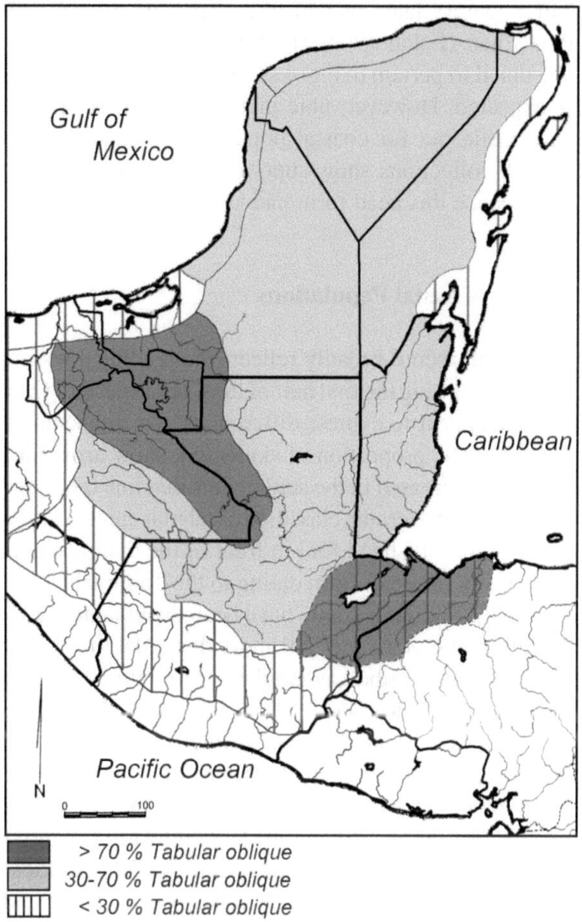

Fig. 8.5 Regional distribution of Maya head elongation vs. head shortening during the Classic period. (Map by V. Tiesler, adapted from Tiesler 2012a)

Note that the continuously uniform kaleidoscope of Classic period head looks differ markedly from the form displayed by Chichén Itzá's populations towards the close of the first millennium. Then, Chichén and its allies grew to dominate political, sociocultural, economic, and ideological life in the northern Maya lowlands, flagship of a new westward-oriented Mesoamerican ideology that roughly correlates in time with the decline and fragmentation of the major centers further south. Chichén's site population itself stages predominantly tabular erect shapes now (Tiesler 2012a). Short and high are also the head profiles represented in its mural paintings and among the monumental sculptures within the huge core areas of Chichén.

Also, the human remains of what were presumably sacrificial victims from the depths of the Sacred Cenote of Chichén Itzá, deserve a special mention for the particular patterns of head forms they disclose. Unfortunately, we cannot affix any secured chronological date but parsimoniously assume that most of them date to the heyday of Chichén's occupation. From the huge sinkhole, a collection of skulls was retrieved, first, during a mission of the Peabody Museum, Harvard University, and half a century later, as part of an excavation led by Mexican archaeologist Román Piña Chan (Folan 1968; Hooton 1940). The majority of these submerged crania, presumed to pertain to those sacrificial victims, displays erect cranial morphologies, as expected. However, note that many of these show the superior compression we have made out for coastal populations. Almost one-third of the evaluable pieces of both collections show superior flattening (29 %; $N = 147$), coming close to the prominence this head form had at El Zapotal, in Veracruz.

8.3.1.2 Coastal Populations

The inland record broadly reflects the predilections in head looks along the shores, especially during the first half of the Classic period. Other coastal expressions of head insignia appear to express differences, which become more evident as the centuries progress. The proportion of skulls that show artificial modeling using compression cradles is increased in the settlements that line the coast of Quintana Roo and Belize, with 75 % of the crania classified as tabular erect. Similar preferences are found on the small Island of Chac Mool in front of the Sian Kaan Biosphere Reserve, presenting exclusively the erect form dating to the Terminal Classic period (Márquez 2006, pp. 224–226). Six more crania have been documented at Colhá, in Belize, located some 20 km off the coastline. All individuals date to the Terminal Classic period. Five of these are described as tabular erect, with a sixth individual showing an oblique form (Massey and Steele 1997). It appears that further down the Belizean shores, head shapes are less abundant during the Classic era. Belizean coastal sites such as Ambergris Cay (Glassman 1995, pp. 74–77), Sarteneja (Kennedy 1983), and Moho Cay (French 2010), display a variation of both tabular erect as well as oblique forms.

The custom of artificial skull modification seems to have lost some of its former popularity among the communities of the Mexican and Belizian Riviera Maya towards the later stages of the Classic period. This trend first appears in sites on Ambergris Cay in northern Belize, where just 10 of 17 evaluable crania (58.82 %) show modification (Glassman 1995, pp. 74–77). Even more reduced is the proportion of modified skulls found in Chac Mool, Quintana Roo. Márquez (2006, pp. 224–226) affirms a frequency of just 26.6 % ($N = 79$) for the Terminal Classic portion of residents. This is still well below the frequencies noted in the Classic sites of Belize, although we wonder whether the results documented by Márquez (or by Glassman for that matter) could be related, in addition to the reduced frequency of modeled skulls, with different classification criteria (see Chap. 11 and Duncan 2009 for a more in-depth discussion).

Towards the second half of the Classic era, the custom of cephalic modeling undergoes noticeable changes along the Maya coastal ranges, in general, which are also significant on the local level, as we will discuss for the coastal settlement of Xcambó on the northern fringes of Yucatán (see Sect. 9.2.2). Those changed preferences appear not just on the east coast, but also on the north and west coasts of the Peninsula. Infant heads are no longer formed to the same degree and in the same number as during the first half of the Classic period. Here, the head splints still seen

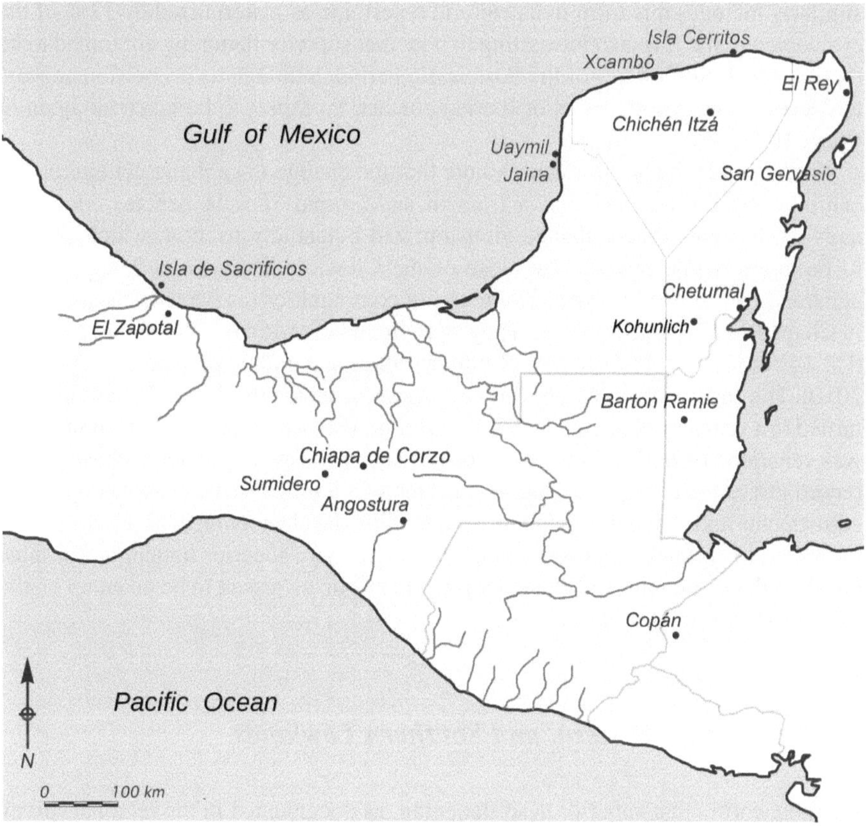

Fig. 8.6 Distribution of documented cases of top-flattened cranial vaults in the skeletal record of Eastern Mesoamerica. (Map by V. Tiesler)

during the first half of the Classic period (with 90.63 % of tabular oblique crania; $N = 32$), gradually give way to cradleboard practices. This change in preferences during the first millennium A.D. is accompanied by a gradual replacement of horizontal constrictions by mid-sagittal grooving.

Also, some new variants in the artificial shapes are revealed in the course of the Late Classic period, thus increasing the diversity in head profiles of their human carriers (Fig. 8.6). Top-flattened heads are now seen along the shores of Yucatán, associated in the material record to abundantly outfitted burials (see Chap. 9). A number of them appear similar to the ones described from El Zapotal; others are inclined and were probably produced by a combination of head devices still in use during the centuries before the Maya collapse. Superior flattening was also reproduced during the Terminal Classic and Early Postclassic periods, with high proportions documented from trader sites, such as Xcambó, San Gervasio, Isla Cerritos, and Chichén Itzá. Top-flattened heads represent 7 % of the port population of Xcambó and 20 % of Isla Cerritos, Chichén Itzá's trader outpost. The population of Jaina in Campeche similarly includes this form in its cultural repertoire, as materialized by 12 % of the evaluable crania. It is also interesting to note that superior flattening continued to be reproduced in later stages of the Postclassic period, although its proportion appears to decrease towards the years of Iberian contact, an aspect to be taken up again in Chap. 10.

We conclude this section with a note that the change in cephalic configurations exhibited by documented crania from in and around Yucatán denotes continuity and—at the same time—change, division, and potentially friction, which appears to be more visible towards the close of the Classic era. Superior flattening was apparently subscribed to by maritime trader communities (as we will discuss again in Chap. 9) with ties to the west. They were life bearers of a new ideological system that accompanied a new network of military allies and trading partners (Tiesler et al. 2010). The early protagonists of this new, forming league would surely have quickly turned to a visual resource to identify themselves with a magical-religious power that was venerated by traders, a sector whose rise becomes evident in the archaeological record just at the end of the Classic era. This was a time that was marked by social adjustments and changes in population, including the abandonment of a large part of the Central Lowlands. As for the individuals who bore superior flattening, we must question their possible ethnic and geographic origin, an aspect to be taken up on the local level in Sect. 9.1.2).

8.3.2 Central, Western, and Southern Lowlands

It is noteworthy that superior head flattening, as documented in the regional survey of this chapter, appears to have been constrained to coastal areas and/or settlements near the coast or connected to it by rivers, such as Kohunlich, Barton Ramie, and Copán. Registries of other inland sites, including abundant skeletal series such as Calakmul ($N = 41$) or the entire collection from the southeast Petén ($N = 91$), do not include examples with this form. A statistical comparison of the coastal Maya series with those of the interior of the Maya world highlights this cultural separation through highly significant differences ($p = 0.002$) in the use of superior flattening, an aspect to be taken up again in Chap. 9 of this book).

In addition to the frequency of this specific head look, we were interested in reviewing the general technical preferences and trends, and in documenting their visible effects on the human carriers among the territories that shared the vast Petén corridor. As previously noted for interior Yucatecans, our attention is also drawn to the diversity of artificial head forms south of the peninsula. Canons of this practice in the northern Petén were closely examined in the sites of Calakmul in Campeche, and Dzibanché and Kohunlich in Quintana Roo (see Tables 8.1 and 8.2). Cultural head modification in this region was present in 94.5 % of the evaluable crania, similar to the frequency along the Usumacinta Basin further west, which culminated in a series frequency 95.1 % greater than in the rest of the Lowlands, especially the series in the eastern corridor of the Petén.

The Belizean watersheds evaluated for this survey were represented by the Baking Pot and the Barton Ramie series, which represented slightly more than half of the total evaluable crania. The proportion of this survey is similar to that documented by Massey and Steele (1997, pp. 69, 70) in their skeletal study of Colha in northern Belize. Note that almost half of the examined crania ($N = 19$) show no modification of any kind, reflecting a lack of usage or low visible expression of this custom in Belize. Unfortunately, the coverage of Belizian skeletal series is too reduced to allow secure comparisons with those of the eastern Maya Mountains and the Belizian coast, where other studies (see for example Glassman 1995, pp. 74–77) would appear to show an equally reduced acceptance of the custom by the population.

Further west, the technical and formal qualities of modified crania in the northern Petén area (and its local diversity) with a proportion of 61.3 % of oblique vs. erect crania, expresses the modalities of the south (with 65.0 %) more than that of the Belizean series (with 70.7 %) or the uniformly oblique shapes of the west. Taken together, the recurrent patterns trace a continuous cultural corridor that extends for more than 1,000 km from the north of the peninsula to the foothills of the Guatemalan Southern Lowlands.

For the central Petén, we may cautiously argue that the customary forms remain constant there when compared with the populations located both to the north and to the south. The results from Nakbé and Uaxactún show that 83.3 % of the residents displayed an artificially modified head, with a ratio of 60–70 % between the oblique and erect forms and a moderate degree of modification on average. These results are similar to the observations of López (1991) on cephalic modification from a second skeletal sample from Uaxactún. Also, the information on head forms from El Perú-Waka', although spurious, is consistent with the trends in our survey (Piehl 2004, 2006, 2008). There, five of six evaluable crania present artificial modification, half of them in the erect form. On the whole, it is patent that the populations of the inner Petén present a penchant for both head elongation and shortening (light gray shaded area in Fig. 8.5). This is expressed by a balanced proportion of both head morphologies. This ratio seems to have remained constant throughout the Classic period. Similar proportions were especially noted in all collections from the Central and Northern Petén that we were able to analyze, showing a continuity and cultural homogeneity for this tradition that echoes the phenotypical population continuity established by analyzing nonmetric dental traits for affinity vs. difference (Cucina and Tiesler 2008).

Yet, cranial forms found in the middle and high Usumacinta Basin to the west are noticeably different from the diversified morphological panorama of the Central Petén corridor. Here, the tabular oblique modification with narrow, reclined profiles dominates both in the skeletal records and in the portraiture of sites, such as Piedras Negras, Bonampak, Palenque, or Yaxchilán. Uniformity is especially prevalent in the cephalic forms exhibited in the territories surrounding the Usumacinta River, where strongly altered crania in the intermediate or extreme tabular oblique form are prominent, including individuals with strong signs of bipolar constriction (pseudocircular). Also, other studies echo the trends found in this survey. Javier Montes (2000) notes that 64 skulls (of $N = 73$) from a skeletal series from Palenque dating to the Late Classic period were reclined, elongated, and constricted, in comparison with just two specimens carrying erect modifications. Andrew Scherer (2006, 2007, 2008) documents seven skulls from the Guatemalan sites of El Kinel and El Tecolote (surrounding area of Yaxchilán). All were artificially altered, and all were reclined backwards. Two series from Piedras Negras recently described by Lori Wright and colleagues (Scherer et al. 1999; Scherer and Wright 2001; Wright 1997; Wright and Witte 1998) and another collection studied by William Coe (1959) similarly confirm the dominant pattern described here. Seven additional elongated skulls are reported for the secondary site of Chinikiha in the Palenque area (Nuñez 2009). These numbers, taken as a whole, provide that 83.9 % of the examined crania manifest the effects of artificial modeling ($N = 31$); and, of these, more than 90 % are oblique, similar to the proportion identified by our own survey for the population of Palenque.

Although less visible than in Palenque, we assume that the collective preference for head backward elongation once extended further up along the Usumacinta river southwards, towards the area of the Río de la Pasión. For example, 90 % of the Classic Era population of Altar de Sacrificios bears an artificially modified head, again displaying a predilection for oblique forms. The proportion of Classic residents of Seibal with artificially modeled heads is similar to a proportion of 93.75 % showing artificial modeling ($N = 32$). Among these, some 70 % show tabular oblique forms. Although less specific, Wright (1990, p. 811) reports that almost all the Dos Pilas' crania under study show the effects of modeling, noting that the tabular oblique shape clearly dominated.

8.3.3 Peripheries

The fringes of the Maya Lowlands are dotted with Classic period settlements west of the Usumacinta River that stretch towards the Grijalva River. Here, the regional capital of Toniná lies, once fierce opponent of the reign of Palenque, and, as it is argued, settled by Mixe-Zoque groups (Ayala 2002). The physical remains of a number of Toniná's inhabitants and those of its satellite villages were excavated by the French Mission during the 1970s (Becquelin and Baudez 1979, 1982; Becquelin and Taladoire 1990; Romano and Jaén 1990). Different from Palenque, broad and shortened head looks predominate here (although not exclusively), many with separated parietal lobes, achieved by using constrictive sagittal bands. These preferences also characterize the Maya Highlands as argued in Sect. 8.1.3. Also further north, in Comalcalco, a similar predominance of erect modeling is observed, although without the use of sagittal bands that was widespread among Maya Highlanders (Tiesler 1999).

The erect cephalic look shown by folk on the outskirts of the Maya area contrasts starkly with the reclined and pseudocircular cephalic shapes found just a few kilometers east in Palenque and in the Usumacinta River basins in general. This difference in head shape must, in life, have been very obvious in the border territories, in fact to the degree that assign cephalic forms relevance as visible ethnic signifiers. This idea is supported by the fact that the boundary between different linguistic groups at that time ran the length of the Chiapas foothills. It also marked the cultural boundary between the Isthmian Mixe-Zoque language groups of Ch'ol (Usumacinta) and Chontal that were spoken on the coastal plains of Tabasco and Veracruz. This boundary was a crossroads of ethnic groups and regional powers as well as a trade corridor that we know gained importance for Maya society near the second half of the Classic period (Justeson et al. 1985, pp. 68–70; Lacadena and Wichman 2002; Pallán 2009; Vargas 2001; Wyllie 2002). We will revisit this idea in Sect. 9.3.

The opposite side of the Maya Lowlands, towards present-day Honduras, was dominated by the centers of Copán and Quiriguá during the Classic period. In addition to some undated skulls from the Ulúa Valley, all of which display a tabular erect modification, regional information on modification was available only from the Valley of Copán (see Sect. 9.2.1). Here, the rural dwellings that surround the city show a preference towards erect head looks when compared to urban Copanecas. Artificial cephalic modeling in the city region was found around three quarters of the examined urban population, with the mimetic tabular oblique variants predominating. It would seem that head-shaping practices were more standardized in urban Copán than in other Maya Lowland centers during the Classic period (see also Saul 1972; Tiesler 1998). The mentioned technique, which generally created moderate morphological changes in head form of local Copanecan folk, probably involved the application of a frontal board with two occipital bands, sometimes complemented, sometimes not, by a sagittal groove. We will return to the Central American peripheries of Mesoamerica in Chap. 9 and take up again the head looks of Copanecans in Sect. 9.2.1.

References

Acosta, María del Rosario, and Gabriela Uruñuela. 1997. Patrones de enterramiento en la Cuenca de Sayula: La fase Amacueca en Atoyac. In El cuerpo humano y su tratamiento mortuorio ed. Elsa Malvido, Gregory Pereira and Vera Tiesler, 161–178. Mexico: Colección Científica, Instituto Nacional de Antropología e Historia.

Alvarado, Luis Adrián. 2013. Tejedores de grandezas. Un análisis de la población de Teopancazco, Teotihuacan, a través de sus actividades ocupacionales. Master diss., Universidad Nacional Autonóma de México.

Ayala, Maricela. 2002. Acerca de los sistemas sociales, políticos y religiosos de Toniná. In La organización social entre los mayas prehispánicos, coloniales y modernos. Memoria de la Tercera Mesa Redonda de Palenque, ed. Vera Tiesler, Rafael Cobos, and Merle Greene, 147–162. Mexico: Instituto Nacional de Antropología e Historia, Universidad Autónoma de Yucatán.

Becquelin, Pierre, and Claude F. Baudez. 1979. Toniná, une cité maya du Chiapas (Mexique). Etudes mesoamericaines. IV, t. 1, C.E.M.C.A., Mexico.

Becquelin, Pierre, and Claude F. Baudez. 1982. Toniná, une cité maya du Chiapas (Mexique). Etudes Mesoamericaines. IV, t. 3, C.E.M.C.A., Paris.

Becquelin, Pierre, and E. Taladoire. 1990. Toniná, une cité maya du Chiapas (Mexique). Etudes Mesoamericaines. IV, t. 4, C.E.M.C.A., Mexico.

Coe, William Robertson. 1959. Piedras Negras archaeology: Artifacts, caches and burials. Museum Monograph 4. Philadelphia: University of Pennsylvania Museum of Archaeology and Anthropology.

Comas, Juan, and Paulette Marquer. 1969. Crânes déformés de l'ile de Sacrificios (État de Veracruz, Mexique). Bulletins et Mémoirs de la Societé d'Anthropologie de Paris 4 (XIIa serie):209–257.

Cucina, Andrea, and Vera Tiesler. 2008. Dinámicas poblacionales y migraciones durante el Clásico y Postclásico en Yucatán: La morfología dental. Los Investigadores de la Cultura Maya 16(2): 165–177.

Domenici, Davide, and Thomas A. Lee. 2009. Periodización y desarrollo cultural del área del río de La Venta, Chiapas. In V Coloquio Pedro Bosch Gimpera, ed. Annick Daneels, 405–433. Mexico: Universidad Nacional Autónoma de Mexico.

Duncan, William N. 2009. Cranial modification among the Maya: Absence of evidence or evidence of absence? In Bioarchaeology and identity in the Americas, ed. Kelly Knudson and Christopher M. Stojanowski, 177–193. Gainesville: University Press of Florida.

Folan, William J. 1968. El cenote sagrado de Chichén Itzá. Mexico: Instituto Nacional de Antropología e Historia.

Franco, María Teresa. 1993. The Huastec and Totonac world. Mexico: Inverlat.

French, Kelley Alyse. 2010. Head shaping of the ancient maya at Wild Cane Cay and Moho Cay, Belize. Master diss., Louisiana State University.

Gervais, Veronique. 1989. Déformations artificielles des crânes préhispaniques au Guatemala et au Mexique. Ph.D diss., Caen University.

Glassman, David M. 1995. Skeletal biology of the prehistoric Maya of Northern Ambergris Caye. In Maya maritime trade, settlement and populations on Ambergris Caye, Belize, ed. Thomas H. Guderjan and James F. Garber, 73–93. Lancaster: Labyrinthos.

González-Licón, Ernesto. 2003. Social inequality at Monte Alban, Oaxaca: Household analysis from Terminal Formative to Early Classic. Ph.D diss., University of Pittsburgh.

González-Miranda, Luis A. 2009. Entierros de Teotihuacán explorados de 1980 a 1982. Mexico: Catalogo del Instituto Nacional de Antropologia e Historia, Instituto Nacional de Antropología e Historia.

Gosse, L.A. 1855. Déformations artificielles du crâ ne, Bailliere, Paris.

Grube, Nikolai, Carlos Pallán, and Antonio Benavides. 2009. La escalinata jeroglífica 1 de Sabana Piletas, Campeche: Nuevos datos sobre el Epiclásico en la región Puuc. Los Investigadores de la Cultura Maya 17(2):93–116.

Hooton, Earnest Albert. 1940. Skeletons from the Cenote of Sacrifice at Chichen Itza. In The Maya and their neighbors: Essays on Middle American anthropology and archaeology, ed. Clarence L. Hay, Ralph L. Linton, Samuel K. Lothrop, Harry L. Shapiro, and George C. Vaillant, 272–280. Nueva York: Appleton-Century.

Justeson, John S., William M. Norman, Lyle Campbell, and Terrence Kaufman. 1985. The foreign impact on lowland Mayan language and script. New Orleans: Middle American Research Institute Series 53, Tulane University.

Kanjou, Youssef, and David Andrade. 2003. Los entierros prehispánicos de Atetelco, Teotihuacán: ritual mortuorio, estructura social y rasgos osteológicos. Estudios de Antropología Biológica 11:773–793.

Kennedy, G. E. 1983. Skeletal remains from Sarteneja, Belize. In Archaeological excavations in Northern Belize, Central America, ed. Raymond V. Sidrys, 353–372. Los Angeles: Monograph 17, University of California.

Lacadena, Alfonso, and Soeren Wichmann. 2002. The distribution of lowland Maya languages in the Classic period. In La organización social entre los mayas prehispánicos, coloniales y modernos. Memoria de la Tercera Mesa Redonda de Palenque, Vol. 2, ed. Vera Tiesler, Rafael Cobos and Merle Greene, 275–353. Mexico: Instituto Nacional de Antropología e Historia, Universidad Autónoma de Yucatán.

Lagunas, Zaíd. 1989. Los antiguos habitantes de Cholula: Prácticas osteoculturales. Notas Americanas 11:8–50.

Lira, Yamile, and Jaime Ortega. 2004. Los entierros de El Tajín, Vercaruz. In Prácticas funerarias en la costa del Golfo de México, ed. Yamile Lira López and Carlos Serrano, 89–116. Mexico: Universidad Veracruzana, Universidad Nacional Autónoma de México, Asociación Mexicana de Antropología Biológica.

López, Nora. 1991. Sistema funerario de Uaxactún: Implicaciones bio-culturales y socio-políticas evidentes en las características de los restos óseos. B.A. diss., Universidad de San Carlos.

Manzanilla, Linda, and Carlos Serrano. 2003. Prácticas funerarias en la Ciudad de los Dioses. Los enterramientos humanos de la antigua Teotihuacán. Mexico: Instituto de Investigaciones Antropológicas, Universidad Nacional Autónoma de México.

Márquez, Lourdes. 2006. Prácticas culturales: Modificación intencional de cráneo y mutilación dentaria. In La población maya costera de Chac Mool. Análisis biocultural y dinámica demográfica en el Clásico Terminal y Posclásico, ed. Lourdes Márquez, Patricia Hernández, and Ernesto González-Licón, 217–231. Mexico: Escuela Nacional de Antropología e Historia, Instituto Nacional de Antropología e Historia.

Martínez, Blanca Lilia. 2007. Estudio de la deformación cefálica intencional tipo Zapotal. B.A. diss., Escuela Nacional de Antropología e Historia, Instituto Nacional de Antropología e Historia.

Martínez, Blanca Lilia. 2009. La deformación cefálica intencional tipo tabular, variante superior, en el Zapotal, Veracruz. Estudios de Antropología Biológica 14(2):489–502.

Martínez, Cira, Marcus Winter, and Pedro Antonio Juárez. 1996. Entierros humanos del proyecto especial Monte Albán. In Entierros humanos en Monte Alban: Dos estudios, ed. Marcus Winter, 79–247. Oaxaca: Centro INAH Oaxaca.

Massey, Virginia K., and D. Gentry Steele. 1997. A Maya skull pit from the Terminal Classic period, Colha, Belize. In Bones of the maya: Studies of ancient skeletons, ed. Stephen L. Whittington and David M. Reed, 62–77. Washington: Smithsonian Institution.

Montiel, Mireya. 2013. Modificaciones corporales en la huaxteca prehispánica. Master diss., Universidad Nacional Autónoma de México.

Montes, Javier. 2000. La práctica de la deformación cefálica intencional entre los pobladores prehispánicos de Palenque, Chiapas. B.A. diss., Escuela Nacional de Antropología e Historia, Instituto Nacional de Antropología e Historia.

Nelson, Greg C., and Felicia C. Mandimenos. 2010. Obelionic cranial deformation in the Puebloan Southwest. American Journal of Physical Anthropology 143:465–472.

Nuñez, Luis Fernando. 2009. Temporada 2008: Informe de actividades. Conjunto habitacional norte. Operaciones 111, 112, 113. In Segundo informe parcial Proyecto Arqueológico Chinikihá Temporada 2008, ed. Rodrigo Liendo, 159–221. Mexico: Universidad Nacional Autónoma de México.

Pallán, Carlos. 2009. Secuencia dinástica, glifos-emblema y topónimos en las inscripciones jeroglíficas de Edzná, Campeche (600–900 d.C.): Implicaciones históricas. Master diss., Universidad Nacional Autónoma de México.

Pereira, Grégory. 1999. Potrero de Guadalupe: anthropologie funéraire d'une communauté pré-tarasque du nord du Michoacán, Mexique. Oxford: British Archaeological Research International Series 816.

Piehl, Jennifer. 2004. Análisis preliminar de los entierros y depósitos con restos humanos de el Perú-Waka' y Chakah. In Proyecto Arqueológico El Perú-Waka': Informe no. 1, temporada 2003, ed. Héctor L. Escobedo and David Freidel, 385–408. Dallas: Southern Methodist University.

Piehl, Jennifer. 2006. Análisis osteológico preliminar de los entierros de El Perú y Chakah, excavados en las temporadas de campo 2004 y 2005. In Proyecto Arqueológico El Perú-Waka': Informe no. 3, temporada 2005, ed. Héctor L. Escobedo and David Freidel, 431–453. Dallas: Southern Methodist University.

Piehl, Jennifer. 2008. Análisis preliminar de los restos humanos de los contextos mortuorios y rituales en Waka' y Chakah. In Proyecto Arqueológico El Perú-Waka': Informe no. 3, temporada 2007, ed. Héctor L. Escobedo and David Freidel, 173–205. Dallas: Southern Methodist University.

Rattray, Evelyn C., and Magalí Civera Cerecedo. 1999. Los entierros del Barrio de los Comerciantes. In Prácticas funerarias en la Ciudad de los Dioses. Los enterramientos humanos de la antigua Teotihuacán, ed. Linda Manzanilla and Carlos Serrano, 149–172. Mexico: Instituto de Investigaciones Antropológicas, Universidad Nacional Autónoma de México.

Rodríguez Betancourt, Felipe, and Octavio R. Sonora Paredes. 1984. Las Banquetas, Chiapas. Mexico: Informe para la Subdirección de Salvamento Arqueológico, Instituto Nacional de Antropología e Historia.
Romano, Arturo. 1973. Deformación cefálica intencional en la población prehispánica de Cholula, Puebla. Comunicaciones Proyecto Puebla Tlaxcala 8:49–51.
Romano, Arturo. 1974. Deformación cefálica intencional. In Antropología física, época prehispánica, ed. Juan Comas, 197–227. Mexico: Instituto Nacional de Antropología e Historia.
Romano, Arturo. 1977. Los cráneos deformados de El Zapotal. In Las esculturas en terracota de El Zapotal, Veracruz, ed. Nelly Gutiérrez and Susan K. Hamilton, 31. Mexico: Instituto de Investigaciones Estéticas, Universidad Nacional Autónoma de México.
Romano, Arturo, and María Teresa Jaén. 1990. El material óseo humano procedente de diversos sitios arqueológicos del valle de Ocosingo, Estado de Chiapas. In Toniná, une cité Maya du Chiapas (Mexique), Vol. 4, ed. Pierre Becquelin and Eric Taladoire, 1661–1687. Mexico: Études Mesoaméricaines 6, Centro de Estudios Mexicanos y Centroamericanos.
Romano, Arturo, Josefina Bautista and María Teresa Jaén. 2011. Análisis craneano de los restos de la cueva Las Banquetas, Chiapas. Mexico: Colección Científica. Instituto Nacional de Antropología e Historia.
Saul, Frank P. 1972. The human skeletal remains of Altar de Sacrificios. Papers of the Peabody Museum of Archaeology and Ethnology 63(2). Cambridge: Harvard University.
Scherer, Andrew K. 2006. Bioarqueología de El Kinel, La Técnica y Zancudero. In Proyecto Regional Arqueológico Sierra del Lacandón: Informe preliminar no. 4, ed. Charles Golden, Andrew K. Scherer and Rosaura Vásquez, 83–94. Guatemala: Dirección General del Patrimonio Cultural y Natural de Guatemala.
Scherer, Andrew K. 2007. Bioarqueología de El Kinel, temporada de campo 2007. In Proyecto Regional Arqueológico Sierra del Lacandón, 2007: Informe de la quinta temporada de campo, ed. Andrew K. Scherer, Charles Golden and Rosaura Vásquez, 186–201. Guatemala: Dirección General del Patrimonio Cultural y Natural de Guatemala.
Scherer, Andrew K. 2008. Bioarqueología de Tecolote, temporada de campo 2004 y 2008. In Proyecto Regional Arqueológico Sierra del Lacandón, 2008: Informe no. 6, ed. Charles Golden, Andrew K. Scherer and Rosaura Vásquez, 160–181. Guatemala: Dirección General del Patrimonio Cultural y Natural de Guatemala.
Scherer, Andrew, and Lori Wright. 2001. Los esqueletos de Piedras Negras: Reporte preliminar no. 4. In Proyecto Arqueológico Piedras Negras: Informe preliminar no. 4, cuarta temporada, 2000, ed. Héctor L. Escobedo and Stephen D. Houston, 553–557. Guatemala: Instituto de Antropología e Historia.
Scherer, Andrew, Cassady Yoder, and Lori Wright. 1999. Los esqueletos de Piedras Negras: Reporte preliminar no. 3. In Proyecto Arqueológico Piedras Negras: Informe preliminar no. 3, tercera temporada, 1999, ed. Héctor L. Escobedo and Stephen Douglas Houston, 387–399. Guatemala: Instituto de Antropología e Historia.
Serrano, Carlos, and Rosa María Ramos. 1984. Perfil bioantropológico de la población prehispánica de San Luis Potosí. Mexico: Universidad Nacional Autónoma de México.
Serrano, Carlos, and Zaíd Lagunas. 1999. Prácticas mortuorias prehispánicas en un Barrio de Artesanos (La Ventilla "B"), Teotihuacán. In Práctica funerarias en la ciudad de los dioses: Los entierros humanos de la antigua Teotihuacán, ed. Linda Manzanilla and Carlos Serrano, 35–79. Mexico: Universidad Nacional Autónoma de México, Instituto de Investigaciones Antropológicas.
Serrano, Carlos, Manuel Rivero de La Calle, and Rosaura Yépez Vázquez. 2003. La deformación cefálica intencional en los habitantes prehispánicos del barrio teotihuacano de La Ventilla. Exploración 1992–1994. In Contextos arqueológicos y osteología del barrio de La Ventilla. Teotihuacán (1992–1994), ed. Carlos Serrano, 103–113. México: Instituto de Investigaciones Antropológicas, Universidad Nacional Autónoma de México.
Sharer, Robert J., and Loa P. Traxler. 2006. The ancient Maya. Stanford: Stanford University.
Sittón, Mair Augusto. 2010. Sacrificios de sangre: El hecho religioso en la civilización del Tajín. Ph.D. diss., Universidad Nacional Autónoma de México.

Stewart, Thomas D. 1953. Skeletal remains from Zaculeu, Guatemala. In The Ruins of Zaculeu, Guatemala, Vol. 1, ed. Richard B. Woodbury and Aubrey S. Trik, 295–311. Richmond: William Byrd.

Tiesler, Vera. 1998. La costumbre de la deformación cefálica entre los antiguos mayas: Aspectos morfológicos y culturales. México: Instituto Nacional de Antropología e Historia.

Tiesler, Vera. 1999. Rasgos bioculturales entre los antiguos mayas: Aspectos arqueológicos y sociales. Ph.D. diss., Universidad Nacional Autónoma de México.

Tiesler, Vera. 2006. ¿Hubo trepanación en la antigua sociedad maya? Una apreciación regional. Antropología Física Latinoamericana 4:169–218.

Tiesler, Vera. 2012a. Transformarse en maya. El modelado cefálico entre los mayas prehispánicos y coloniales. Mexico: Universidad Autónoma de Yucatán, Universidad Nacional Autónoma de México.

Tiesler, Vera. 2012b. Studying cranial vault modifications in ancient Mesoamerica. Journal of Anthropological Sciences 90:1–26.

Tiesler, Vera. 2012c. Formas cefálicas, etnicidad y cambio social en las franjas costeras del Golfo. Una mirada peninsular. In Haciendo arqueología. Teoría, métodos y técnicas, ed. Sara Ladrón de Guevara, Lourdes Budar, and Roberto Lunagómez, 95–110. Xalapa: Universidad Veracruzana.

Tiesler, Vera, Arturo Romano, and Carlos Pallán. 2010. Las formas cefálicas en las vísperas del periodo Clásico: Implicaciones para el cambio social en el área maya. Los Investigadores de la Cultura Maya 18(1):83–96.

Tiesler, Vera, Arturo Romano, Jorge Gómez-Valdés, and Annick Daneels. 2013. Posthumous body manipulation in the Classic period Mixtequilla: Reevaluating the human remains of Ossuary I from El Zapotal, Veracruz. Latin American Antiquity 24:47–71.

Vargas, Ernesto. 2001. Itzamkanac y Acalan: Tiempos de crisis anticipando el futuro. Mexico: Instituto de Investigaciones Antropológicas, Universidad Nacional Autónoma de México.

Winter, Marcus, William O. Autry Jr., Richard G. Wilkinson, and Cira Martínez. 1996. Entierros humanos en un área residencial de Monte Albán: Temporadas 1972–1973. In Entierros humanos de Monte Albán: Dos estudios, 11–78. Oaxaca: Centro INAH Oaxaca.

Wright, Lori E. 1990. Estudios mortuorios y osteológicos en la región de Petexpatún: Un reporte interino. In Proyecto Arqueológico Regional de Petexbatún, Informe preliminar no. 2, ed. Arthur Demarest and Stephen Douglas Houston, 807–812. Guatemala: Instituto de Antropología e Historia.

Wright, Lori E. 1997. Los restos óseos humanos de Piedras Negras: Un reporte preliminar. In Proyecto Arqueológico Piedras Negras: Informe Preliminar no. 1, primera temporada, 1997, ed. Héctor L. Escobedo and Stephen D. Houston, 213–218. Guatemala: Instituto de Antropología e Historia.

Wright, Lori E., and Amy M. Witte. 1998. Observaciones sobre la osteología humana de Piedras Negras: Reporte preliminar no 2. In Proyecto Arqueológico Piedras Negras: Informe preliminar no. 2, segunda temporada, 1998, ed. Héctor L. Escobedo and Stephen Douglas Houston, 393–397. Guatemala: Instituto de Antropología e Historia.

Wyllie, Cherra E. 2002. Signs, symbols, and hieroglyphs of ancient Veracruz: Classic to Postclassic transition, Ph.D. diss., Yale University.

Yépez, Rosaura. 2001. El modelado cefálico intencional en los pobladores prehispánicos del barrio teotihuacano de La Ventilla. Exploración 1992–1994. Master diss., Universidad Nacional Autónoma de México.

Stewart, Thomas D. 1953. Skeletal remains from Zaculeu, Guatemala. In The Ruins of Zaculeu, Guatemala, vol. 1, ed. Richard B. Woodbury and Aubrey S. Trik, 295–311. Richmond: William Byrd.

[illegible]

Chapter 9
Growing Up Maya. Gender, Identity and Dynasty

The previous chapter has set out some regional trends in Mesoamerican modeling techniques and head shapes during the Classic period. By themselves, these do not, however, grant any further insights into any internal, familiar, emblematic, and daily roles, as conceptualized in Chap. 6. In the following, these aspects are examined specifically for the ancient Classic Maya, who witnessed a growing hierarchy. Since the Late Preclassic period, long-standing Maya forms of organization evolved into a progressively institutionalized state system, reinforced by an ideological, administrative, and coercive apparatus that had its seat in capital cities and was regionally instrumented at the level of city states (Houston and Inomata 2009). In the course of the Classic period, a dynamic mosaic of territorial units evolved, led by local aristocrats with changing political and family ties (Martin and Grube 2008). It appears as if the regions maintained a certain cultural independence despite apparently feeling an ethnic and language cohesion, at least among the major sectors of Maya territories.

This chapter seeks to exemplify local head attributions and embedded ideological meanings for the Classic period Maya, expressly their importance within the nuclear family, the role of the practitioners, and the infants whose heads they molded. These aspects confer importance on head-modeling traditions at the community and settlement level, to be explored in two case studies in this chapter. With its rich legacy of writing, art, and abundance of material vestiges, the data-rich Maya area, a hub of Mesoamerican research, is especially well suited to such an endeavor. More than most other Mesoamerican cultural settings, this study environment (and the bulk of published Mayanist research available) concedes more nuanced and culturally embedded interpretations. For the purposes of this volume, I will refer to my own systematic cranial survey of some 2,000 crania from the Maya area, which includes 1,200 individuals dated to the Classic period (Table 9.1). The latter represent 91 site collections that span over southern and southeastern Mexico, Guatemala, Belize, and Honduras (Tiesler 2012).

The documented distribution of head forms over their cultural territories roughly follows some of the Mayan vernacular language divisions during the Classic period and points to their significance as visible emblems of group affiliation. This will be explored here also for regional adoptions of obelionic flattening, a compression of the parietals that has been discussed already in Chap. 8 for Mixtequilla populations

V. Tiesler, *The Bioarchaeology of Artificial Cranial Modifications*,
Interdisciplinary Contributions to Archaeology 7, DOI 10.1007/978-1-4614-8760-9_9,

Table 9.1 Frequency of artificial cranial-vault modifications in the Maya area according to time periods and phases ($N = 1{,}918$). (Tiesler 2012, p. 107)

Period/phase[a]	Absence (0–0.25)[a]	Presence (≥ 0.5)[a]	N (% modified)[b]	Average degree of modification in modeled adult crania (0.5–4)
Middle Preclassic (B.C. 1,000–300)	6	11	17 (64.71)	1.69
Late Preclassic (B.C. 300–A.D. 100)	6	16	22 (72.27)	1.50
Terminal Preclassic (A.D. 100–250)	4	24	28 (85.71)	2.19
Early Classic (A.D. 250–550)	19	107	126 (84.25)	1.80
Middle-Late Classic (A.D. 550–800)	140	530	670 (79.10)	1.90
Terminal Classic (A.D. 800–900)	10	117	127 (92.13)	2.17
Early Postclassic (A.D. 900–1200)	6	64	70 (91.42)	2.16
Middle to Late Postclassic (A.D. 1200–1521)[c]	6	98	127 (92.91)	2.01
Colonial/postcolonial (A.D. 1521–1900)	53	24	77 (31.17)	1.67
Total	250	992	1,242	

[a]Only dated individuals included; included in each category were individuals with dates that span toward the subsequent phase
[b]Percentage of total number of cases in each category
[c]The sinkhole series of San José Mayapán, Yucatán, was excluded from this analysis for potentially dating to colonial times

of coastal Veracruz. Apart from integrative roles, head shapes did not seem to decode any meanings of vertical distinction among the ancient Maya. The aristocracy did not appear to perform head modeling to proclaim their noble status, but instead displayed analogous head forms to their commoner underlings. The regional survey of the Maya is balanced and fleshed out by local glimpses of head-modeling customs.

9.1 Maya Practitioners and Their Infants

One of the inherent qualities of head modification is that it can be practiced only during the first year of infancy and to a lesser degree during the second, thereby demarcating biologically both the maximum duration of the compression and the corresponding age range (see also Chap. 3). Regarding the Maya, historical accounts remain rather vague as to the real duration of the compression routine. According to most early colonial Maya references, compression was initiated just days after birth. Friar Diego de Landa states that among native Yucatecans it was custom to start head compression "four or five days after the infant was born, [and continued] ... until at the end of several days, the head remained flat and molded" (Tozzer 1941, p. 125). The Friar further specifies that "once through with the torment of flattening their forehead and heads, they went with them to the priests, so that they might see his destiny and tell the profession which he was to pursue, to give the name which he was to bear during his childhood ..." (Tozzer 1941, p. 129). If the compression of infant heads really ended after only days or weeks, we may assume that subsequent natural cranial growth would soon neutralize the effects of compression completely, or at least mitigate the changes to a substantial degree. It is significant that Postclassic and Classic period figurines from the Maya area include representations of infants with headboards in upright sitting posture, while carried by their female caretaker. The capability of sitting upright leads us to conclude that still older babies were continuing to wear compression devices. These age groups, well into the second half of the first year, or maybe beyond the completion of the first year, definitely surpass the perinatal period that colonial testimonies associate with active head compression. If the written and figurative information from the Maya region is simply vague or if it is an expression of the diversity in its regional and local enactment, we cannot be certain. The Classic and Postclassic Maya skeletal record probably supports the latter possibility, with degrees of artificial modifications ranging between absence to severe modifications within many archaeological site series.

The presence and expression of postcoronal grooves behind *bregma*, a spot just behind the capillary widow's peak on the forehead, provides additional clues as to the duration of the compression process in artificially modified skulls. As argued in Chap. 3, continuous postcoronal grooves occur when the compression of the baby's calotte continues past the closing of the anterior fontanel (see also Tiesler 1999). As the fontanel is obliterated in around 96 % of the infants at the conclusion of the second year of life (Scheuer and Black 2000), the presence of the postbregmatic sulcus suggests that the process was prolonged beyond this age. When comparing the average expressions of postcoronary grooves around *bregma* (measured on a scale ranging from 0 to 3),[1] our Maya regional cranial survey confirms a shift in its expression in the dated individuals. This shift denoted a gradual increase in visibility of the postcoronary groove between the Preclassic (0.35), then the Classic (0.53) and finally the Postclassic era (0.67). This increase, which in other elaborations has shown to be relatively independent from the specific compression technique used, appears to suggest a gradual prolongation of head-compressor use among the ancient Maya.

The profiling of the practitioners of the body modification is also of interest, as it addresses the potential role of skull transformation as a visible gender expression, a female way to pass on family or community membership and identity. The fact that it was women who were in charge of crafting the desired shape in their children's heads illuminates the collective ritual enactment from an angle that has been little explored until now: the perspective of females and their participation in the ideological re-creation of group identity and accouchement of ethnogenesis and social change. Pre-Hispanic imagery and the ethnohistoric record appear to identify the female caretakers who enacted the daily compression routine as mothers, midwives, and related female kin. Some of these women were well beyond child-bearing age. The elderly practitioners are portrayed with warts, as toothless and wrinkled (see Chaps. 4 and 5; Fig. 9.1). From what is known of the pre-Hispanic cycling of life, we may infer therefore that apart from the mother herself, there were a lost of helpers, like midwives, godmothers, grandmothers, and other kin, many allegedly seniors, who were responsible for teaching or actually performing head swaddling and compression techniques.

[1] For reasons of consistency, only tabular erect flattening was considered for this score.

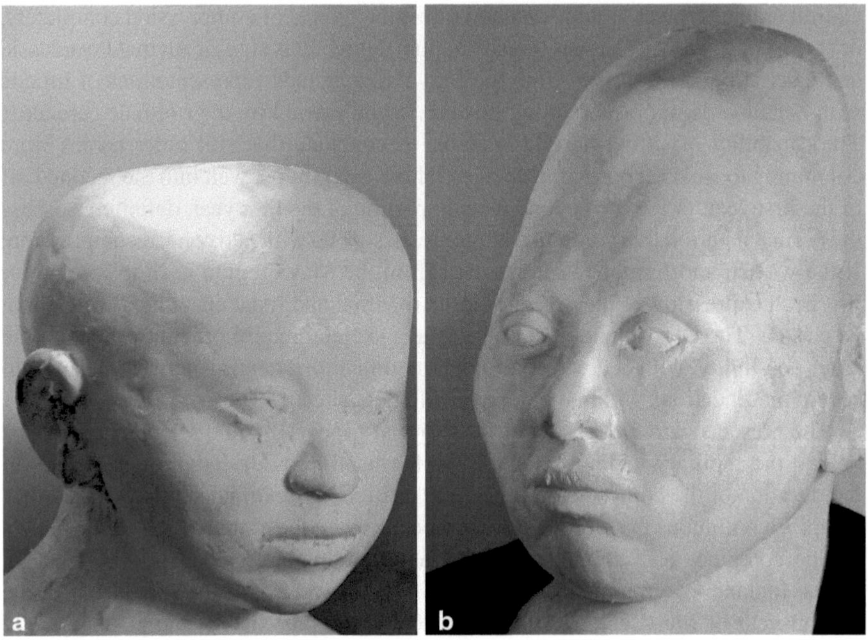

Fig. 9.1 Silicon molds of two head sculptures reconstructed on skulls, using forensic methods. The artistic renderings express different head looks from northern Yucatán. The sculpture on the *left* (**a**) was reconstructed from the cranium of a child recovered from the Sacred Cenote of Chichén Itzá, Yucatán (DAF/INAH). It displays strong superior flattening, resulting in a characteristic broadening of the head. The individual on the *right* (**b**) was reconstructed from a cranium deposited at the site of Yaxuná, Yucatán, which dates to the Early Classic period (Proyecto Arqueológico Yaxuná/UDLA). It shows an extreme pseudocircular form of head elongation, which appears prominently represented in Classic portraiture. (Facial reconstruction and replication by M. Sánchez and J. Chi; Laboratorio de Bioarqueología, University of Yucatán, Mérida; photo by V. Tiesler)

It is interesting that the majority of those Maya figurines that represent pairs of adult practitioners with their infants are modeled on hollowed and perforated clay tubes, crafted to serve as whistles or flutes. This function makes me believe that these objects had been used before either as child toys or might have intonated infant transition ceremonies (Figs. 4.11, 4.12, and 6.3; Tiesler 1999, 2012). Unfortunately, the lack of contextual information on the exact origin of these statuettes, many of which are curated at the Popol Vuh Museum of Guatemala, limits the real possibilities of inferring their use contextually. Conspicuously, whistles and flutes also equipped many Maya infant burials, again emphasizing their close association with children. It is probably no coincidence that seven of eight whistle-equipped burials from systematically scored coastal Maya burials—concretely of the mortuary records of Jaina and Xcambó—pertained to subadults. This aspect still awaits systematic study.

Apart from the portraiture of female head modelers, the produced head forms also grant crucial glimpses of the ways how the ancient practitioners enacted head flattening and their potential purposes (Fig. 9.1, see also Fig. 9.2).

Fig. 9.2 Geographic location of Copán and Xcambó, Yaxuná and Chichén Itzá. (Map by V. Tiesler)

A comparison of cephalic models between males and females has the potential of expressing gender differences in maternal treatment of babies. We might wonder, for example, if Maya mothers employed the same techniques and modeling devices on the heads of their infant daughters as they did on their infant sons. Even though sexing subadult crania is problematic due to the lack of morphological dimorphism in infant skeletal formations, the individuals who survived beyond infancy still retained the artificial head shape conferred on them after birth and therefore can be studied and compared just as well. To this end, we confronted the distribution of presence, formal characteristics, and head compression techniques between adult males and females.

On the whole, the scores of the Maya crania indicate that the custom was performed similarly on male and female offspring. Boys acquired the same head shapes as girls. These, we assume, reflected regional and local preferences during different stages of Maya cultural development. Also, the overall proportion of female crania which are not artificially altered appears similar to that of males. Only the female degree of head modification (scored as 1.99 on average on a scale of 0 to 4) was

Table 9.2 Average degree of cranial modification in pre-Hispanic male and female Maya (0–4)

Type of modification	0	0.25–1	1.25–2	2.25–3	3.25–4
Presence in males (% of total)	33 (55.00)	78 (56.12)	76 (50.33)	94 (55.95)	6 (33.33)
Presence in females (% of total)	27 (45.00)	61 (43.88)	75 (49.67)	74 (44.05)	12 (66.67)
Total	60 (100)	139 (100)	151 (100)	168 (100)	18 (100)

Table 9.3 Proportions of cranial modification in pre-Hispanic male and female Maya

Type of modification	Tabular erect	Tabular oblique	Total
Presence in males (% of total)	167 (57.99)	32 (52.46)	241 (54.90)
Presence in females (% of total)	121 (42.01)	29 (47.54)	198 (45.10)
Total	288 (100)	61 (100)	439 (100)

slightly but not significantly more pronounced than male skulls of our series (1.92 on average; Tiesler 1998, pp. 122–123; Tables 9.2 and 9.3). However, this slight discrepancy might just as well be the byproduct of scoring dimorphic male and female cranial morphology. Note that there are subtle gender-based shifts in some areas and site series. These shall be explored further for urban Copán, Honduras, and the coastal settlement of Xcambó (see also Tiesler and Cucina (2008) for the Southeast Petén area).

The apparent uniformity of male and female models points to an absence of any gender preferences, distinction, or discrimination in this body tradition, at least in the general regional sample scored. It appears that the female child's head was protected in the same way as that of the boys'; the same methods would have been used in modifying the little calottes. This conclusion, at first glance surprising, may find its explanation both in the idea that females treated their offspring uniformly and in the fact that the individuals subjected to the practice were still infants. Once childhood was reached, the gender distinctions acquired greater weight as the individual grew, matured, and as puberty drew near. This argument should also hold true for other cultural areas of Mesoamerica (see Chap. 6).

9.2 Local Head-Shaping Practices Among the Maya

This section explores the distributions of archaeologically contextualized head shapes within archaeological sites. The site-internal profiling of different cranial-vault modifications provides important clues on daily life within residences and particular family, and perhaps lineage, traditions. The culturally transmitted emblems furthermore express the residence, permanence, and sometimes mobility of their human carriers (Tiesler 1998, 1999). Naturally, the analytical possibilities of comparing cephalic modification among different neighborhoods within settlements depend directly on burial patterning, numbers of burials, and wealth of contextual information. These conditions were met relatively well at the Classic period sites of Copán and Xcambó, inasmuch as both harbor broad, well- documented and dated skeletal collections (Fig. 9.2). The balanced distribution between sexes and age ratios in both samples likewise suggests a suitable representation of their residents, and, in the case of urban Copán, of its social sectors and neighborhoods.

9.2.1 *Displaying Head Models in the Neighborhoods of Copán, Honduras*

The archaeological ruins of Copán, located in western Honduras near the Guatemalan border, once functioned as an important capital on the southeastern edge of a dynamic mosaic of interconnected Maya regional states. These were Maya borderlands fringed with non-Maya populations that a number of scholars identify with Lenca cultural heritage. Here, small farming hamlets started to grow during the second millennium B.C. and monumental construction ensued toward the Early Classic period (ca. A.D. 200–400). During the fifth century A.D., a Maya royal dynasty was founded at Copán by a foreign-born individual known as K'inich Yax K'uk' Mo'. From there, Maya aristocracy ruled the area for several centuries, until ca. A.D. 822 (Bell et al. 2004; Martin and Grube 2008; Price et al. 2009; Webster et al. 2000). At its peak, Copán's population probably numbered between 20,000 and 30,000 individuals (Webster et al. 1992). The architectonic vestiges of the ancient city extend over an area of approximately $16\,km^2$ in the Copán Valley. Elite and commoner residential compounds surround a central core area, including an acropolis. Their architectural design ranges from large, masonry palaces to low earthen mounds that once supported pole-and-thatch houses.

The skeletal population from the Copán Valley available for this survey, includes is comprised of 478 skeletons that were recovered during Copán Project Phases I and II in Copán, Honduras (*Proyecto Arqueológico Copán,* PAC I (1977–1980) and PAC II (1981–1984); Operations III-XXVII; Tiesler 1997, 1999, 2005). The Copán skeletal series is one of the largest series from the Maya Lowlands and one that has been extensively studied over the years (Storey 2005; Whittington 1989). The PAC I and PAC II samples include primary burials from residential and ceremonial contexts in the center of the city, as well as secondary burials and isolated sets of bones. This varied mortuary record materializes the diverse domestic and public activities of the city, providing important archaeological data on key aspects of life and death for the Maya of the Classic period. Additional skeletal collections, included here, had been recovered in and around Copán during the explorations conducted by Harvard University and are now curated at the Peabody Museum, Cambridge.

It is noteworthy that only one-third of the total number of skeletons scored from the Copán Valley ($N = 583$) could be evaluated in terms of cephalic modification, mainly due to deterioration, especially for the earlier phases of occupation of the site (Fig. 9.3). None of the skulls that were dated to the Middle Preclassic (Gordon Phase) or Bijac Phase (100–400 A.D.) was sufficiently preserved to allow a determination of cultural skull modification. The earliest preserved crania that showed signs of cephalic modeling date to the Acbi Phase (400–700 A.D.). By contrast, funeral contexts from the Coner Phase of the Late Classic period (700–900 A.D.) constitute the largest portion of the sample, while only three skulls from later phases

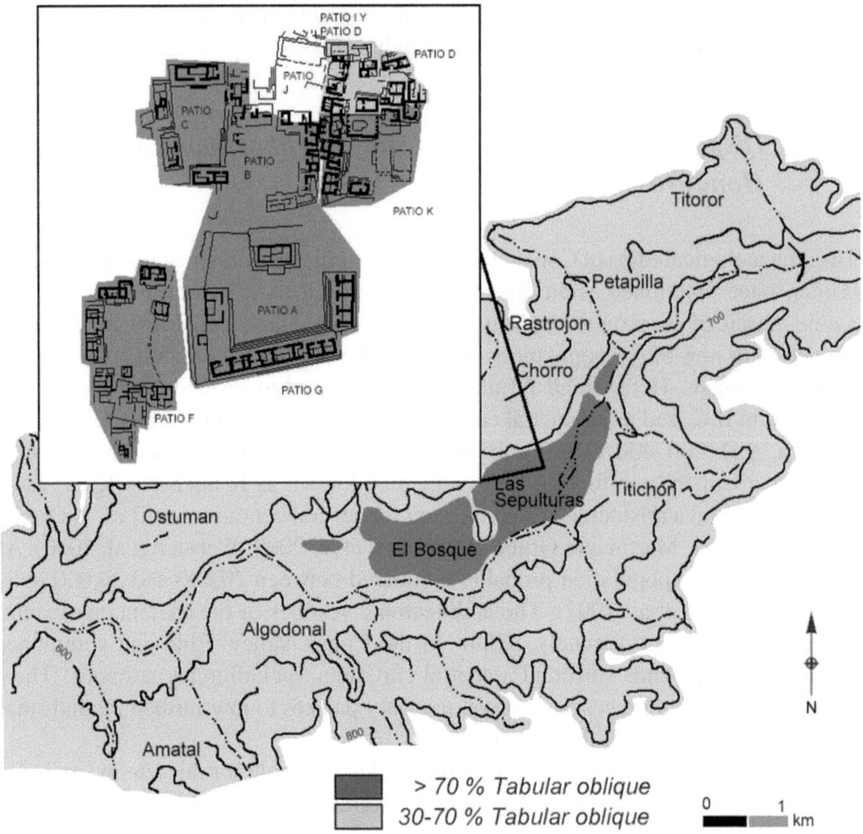

Fig. 9.3 Distribution of head shapes in and around Copán. *Inset*: Distribution at the multifamily residential compound 9N8, Las Sepulturas, Copán. (Redrawn by B. Ceballos)

were available for study. For purposes of methodological consistency, we present below only the results of the series dated to the Classic period, of which artificial modification was noted in 119 of 154 scored skulls (77.3 %), most of which were dated to the Coner Phase (see also Tiesler and Cucina 2010). In additional four cases, shaping was suspected but too slight to be scored as present. The overall proportion in the series from the Copán Valley is roughly in line with other frequencies around the Maya area (see Chap. 8).

In a nutshell, our findings show that Classic period Copanecans had a strong penchant for tabular oblique modifications in mimetic expressions (implying two separate posterior planes). They performed this kind of modification mainly with head splints that were applied on the head solely, or by adding circular constriction as was documented in 69.14 % of the modified crania in our series that could be classified by type ($N = 81$). When their orientation is averaged, the two flattened surfaces on the back of the head make one combined plane that appears approximately parallel to the profile of the forehead. There is a general rearward inclination of the head, especially the forehead, an appearance that confirms the use of a cephalic device.

An additional 10 % of the examined specimens from Copán held other tabular oblique shapes; namely, extreme, intermediate, occipital curve or top-flattened variants. Two of these reclined crania, classified as extreme, show by far the strongest modifications. Another 22 crania displayed tabular erect shapes, produced in classic compression in cribs. They only make up 27.2 % of the classifiable specimens, compared to 72.84 % of artificially reclined (oblique) crania. Local crib use normally produced alterations that were much less severe in comparison with those resulting from head splinting. Many of the mimetic variants and the erect skulls show the impression of a sagittal sulcus in the superior part of the cranium, just where the anterior and posterior boards should have been held together on the top of the head. In this collection, a (pseudocircular) constriction band was used only in combination with head splints (tabular oblique forms). Upon comparing the presence of each technique, the resulting modification patterns are similar to those found in other sites of the Classic period Maya Lowlands, but differ noticeably from the head shapes staged later, during the Postclassic period.

As we confront urban head profiles with those from the satellite villages, differences in looks are immediately apparent. Some 76.5 % of Central Copanecans ($N = 68$) display an elongated, oblique head morphology, whereas in the outskirts of the city, head modification appears less expressed and, when practiced, was enacted in a more diverse way, including the use of compression cribs. Here, tabular oblique shapes are less frequent than inside the city (58.3 %). Hinterland dwellers also display less severe cranial modifications, which are on average slighter (0.89 on average [0–4]) than among central urban residents (1.35 on average [0–4]). The latter express a strong preference for mimetic forms, produced by head splinting. This was in all probability the common local pattern at Copán, at least during the Late Classic period.

The difference between urban and rural Copán's material culture has been the subject of mainstream archaeological discussions. These revolve around the possibly multiethnic nature of the population that occupied the Copán Pocket during the Classic period, with ethnic Maya urban folk surrounded by possibly non-Maya populations. Given the patent differences in head form, we ask ourselves if the old rural stock surrounding Copán distinguished themselves ostensibly by their shortened heads? And did the Mayan-speaking urban Copanecans take pride in their elongated head profiles and their reclined foreheads?

In order to gain a deeper understanding of the distribution patterns of different head styles within and between the urban households of Copán, we zoomed in on their presence among the patio compounds of Las Sepulturas, a large residential area to the north of Copán's acropolis. The distribution of artificial skull modification within Copán's residential areas helps to reevaluate its emblematic role in the residential and family ambits, where differences in head shapes are evident on the scale of many household compounds. To this end, head shapes were scored according to residence type, location, and, in the case of Group 9N-8, also according to patio number.

Within the multipatio compound of Group 9N-8, within the neighborhood of Las Sepulturas, most of its patio compounds show a marked preference for the mimetic forms of the oblique type; here the erect type is practically absent, at least in our study

Table 9.4 Different expressions of artificial head modeling at Classic Copán, according to sex and age, with chi-square values

	Infants (0–10 years)		Adolescents and adults (> 10 years)			Females		Males		
	%	*N*	%	*N*	*p*	%	*N*	%	*N*	*p*
Frequency	82.1	28	76.2	126	0.4966	70.1	67	83.3	48	0.1043
Frequency of circular wrap	0.0	5	18.8	69	0.2851	18.4	38	20.7	29	0.8160
Frequency of sagittal groove	0.0	4	32.9	79	0.1662	34.1	44	31.3	32	0.7947
Tabular oblique/erect	100.0	9	69.4	72	*0.0520*	72.2	36	64.7	34	0.4984
Mimetic/rest	75.0	8	68.1	72	0.6877	73.7	38	58.1	31	0.1710
Superior flattening/rest	0.0	8	6.9	72	0.4414	7.9	38	6.5	31	0.8181

Significant differences appear in italics

sample. This strong preference for inclined head styles contrasts markedly with the more diverse head forms observed in Patio D, whose dwellers appear to have had a penchant for erect head shapes. Although no statistical validation was intended due to insufficient sample size, the trend traced by cranial forms from Patio D is echoed by other findings on architectural style and ceramic wares (Gerstle 1985; Diamanti 1991). Melissa Diamanti (1991), Andrea Gerstle (1985), and Julia Hendon (1987) infer from the material evidence that Patio D must have been occupied by folk that were culturally distinct from the neighborhood of Group 9N-8, perhaps ethnically related to Lenca stock from the interior of present-day Honduras.

The degree of in-migration and internal heterogeneity becomes also apparent upon confronting the adult frequencies of the different models with those of minors, who did not live long enough to reach adolescence (Tiesler and Cucina 2010; Table 9.4). This age group comparison provides relevant clues on local residence vs. population mobility. Its underlying idea is that, given their young ages, infants and children (here defined as age 10 years and below) have lived less than grown-ups and are therefore less likely to have moved from one place to another. It follows that children, and especially babies, must express the local population and their longstanding modeling preferences much more closely than grown-ups. Adults, because of their longer life span, are more prone to have moved in the course of their lives; accordingly, their head forms are more inclined to differ from the local modality at the settlement where they die and are buried, especially when having originated in a place that did not share the head-shaping traditions at the place of death and burial.

Table 9.4 expresses different qualities of artificial cranial modifications, such as its popularity (as manifested by the proportions of shaped heads), the uses of sagittal and circular wraps along with the preferences of modification types and diagnostic varieties (superior flattening and mimetic forms). Our findings underline differences between the enactment of shaping in minors and in grown-ups, which is especially evident when comparing the ratio of oblique as opposed to erect head forms ($p = 0.052$) None of the babies and infants displayed the latter, which was present in every one in three adults. Following our conjecture, this discrepancy implies that infant cradleboarding did not count among the local head traditions among Copán's urban neighborhoods and that incoming folk were to adopt the more common splinting customs soon after arrival.

Note deserves also that the differences in head shapes between men and women are less noticeable than those between adults and subadults, as expressed again in Table 9.4. Comparing the artificial head forms among sexed adults, no clear preference was noted on the overall sample in terms of technique or type of molding, although men appear to have been slightly more subject to undergoing the procedure during infancy in comparison with women (83 % vs. 70 %, $p = 0.1043$; Table 9.4). Although small sample sizes constrain statistical validation, the frequencies suggest a slight male preference for this custom or, alternatively, a higher proportion of incoming women from areas where it was not as popular as in the Copán pocket itself. The fact that the global series of sexed adults showed similar diversity and prevalence in head deformation types may have also important implications for understanding mobility migration, suggesting that incoming families, not individuals of one or the other sex, arrived and settled at Copán.

Regarding the residential organization within the urban spaces, we identify the coexistence of various shaping traditions. Presuming the techniques were passed on through the female line in a society, which we know apportioned chores and occupations according to gender, the recurrent presence of different modification techniques among females in each housing complex (and given a continuity between living and mortuary spaces), could well be associated with the coresidence of females from different kin groups. This would suggest a patrilocal more than matrilocal organization of extended families inside Copán, an issue that has also been explored from other angles (see, for example, Diamanti 1991).

In synthesis, the patterning of cranial modifications adds to the general research on Copán's population and its vibrant social and multiethnic makeup on the fringes of Maya civilization. The outcome of this analysis is consistent with other sets of data regarding the cultural differences between Copán's residential areas and its surrounding hinterland (Hendon 1987; Hendon 1991; Diamanti 1991; Fash and Agurcia 1991). Although not necessarily related to any ascribed distinctions, these observations reveal artificial head shapes in residential contexts to be powerful (bio)archaeological indications of everyday social and ideological reproduction, especially if we as some that the burial places were associated with places of residence and, therefore, with the spheres of family and interfamily interaction, perhaps also lineage, descent, cultural, and ethnic affiliation.

9.2.2 Head Styles Among the Merchants of Coastal Xcambó, Yucatán

Still more numerous than the cranial series from Copán is the collection from the small settlement of Xcambó, nestled in the off-shore marshlands of northern Yucatán (Fig. 9.4). The site occupies a 700 m east–west by 150 m north–south area on top of a natural mound that was artificially expanded and raised above the sea level by

Fig. 9.4 Center of Xcambó, Yucatán, with view toward the coastline of northern Yucatán, Mexico. (Photo by V. Tiesler)

its Classic period settlers. The 564 skeletal remains, which were recovered between 1996 and 2000, are mostly well preserved, contextualized, and dated, spanning the Early (300–600 A.D.) and Late Classic period (600–750 A.D.). They belong to mostly well-equipped burials placed in simple or cist graves. Many of the graves contained imported ceramic vessels and figurines along with diverse local or imported items of personal adornment (Medrano Chan 2005; Sierra 2004). Under the direction of archaeologist Thelma Sierra (2004), the Proyecto Arqueológico Xcambó (INAH) has combined various lines of research and data-sets to explore the trajectory of the settlement and its role in the shifting coastal Maya networks during the Classic period. The ongoing interdisciplinary efforts prominently embrace bioarchaeological data sets (Ceballos 2003; Cetina and Sierra 2005; Jiménez 2002; Maggiano et al. 2008; Méndez et al. 2009; Peña and Sierra 2004; Quintal 2000; Sierra 1999a, b, 2001, 2004; Suzuki et al. 2009; Tiesler et al. 2002, 2004, 2005; Tiesler and Cucina 2010; Wanner et al. 2007).

During its 500 years of formal settlement occupation, Xcambó functioned as a salt production center and port, maintaining long-reaching ties with other parts of the Maya world and Veracruz. This underlines Xcambó's active interaction with other parts of the Maya world and toward Veracruz. Specifically, the western coastal communication route is critical for understanding the broader social and economic dynamics during the centuries that anticipated the so-called "collapse" of inland Classic hegemonies which on the peninsula went in tandem with a reorientation of political and economic networks from the coast.

Toward the Late Classic period, Xcambó's exchange routes and connections shifted and expanded (Sierra 2004). Xcambó's settlers must have experienced new prosperity, which was expressed by growth and architectural transformation with all its earlier structures being built over. Overbuilt and densely Packed residential spaces, which surrounded two public plazas, replaced earlier ceremonial architecture and the former storage facilities. The central plaza was surrounded by new monumental structures, built with fine carved stones (Sierra 1999a, b). Civic, religious, and administrative functions were likely carried out in this plaza and the adjacent north pier of the elevated island settlement with embarking facilities. A swift end of Xcambó's occupation follows soon after A.D. 700 and has been associated with the rise of a new order of coastal traders toward the end of the Classic, watched over by Chichén Itzá and its coastal outpost Isla Cerritos.

For the purposes of this study, we used 371 skulls that were preserved enough to be scrutinized at least in terms of presence or absence of head modification. The cephalic modification was present in 80.1 % of this overall series. Among those crania that could be classified according to the type of modification, mimetic tabular oblique profiles prevail (in 112 cases), evidencing a marked preference for this technique, in fact, quite similar in fact to urban Copanecan head looks. Other tabular oblique forms, and still others, catalogued as tabular erect, are present in smaller proportions among Xcambó's neighbors (22.2 %). Here, oblique modifications often result in severe to extreme morphological change. While the dominant mimetic look is the outcome of specific forms of head splinting and wrapping, the erect configurations were produced, by definition, solely by compression cribs. The almost complete absence of supra-inial lesions is noticeable at Xcambó and is probably due to the reduced use of compression cribs at Xcambó. Sagittal bands are evident in 44 of 169 examined individuals (26.05 %).

The scarcity of erect models and the patent preference for oblique shapes in the practice of cultural modeling is manifested above all during the first phase of occupation. Elongated, reclined head forms also characterize other folk that lined the western and northern shorelines of the peninsula during the Classic period (Tiesler 2012; see also Sect. 8.3). This predilection for oblique variants persisted in Xcambó during the Late Classic, although to a lesser degree. Probably even more than in Copán, mimetic shaping constituted a standardized local tradition among Xcamboans and sets it apart from coeval inland series. Further south and far from the coast, the Maya engaged in various different shaping practices, especially the larger inland centers of Yucatán and the Central Petén, such as Dzibilchaltún and Yaxuná, Calakmul, Ixtonton or Dzibanché (Tiesler 1999, 2013).

The degree of local standardization becomes even more apparent upon confronting the adult frequencies of the different models with those of minors, who did not live long enough to reach adolescence (Tiesler and Cucina 2012; Tables 9.4 and 9.5). Our findings show that, during the Early Classic, some 94 % of Xcambó's population still displayed an artificially modified head. Back then, the local mimetic oblique form was still crafted in over 90 % of the modified individuals (Fig. 9.5; Table 9.5). Their reclined forehead was backed up by an elongated oblique skull vault with two flattened areas on its back. Note that during Xcambó's early occupational phase, this style is similarly dominant in adults, children, and infants. According to our proposed scenario of population mobility vs. permanence, the uniformity of head looks and the similarity of head form between the different age segments, come to confirm

Table 9.5 Different frequencies of "diagnostic" features of artificial head modeling among the Early and Late Classic Maya settlers of Xcambó, Yucatán, according to sex and age

	Subadults (0–10 years)		Adolescents/ adults (> 10 years)			Females		Males		
	%	*N*	%	*N*	*p*	%	*N*	%	*N*	*p*
Early Classic										
Frequency	94.4	18	94.4	36	1.000	100.0	12	95.0	20	0.4313
Frequency of circular wrap	55.6	9	64.7	17	0.6482	62.5	8	66.7	9	0.8576
Frequency of sagittal groove	0.0	7	26.3	19	0.1310	37.5	8	22.2	9	0.4902
Tabular oblique/tabular erect	90.9	11	94.7	19	0.6855	100.0	7	90.9	11	0.4117
Mimetic/rest	100.0	12	84.2	19	0.1475	66.7	9	100.0	9	*0.0578*
Superior flattening/rest	0.0	12	0.0	19	1.000	0.0	9	0.0	9	1.0000
Late Classic										
Frequency	89.5	76	73.2	164	*0.0043*	70.8	72	75.6	78	0.5059
Frequency of circular wrap	31.8	22	7.9	76	*0.0036*	5.4	37	8.6	35	0.5974
Frequency of sagittal groove	13.0	23	33.3	87	*0.0567*	52.6	38	18.6	43	*0.0013*
Tabular oblique/erect	97.3	37	72.6	73	*0.0019*	56.3	32	81.3	32	*0.0310*
Mimetic/rest	68.8	32	73.5	68	0.6195	43.3	30	55.2	29	0.3632
Superior flattening/rest	0.0	32	10.3	68	*0.0598*	13.3	30	10.3	29	0.7227

Significant chi-square values in italics

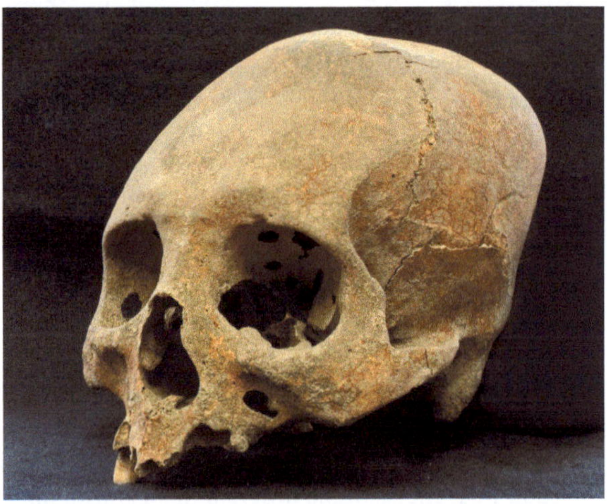

Fig. 9.5 Lateral frontal view of tabular oblique flattening at Xcambó, Yucatán, Mexico. Note the elimination of the nasal root. (Proyecto Arqueológico Xcambó, Yucatán/INAH; photo by S. Suzuki)

the high standardization of head devices among local Early Classic practitioners and identifies Early Xcamboans as folk who were firmly rooted to their native soil.

The distribution of head looks changes noticeably toward the Late Classic era. Table 9.6 presents direct values from the comparison of the Early and Late Classic periods, many of which are statistically significant. The overall frequency of the practice has decreased, by them as both groups of adolescents/adults, female and male, show a significant or almost significant decrease in the proportion of shaped

Table 9.6 Differences in head-shaping practices between the Early and Late Classic at Xcambó, Yucatán, according to sex and age

	Subadults	Adolescents/ adults (> 10 years)	Females	Males
Frequency	$p=0.5193$	$p=0.0060$	*$p=0.0308$*	*$p=0.0553$*
Frequency of circular wrap	$p=0.2181$	*$p=0.0000$*	*$p=0.0001$*	*$p=0.0001$*
Frequency of sagittal groove	$p=0.3138$	$p=0.0622$	$p=0.4366$	$p=0.8023$
Tabular oblique/tabular erect	$p=0.3519$	*$p=0.0406$*	*$p=0.0288$*	$p=0.4541$
Mimetic/rest	*$p=0.0276$*	$p=0.3361$	$p=0.2193$	*$p=0.0133$*
Superior flattening/rest	$p=1.000$	$p=0.1447$	$p=0.2475$	$p=0.3147$

Significant chi-square values in italics

heads. This later stage of occupation witnesses a significant drop in artificially modified heads ($p=0.006$), which now make up only 73.2 % of the adult segment of the series, compared to 94 % during the Early Classic (Tables 9.5 and 9.6). This later phase also displays a greater variety in head shapes and a trend toward erect forms, which again is significant for the adult segment ($p=0.0406$). Now, superior flattening made its debut in the heads of Xcambó's settlers. Here, some 10.3 % of Late Classic adolescents and adults show this distinct form, which has also been described for other coastal Maya populations past A.D. 500 (Tiesler et al. 2010), and for Mixtequilla sites, where they were identified as "El Zapotal" modification by Arturo Romano (1977) who described these distinct vault shapes for the first time 40 years ago (see Chap. 8).

Also, when Late Classic period adult and subadult head forms are compared, the results show significant differences in almost all of the examined criteria (Table 9.6). While Late Classic subadults keep presenting the local oblique shapes (97.3 %), adults show a much broader range of head looks, also when compared to the previous occupational phase and when compared to children. Our chronological comparison shows statistically significant differences (chi-squared) between the frequencies of forms found during the Early and Late Classic periods. This difference is also significant when comparing the population of children (up to 10 years of age at death) with the adolescent and adult segment of its population. In addition to age group discrepancies, there are also differences now, although less marked, between the head shapes that men and women exhibit (with $p=0.05$; Table 9.6; see also Tiesler and Cucina 2011).

If we follow our age-group conjecture, the above elaborations affirm the residential permanence of Xcambó's settlers during the Early Classic and of a higher mobility during the Late Classic. Apparently, foreign individuals (adults, perhaps young families without children) with a different head look were arriving at Xcambó to live and die there, but still expressing their cultural and geographic origin by their distinctive head shape. Evidence from strontium isotope ratios, performed on 131 individuals from Xcambó appears to confirm the increase in mobility among its residents during the Late Classic period and is described in detail in other work (Sierra et al. 2013). For our topic, it is relevant that the individuals born in the inland areas to the south, and more so, those residents who have been presumed to have emigrated to Xcambó from the shores several hundred miles away, were buried predominantly in the administrative core of the settlement, adjacent to the site's main plaza. As expected, this segment stages a higher variety in head looks when compared to coeval locals.

Especially noteworthy is the high number of women with erect head shapes among the Late Classic folk. These represent over 40 % of the population in comparison with only 20 % of men displaying this head form during the same occupation phase (significant difference; Tables 9.5 and 9.6). The difference between adult female head shapes and that of the minors, who still largely exhibited elongated heads, comes as a surprise if we recall that it was the women who were in charge of modeling the heads of their little ones. Our findings suggest instead that incoming prospective mothers, who had been cradleboarded as girls elsewhere, did not reproduce this same technique in their children, but must have swiftly adopted the treatment that was the local custom. It is difficult at this point to infer any specific mechanism or circumstance that might have led to said adoption but we may assume a medianism of local assimilation. We may speculate as to whether the incoming women were part of extensive family networks, some of whose members had already settled at Xcambó. Had they been initiated by Xcamboan midwives and local relatives on the suitable forms of local cephalic modeling?

9.3 Classic Maya Head Shapes and Ethnicity

9.3.1 Regional Diversity

Especially during the Classic period, artificially produced head shapes must have constituted a widespread and highly visible body emblem of group identification, as we argue also in other work (García and Tiesler 2011; Tiesler 2010, 2011, 2012; Tiesler and Cucina 2010). This is when culturally modified head forms reached a peak both in popularity and diversity. This section attempts to trace some of the deeper cultural undercurrents that could explain the vibrant mosaic of head forms and the shifts of cranial-vault modifications across the cultural geography of the Maya territories. Here, a noticeable predilection for erect head shapes is observed across the southern mountain chain that connects the Highlands of Guatemala to the Central depression of Chiapas and down the isthmic strait that extends toward Oaxaca, still further west. These later areas were the lands of Mixe-Zoque-speaking populations, who were arguably non-Maya (Sharer and Traxler 2006; Wichmann 1999).

Toward the north, the mountains open toward the vast Maya Lowland and the Petén corridor. Toward the Lowlands, the erect head forms are gradually replaced by a diversity in head profiles, a pattern staging different degrees both of broad and short, elongated, and artificially narrowed cephalic models. This distribution is also reproduced by the Classic communities lining the western and northern shores of Yucatán.

Toward the later stages of the Classic period, however, changes in head form become patent along these coastal stretches, most noticeably on the east coast of Yucatán with a shift toward cradling traditions and the appearance of superior flattening, to be discussed in detail in the last segment of this chapter. Apparently, this shift is not paralleled by peninsular inland head traditions, which remained practically unchanged over the Classic period (statistically reduced difference in a chi square analysis, with $p = 0.611$). This shift advocates a segregation of some sort—being populational, merely cultural, or social—between coastal Maya folk and inhabitants of the interior. This divide becomes more pronounced in the second half of the first millennium A.D., as is statistically expressed in a comparison of tabular erect to tabular oblique forms in the coastal series with those from inland sites ($p = 0.000$; highly significant).

Also further south, in the Central Petén, the inland canons of cephalic modification continued to be practiced toward the close of the Classic period with fewer changes than along the coast. The skeletal series from this area conveys the notion that each population echoed others in terms of preferences, shapes, and general enactment of the custom. Here, different tabular oblique and erect shapes appeared in a balanced proportion that remains virtually unaltered throughout the first millennium. The Central Petén and Northern Petén, in particular, show similar percentages in all evaluable collections, communicating a cultural uniformity within the Petén corridor that simultaneously emphasizes, although indirectly, the importance of family ties and horizontal relations that must have prevailed between the groups, an aspect to be revisited later in this work. Patterns found in dental morphology (Cucina and Tiesler 2008; Tiesler and Cucina 2012) similarly give testimony of an open population dynamic and continuous and stable occupation in the Petén corridor before the collapse. This vision complements views on the dynamic, essentially unstable nature of the political landscape of those hegemonic networks that dominated the Lowlands over the centuries (Martin and Grube 2008).

Our regional survey of head forms also suggests that the head shapes preferred in the Petén zones differ noticeably from those found further west, along the banks of the Usumacinta River. This observation once again gains importance as we recognize the cephalic forms that were most frequently represented in images of the pantheon of Maya gods. We wonder if these discrepancies in preferences between the western, eastern, and northern stretches of the Maya Lowlands could have expressed a deeper cultural divergence. This perhaps linked linguistic and/or ideological differences, an interpretation that seems feasible considering the diversity and extent of the cultural geography represented by the Maya Lowlands during the Classic period.

Our idea of cultural separation also finds support through less tangible cultural expressions such as speech and style conventions, whose geographic distributions follow those of the type of cephalic modeling (see for example Kettunen 2008, pp. 182–186). Alfonso Lacadena and Soeren Wichmann (2002) recently inferred a linguistic line that divides the eastern and western territories of spoken Ch'olan. This linguistic boundary, which runs parallel to the Usumacinta River, must have been located somewhere west of the Petexbatún region, erasing the differences in speech along both sides of the Río de la Pasión watershed to the south. Correlated with the distribution of head forms, the linguistic demarcation between different versions of Ch'olan roughly follows that of preferences in head shapes. This is probably no coincidence and adds ethnic value to cephalic modification and its potential for tracing and interpreting populational and cultural dynamics that at one time or another grounded the evolvement of ancient society.

Among the living, the differences in cranial morphology must have been most visible in and around the Usumacinta basin. In the multiethnic Maya fringes, territories of contact and exchange, Western Ch'ol-speaking folk from Palenque, Yaxchilán, or Bonampak, would be readily recognized as such by their extremely reclined and tubular head form. Their western neighbors toward Toniná and Chiapa de Corzo would not share these looks. Many of their mostly short and artificially broadened heads appear to be opposite to western Cholan styles, still enhanced by strong sagittal grooving, which divided both parietal lobes and thereby increased the bilateral expansion still further.

9.3.2 *Families and Communities*

Apart from regional ethnic attributions, we believe that the different Maya head modifications must also have held ideological value for their bearers on the local level, as proposed by us in Chap. 6. As a number of scholars argue (García and Tiesler 2011; Houston et al. 2009; Sotelo and Valverde 1992), head shape, per se, would have acted as a visible sign of outer and inner beauty; the latter was identified with sacred powers, an attempt perhaps to emulate the gods that protected the families or communities or those that were venerated by different sectors of the community. This connection was established already during the 1980s by Arturo Romano, who affirmed that the "variety of artificial head forms among different Maya groups [. . .] should reflect to an important degree mythical, magical and religious thought . . . " (Romano 1987, p. 25). The attributes of head morphology in the portraiture of Maya gods provides a promising point of departure to gain a deeper understanding of the aggregate religious meanings that the ancient Maya practitioners might have reproduced on the heads of their infants. Other published work, conducted together with my colleague, Ana García Barrios (García and Tiesler 2011), focuses on this aspect by surveying systematically Classic Maya anthropomorphic portraits of the supernatural. For this purpose, we revert to a database of some 300 images representing sacred forces. In this sample, head form was inferred categorically as either indistinguishable, unshaped, short, flattened on top, or artificially narrowed and elongated.

Our findings show that Classic period Maya artists did not assign head attributions by chance. They delineated each deity preferrable with one specific head configuration, although there is some variation in the conventions, which adhered to the Maya cartography of expressing the sacred. In general, we found the elderly gods were displayed with erect profiles, while young deities appeared with elongated and reclined heads (García and Tiesler 2011). The young God E, especially, of the Schellhas pantheon is represented in this fashion. As an image of the fertile earth and life-sustaining maize, he is figuratively converted into a husk of corn. His head is drawn with an elongated, reclined profile, quite similar to the rendering of the affiliated Moon

Fig. 9.6 Representation of the foliated young Maize God, showing strong head reclination and elongation. Note the visible postcoronary groove above the forehead, a secondary effect of prolonged compression. (Adapted from Taube 1992, Fig. 21b, p. 49; drawing by M. Sánchez)

Goddess (Taube 1992, pp. 64–69). The identification of these, complemented by their equal head forms, are repeatedly expressed by the artists of the first millennium A.D. Stephen Houston and his colleagues (2006, p. 45) supplement with insight on this form was accepted as a beauty ideal and by the female practitioners as the desired form to imprint on their offspring (Fig. 9.6; see also Taube 1992, pp. 46–50).

Other gods are drawn consistently with erect heads, which are distinct from the Maize God's rendering. Such is the case of Chaahk, the God of Rain, and, K'awil, the Scepter God. Also, the Solar God G, the elderly Goddess O, and gods N and A, the Lord of Death, are consistently delineated in this fashion (see also Taube 1992). The latter sometimes also appear with natural head profiles in Classic period portraiture, showing a visibly rounded, protruding occiput (Fig. 9.7). Lastly, God L, who was venerated by merchant folk, is also represented as a Muan bird in native iconography (Taube 1992, pp. 79–88). The anthropomorphic rendering of this trader deity appears in many portraits with a strong superior flattening when not covered by his wide-brimmed head.

Naturally, it is problematic to establish categorical associations between the head attributions of the Maya pantheon of gods and more so of their potential emulation among Classic Maya families and communities. The exception to this is, perhaps,

Fig. 9.7 God N in a palace scene (left). His head does not show signs of artificial head reclination or elongation. Note the visibly rounded occipital bulge behind the ears. (Kerr Archive 4113; redrawn by B. Ceballos)

of some extremely elongated head forms that strongly recall the look of the young Maize God, and superior parietal flattening, which is reminiscent of God L, as we have argued. It is probably also premature to speak with certainty on the exact ideas and mechanisms that among the living practitioners justified the selection of cephalic shapes, an area that still awaits examination in specific contexts of the Maya world or for Mesoamerica in general. For I shall take up the aspect of supernatural emulation for specific segments of society; namely, the royal upper crust, which is known to have reverted to a myriad of visible recourses to underline their divine right to authority and rulership.

9.4 Ascribing Social Roles to Maya Head-Shaping Traditions

9.4.1 The Head Forms of Classic Maya Aristocracy

Human interactions occurred at many different levels and among different segments of society. In this vein, the aristocracy that once protagonized Classic Maya society—and the personal cult that was celebrated on their behalf—are food for thoughts about the possible aggrandizing roles that cephalic models once held among courtiers (McAnany 1995). This function may have marked exclusivity, the exhibiting of certain head shapes by some but not others, as was common among the Inca aristocracy, for instance (Chap. 5). Specifically, head shaping lends itself as an ostentatious means of distinction, maybe by forcing an exceptional look on the head or by visibly emulating sacred forces. In this section, I wish to explore potential symbolic transposition and the role of cephalic modeling in the appropriation for self-aggrandizement. I will go about this idea by comparing the cephalic looks of the Classic Maya elite with those of the mass of society and use tandem information from the iconographic record.

In fact, the idea of visual self-enhancement and identification with the sacred among aristocrats has prominently engaged the minds of Mayanist scholarship who have explored this aspect mainly from the perspective of Classic period imagery (Houston et al. 2006). An iconographic study of head modification, anchored in the portraiture of Palenque's upper crust (Romano 1987), suggests that the tabular oblique modification predominated among members of the ruling elite. Also, the skeletal record has been consulted to establish potential associations with the ancient elite. Haviland and Moholy-Nagy (1992, p. 56) conclude that compression cribs were used exclusively by members of the aristocracy of the city of Tikal and that the initial enactment of cranial modification was ascribed to elite contexts.

However, more recent regional surveys of head looks among tomb occupants of the most exquisite Classic Maya funeral contexts have come to deny any elite associations of head shaping (Tiesler and Benedict 2001; Tiesler 2012). Namely, a series of 25 richly attired funeral contexts (with status markers of 4 and 5 (0–5); $N = 25$; Table 9.7) was consulted to examine diagnostic attributes of modeling. These dignitaries come principally from urban centers of the Central Petén and the Usumacinta areas; most of them date to the Late Classic period. This list includes the famous king Janaab' Pakal, of Palenque and his fernale consort. From Calakmul comes the body of who is presumed to be Lord Jaguar Paw, fierce rival of Tikal's powerful league (Martin and Grube 2008; Tiesler 2004), and who may be Lord Sky Witness, from Dzibanché. Other political heavy weights come from primary and secondary Maya centers of hegemonic power, like Copán in Honduras, Holmul in Guatemala, as well as Toniná and Chiapa de Corzo from Chiapas, Mexico, and Xuenkal and Oxkintok from Yucatán, Mexico.

As expected, the majority of these personages are males, although the group does include women. One of the prominent females who led Palenque's high society was Lady Tza'k bu' Ahaw whom in life was married to the decorated ruler Janaab' Pakal of Palenque (Tiesler et al. 2004; Fig. 9.8). She has been known in popular literature as the Red Queen, given the lack of inscriptions on her tomb and the thick layer of red cinnabar that covered her skeleton. As with Janaab' Pakal, her cephalic physiognomy is determined by a pronounced tabular oblique modeling in its pseudocircular variant, a type of modification that is shared by the majority of the population buried in and around Palenque (Montes 2000; Tiesler 1999). It was achieved by a prolonged anteroposterior compression of the head by use of cephalic splints, which were reinforced with tight constricting bands, which reduced the bilateral cephalic expansion. As a result, the head of Lady Tza'k bu' Ahaw was lengthened and her forehead severely reclined. As with other extreme forms of cranial constriction, her facial profile looks outthrusting, dominated by an aquiline nose with no visible nasal root and a buccal protrusion (Fig. 9.8).

Table 9.7 Artificial head forms among Classic Maya aristocrats

Burial/name	Archaeological site	Sex	Age range (years)	Status score (0–5)	Type of modification	Variant	Degree/ presence
16-1-3a	Xuenkal, Yucatán, Mexico	M	Young to middle aged adult (25–45)	4	Tabular erect	Intermediate	1.5?
16-1-3b	Xuenkal, Yucatán, Mexico		Young adult (15–25)	4			2
E. 2 AC/CA-3A	Oxkintok, Yucatán, Mexico	F	Adult (no range)	4	Tabular oblique	Mimetic	Present
T. XIII-3 "Tzak' Bu Ahaw"	Palenque, Chiapas, Mexico	F?	Mature adult (50–60)	5	Tabular oblique	Intermediate	3
1 TI "Janaab' Pakal"	Palenque, Chiapas, Mexico	M	Old adult (> 55 y.)	5	Tabular oblique		Present
XVIII-1 (1956) A/B	Palenque, Chiapas, Mexico		Adult (no range)	4	Tabular oblique		Present
XVIII-T.3	Palenque, Chiapas, Mexico	F	Young adult (25–30)	5	Irregular		0.25?
IV-6 (1)	Toniná, Chiapas, Mexico	F	Middle aged adult (35–45)	4	Tabular erect	Intermediate	2.25
121-1		M	Young to middle aged adult (25–45)		Tabular erect	Superior flattening	2.25
121a	Chiapa de Corzo, Chiapas, Mexico	F	Adult (no range)	4	Tabular erect	Intermediate	1.25
E.8 (27-2/2000) "Our Lord"	Kohunlich Quintana Roo, Mexico	M?	Mature to old adult (> 45 y.)	4			Present
3004 (D/B-2)	Dzibanché, Quintana Roo, Mexico	M	Middle aged adult (40–45)	4	Tabular oblique	Intermediate	3?
3005 (D/A-1)	Dzibanché, Quintana Roo, Mexico	F	Middle aged to mature adult (35–55)	5	Tabular oblique	Mimetic	1.25
3009 (K/A-1)	Dzibanché, Quintana Roo, Mexico	M	Young adult (25–30)	5			0
3010 (K/A-1)	Dzibanché, Quintana Roo, Mexico	M	Mature to old adult (> 45)	5	Tabular erect	Intermediate	0.5
E-2 (north)	Dzibanché, Quintana Roo, Mexico	M	Young to middle aged adult (25–45 y.)	4			Present
E-2 (south) "Lord Sky Witness"	Dzibanché, Quintana Roo, Mexico	M	Young to middle aged adult (25–35)	5	Tabular oblique	Mimetic	2
XIII-1	Calakmul, Campeche, Mexico	F?	Young adult (20–25)	4	Tabular oblique	Intermediate	2.5
XV-1	Calakmul, Campeche, Mexico	F	Mature to old adult (45–60)	5	Tabular	Mimetic?	
III-t.5 (9)	Calakmul, Campeche, Mexico	M	Middle aged adult (35–45)	5	Tabular oblique	Mimetic	3?
VII-t.1 (1)	Calakmul, Campeche, Mexico	M	Middle aged to mature adult (35–55)	5			Present
II-4a (97) "Lord Jaguar Paw"	Calakmul, Campeche, Mexico	M	Mature to old adult (50–60)	5			0?
II-B-5A	Holmul, Belize	M	Middle aged adult (35–45)	4			Present
36	Copán, Honduras	M?	Middle aged adult (30–35)	5	Tabular erect	Occipital plane	0.5
V-4	Copán, Honduras	F?	Middle aged to mature adult (35–55)	5			Present

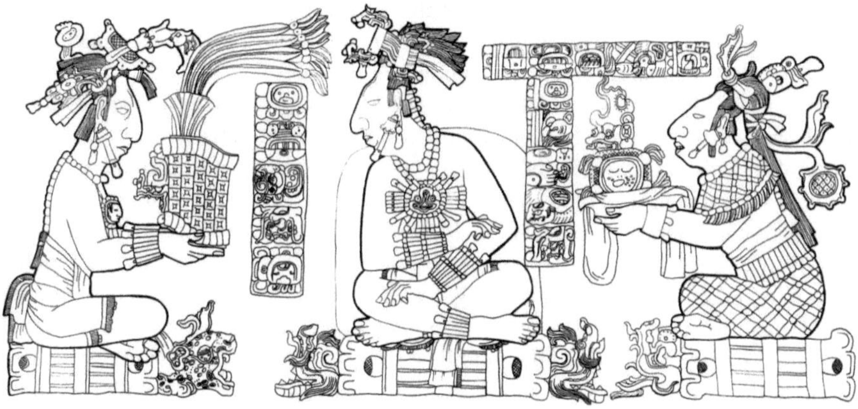

Fig. 9.8 Royals interacting in palace scene at Palenque, Chiapas, Mexico; all of them display strongly reclined head profile, as is common in the Usumacinta area. On the *left*, Janaab' Pakal is shown, handing the attributes of royal power to his son. To the *right*, Pakal's consort, Lady Tza'k bu' Ahaw is participating in the ceremony. She is presumed to be the unidentified dignitary, discovered in 1994 in Palenque's Temple XIII-sub. (Adapted from Greene 1991; drawing by M. Sánchez)

More diverse than at Palenque were the head looks displayed among the elites of the opposite side of the Classic Maya cultural sphere. At Copán's acropolis, recent research (Buikstra et al. 2004, pp. 194–195) has been conducted on a skeletal series from the Early Acropolis. The findings offer valuable insights into the role of cephalic modification in the looks of Copán's early aristocrats. From here comes the "Hunal" Tomb whose occupant was accredited with founding the Copán dynasty and in life bore the title of K'inich Yax K'uk' Mo'. This man died at an advanced age and in the years of his youth seems to have migrated to Copán from some place in the Petén (Price et al. 2009). Adjacent to his funeral chamber is another tomb which archaeologists colloquially call "Margarita." This tomb contained the remains of a woman who was also buried with abundant grave goods. She was a resident of Copán or its surroundings and must have lived shortly after Yax K'uk' Mo' (Price et al. 2009). For the purposes of this review, it is significant that, unlike the majority of later Classic Copanecans (with mostly mimetic oblique modifications), both dignitaries' heads display a tabular erect form.

In a nutshell, the results detailed in Table 9.7 indicate categorically that aristocratic head modifications were not patently different from those modifications that the mass of local society displayed. Tabular erect and oblique shapes are equally represented in the series; the latter appear somewhat below the general ratio for the Classic period. The proportion of artificially shaped crania ascends to 88 %, with a moderate expression on average of 1.928 (in a scale of 0–4), again being equivalent to the patterns exhibited by the popular sectors. Consequently, we can affirm that prevalence and choice of head form is similar to the models established for the remaining population of its time and area. Therefore, there appears to be no indication that could argue for any exclusive head form or any specific enactment of these body modifications among nobles.

In a second approach, we compared the average status index for each type of cephalic modification to distinguish if any particular form appeared more frequently in richly attired funerary contexts. While tabular erect forms marked an average score of 0.982 ($N = 257$), tabular oblique shapes, which the literature identifies with dignitaries of the Classic Era Maya courts, are associated with a score of 0.899 ($N = 449$). This result is still below the average for erect modifications and again discounts any assignment of exclusivity to reclined and tabular heads from the Classic era. Also, the criterion of presence (0.913, $N = 1{,}000$) vs. absence of modification (0.875, $N = 206$) does not denote any privilege or wealth either for the practice or lack of practice. Likewise, the degrees of modification offer no elements that could establish correspondence between status and the visibility of modification (such as the more pronounced the modification, the higher the status). As previously discussed, presence, degrees, and forms were not applied preferentially, let alone exclusively, among members of different social sectors among Classic period Maya. This observation is also confirmed for specific site series from our database. In the skeletal series of Copán, Honduras, for example, which includes individuals of different social groups, no significant difference was documented when presence and type of modification were compared between high and low status groups ($p = 0.936$ and $p = 0.873$, respectively).

In view of the lack of any tendency, there is no evidence that would lead us to suppose that cephalic modification, in itself or its generic types, performed a role as a marker of social position, a conclusion that confirms the results of previous phases of my regional survey of the Maya area (Tiesler 1997, 1999). The assimilation of popular cephalic forms by the Maya aristocracy differs somewhat from Andean head practices, for example (Torres-Rouff 2002; Yépez 2006, 2009). Cognizant of the importance of other body attributes to denote aristocratic exclusivity and even divinity, the apparent social equality communicated by the enactment of infant head modeling comes as a surprise. Explanations could lie in the fact that the practitioners were female and the idea that those who experienced the head procedure were too young to acquire leading roles and authority. Among central Petén elites, we note that their real artificial head forms do not necessarily coincide with their head rendering in royal portraiture, which tends to stick to the reclined aesthetic ideal.

There is a single exception that contrasts noticeably with the panorama described earlier in the regional survey. It identifies the superior head flattening, which is achieved either by using cradleboards or cephalic devices (Chap. 4). This form, which in funerary contexts is associated with richly attired burials, proliferated in the Maya area during the latter half of the Classic period. The elite connotation of this modification is expressed in markers of status, which scores an average of 1.90 in our series, well above the remainder of the series (0.83). Although modifications with flattening also appear in women and children, the majority of the individuals who show superior flattening were men, constituting 71.15 % of those individuals for whom it was possible to determine the sex ($N = 52$). Who were the people who showed their head flattened on top? What was their role in the ancient social fabric? Why was this artifice introduced only in the Late Classic period in the area? Why was it not used as was the vast majority of the forms centuries or even millennia before? We will explore these and other questions in the following section.

9.4.2 The Social Signification of Superior Flattening

We recall that a specific form was added to the Mesoamerican caleidoscope of cephalic shapes during the Classic period (Chap. 8). This model, called parallelepiped, superior, parietal or obelionic flattening, or simply El Zapotal type (Romano 1973), displays a visible compression plane in the upper portion of the neurocranium. The cultural origin of this cranial configuration appears to lie in southern-central Veracruz. At least here, the practice found its most visible expression during the Classic period (Chap. 8).

Apart from other cultural areas within the Mesoamerican sphere, which we have already documented in the previous chapter, superior head flattening also came to be known and applied in the Maya area and vicinities. A total of 74 shapes of the type were registered in the recently expanded survey of the Maya area. These prominently identify males from rich funerary contexts, suggesting they were wealthier than most other Maya and/or enjoyed a higher level of prestige. Superior flattening seems to have appeared in the circum-Maya area during the Middle Classic period. Here, skulls bearing this form are recorded from the Mixe-Zoque region around Chiapa de Corzo. Shortly afterward, the people bearing top-flattened head forms appear to have breached immense distances along the coastline of Yucatán toward Honduras and adjacent settlements connected to maritime trade. Also, a few inhabitants of the Copán Pocket, on the other side of the peninsula, exhibit superior flattening. It is noteworthy that superior flattening appears to have been practiced among coastal or nearby communities, that is, by those with direct or indirect access to the sea such as Kohunlich, Barton Ramie, and Copán. Registries of other inland sites, including numerous skeletal series such as Calakmul ($N=41$) or the entire collection from the southeastern Petén ($N=91$), do not include a single remain with this form. A statistical comparison of the presence of superior flattening among coastal Maya series with the interior areas of the Maya world highlights this cultural separation ($p=0.002$).

Heads with their top portion flattened, especially in the ports of Xcambó and Isla Cerritos, are visible in 7 and 20 % of the crania, respectively. The population of Jaina in Campeche likewise includes this form in its cultural repertoire. Here, it is apparent in 12 % of the evaluable crania. Note that this style continued to be crafted in infant heads up to the Early and Late Postclassic period, although in a smaller proportion. It is still either coastal populations that display it or those from the Chiapas Mixe-Zoque peripheries of the Maya area. The configuration is expressed by 2.7 % of the evaluable population for the Postclassic site of El Rey, and still makes up 6.3 % of all evaluable crania from the island site of San Gervasio, most of which date to the Early Postclassic period. Conversely, we could not document any cases of superior flattening in the skeletal populations of Tulum or Champotón, which date to the Late Postclassic period (Vargas 1997; Folan et al. 2003).

Although we cannot confirm exact chronological ranges for the human remains of what were presumably sacrificial victims, recovered from the depth of the Sacred Cenote of Chichén Itzá, these deserve special mention, as they represent the vast majority of the Maya crania documented with superior compression so far. This trend is noted both in the series of skulls found in the Peabody Museum, Harvard University, and in another more recent series recovered by Piña Chan in the 1960s. Almost one third of the examined pieces show superior flattening (29 %; $N = 147$), which constitutes almost the same level of acceptance noted centuries before among the residents of El Zapotal in Veracruz. Given the strong pan-Mesoamerican trader network controlled by Chichén Itzá (Vargas 2001; Wyllie 2002), the skulls with cranial forms reminiscent of merchant folklore seem to decode the cultural and possibly ethnic pertinence of the individuals presumably offered for ritual purposes by Chichén's hegemonic rulers (see Chaps. 8 and 10).

In synthesis, the fact that folk with top-flattened heads appear to have been propagated along the coast, and given the similarity found with head forms further west, toward Veracruz, invites reflection on their possible meanings as visible signifiers of ethnicity, social role, and ideological association. As for the individuals who bore superior flattening, we must engage questions on their possible ethnic and geographic origin. Recalling the similarity of this head form with that of the Maya Merchant God, we cautiously infer that this model emulated this magical–religious power venerated by traders. The cult of God L, enacted by Maya merchant folk from the overarching networks of ever more Mesoamerican maritime and terrestrial trade routes, gained strength in the area toward the Late to Terminal Classic period (Wyllie 2002; see also Tiesler et al. 2010). The geographic and chronological distribution of those Maya individuals who display superior flattening, along with the richness of their funerary clothing, appears to reaffirm their collective identity and affiliation with traders. Ideological elements introduced, flows of materials such as obsidian, and the presence of foreign personages, likewise suggest strong influences that emanate from the west along the Gulf coast (Grube et al. 2009; Pallán Gayol 2009). In this line of thought, the coastal Maya head shapes with their similarity to forms common in Veracruz would therefore have constituted a visible expression of the new sociocultural dynamics that were beginning to be felt in the receiving Maya territories, including centuries before the so-called "collapse."

We conclude this section by musing on the role of the top-flattened head looks and its propagation in collective group ideology and visible representation. Broader connotations relate to the nature of trade and the mobility of traders. If the superior flattening indeed denotes merchant identity, then Late and Terminal Classic trade was conducted by people who breached large distances along the shores that surrounded the Maya area to the east and north. They did not only move and introduce objects and styles but also made themselves conspicuous by showing a distinctive head form. Set in the historical context of precollapse Maya society, heads with superior flattening somehow appear as living bearers of a new ideological system that accompanied a new network of military allies and trading partners. Its initial expansion along the Gulf coast would soon reach the coastal borders and territories on the other side of the Peninsula (Tiesler et al. 2010). The early protagonists of this new order could well have turned to the visual resource of head shape to identify themselves with a magical–religious power that they venerated as traders, a sector whose rise was consolidated around the end of the Classic era. This was a time marked by dramatic social adjustments and changes in population, including the abandonment of a large part of the Central Lowlands and the rise of new powerful hegemonies, the first of which was Chichén Itzá in the northern Peninsula.

References

Aguilar, Ana Cristina. 2012. Género en los contextos funerarios de una aldea de los períodos Sapoá y Ometepe (800–1550 D.C.) en la Bahía de Culebra. Master diss., Universidad de Costa Rica.

Bell, E. Ellen, Robert J. Sharer, Loa P. Traxler, David W. Sedat, Christine W. Carelli, and Lynn A. Grant. 2004. Tombs and burials within the Early Classic Copan Acropolis. In Understanding early Classic Copan, ed. Ellen E. Bell, Marcello A. Canuto, and Robert J. Sharer, 131–158. Philadelphia: Museum of Archaeology and Anthropology, University of Pennsylvania.

Buikstra, Jane E., T. Douglas Price, Lori E. Wright, and James A. Burton. 2004. Tombs from Copan's Acropolis: A life history approach. In Understanding Early Classic Copan, ed. Ellen E. Bell, Marcello A. Canuto and Robert J. Sharer, 191–212. Philadelphia: Museo de la Universidad de Pennsylvania.

Ceballos, Teresa Noemi. 2003. La cronología cerámica del puerto maya de Xcambó, costa norte de Yucatán: Complejo Xtampú. B.A. diss., Universidad Autónoma de Yucatán.

Cetina, Aleida, and Thelma Sierra. 2005. Condiciones de vida y nutrición de los antiguos habitantes de Xcambó, Yucatán. Estudios de Antropología Biológica 12:661–678.

Cucina, Andrea, and Vera Tiesler. 2008. Dinámicas poblacionales y migraciones durante el Clásico y Postclásico en Yucatán: La morfología dental. Los Investigadores de la Cultura Maya 16(2):165–177.

Diamanti, Melissa. 1991. Domestic organization at Copán: Reconstruction of elite maya household through ethnographic models. Ph.D. diss., Penn State University.

Fash, William L., and Ricardo Agurcia. 1991. Scribes, warriors and kings. The city of Copán and the ancient Maya. London: Thames and Hudson.

Folan, William, Lynda Florey, Abel Morales, Raymundo González, José Hernández, Rosario Domínguez, Vera Tiesler, David Bolles, Roberto Ruiz, and Joel D. Gunn. 2003. Champotón, Campeche: Su presencia en el desarrollo cultural del Golfo de México y su corredor eco-arqueológico. Los Investigadores de la Cultura Maya 11:64–71.

García, Ana, and Vera Tiesler. 2011. El aspecto físico de los dioses mayas. Arqueología Mexicana 112:59–63.

Gerstle, Andrea I. 1985. La arquitectura ceremonial de las Sepulturas, Copán. Yaxkin 7(1–2): 99–110.

Greene, Merle. 1991. The Sculpture of Palenque, Vol. 4, The Cross Group. Princeton: Princeton University.

Grube, Nikolai, Carlos Pallán, and Antonio Benavides. 2009. La escalinata jeroglífica 1 de Sabana Piletas, Campeche: Nuevos datos sobre el Epiclásico en la región Puuc. Los Investigadores de la Cultura Maya 17(2):93–116.

Haviland, William, and Hattula Moholy-Nagy. 1992. Distinguishing the high from the mighty from the hoi polloi at Tikal, Guatemala. In Mesoamerican elites: An archaeological assessment, ed. Arlen Chase and Diane Chase, 50–60. Norman: Universidad de Oklahoma.

Hendon, Julia A. 1987. The uses of Maya structures: A study of architecture and artifact distribution at Sepulturas, Copán, Honduras. Ph.D. diss., Harvard University.

Hendon, Julia A. 1991. Status and power in Classic Maya society: An archaeological study. American Anthropologist 93(4):894–918.

Houston, Stephen D., Takeshi Inomata. 2009. The Classic Maya. New York: Cambridge University Press.

Houston, Stephen, David Stuart, and Karl A. Taube. 2006. The memory of bones: Body, being and experience among the Classic Maya. Austin: University of Texas.

Houston, Stephen, Claudia Brittenham, Casandra Mesick, Alexandre Tokovinine, Christina Warinner. 2009. Veiled Brichtness, a History of Ancient Maya Color. Austin: University of Texas Press.
Jiménez, Socorro del Pilar. 2002. La cronología cerámica del puerto maya de Xcambó, costa norte de Yucatán: Complejo cerámico Xcambó y complejo cerámico Cayalac. B.A. diss. Mérida: Universidad Autónoma de Yucatán.
Kettunen, Harri. 2005. Nasal motifs in Maya iconography. Ph.D. diss., University of Helsinki.
Lacadena, Alfonso, and Soeren Wichmann. 2002. The distribution of Lowland Maya languages in the Classic period. In La organización social entre los mayas prehispánicos, coloniales y modernos: Memoria de la Tercera Mesa Redonda de Palenque, Vol. 2, ed. Vera Tiesler, Rafael Cobos, and Merle Greene, 275–353. Mexico: Instituto Nacional de Antropología e Historia, Universidad Autónoma de Yucatán.
Maggiano, Isabel S., Michael Schultz, Horst Kierdorf, Thelma Sierra, Corey M. Maggiano, and Vera Tiesler. 2008. Cross-sectional analysis of long bones, occupational activities and long-distance trade of the Classic Maya from Xcambó, Archaeological and osteological evidence. American Journal of Physical Anthropology 136:470–484.
Martin, Simon, and Nikolai Grube. 2008. Chronicle of the Maya kings and queens. New York: Thames and Hudson.
McAnany, Patricia. 1995. Living with the ancestors. Kinship and kingship in ancient Maya society. Austin: University of Texas.
Medrano Chan, Lucy. 2005. Tratamientos póstumos del cuerpo humano en la tradición funeraria de Xcambó, Yucatán. B.A. diss., Universidad Autónoma de Yucatán.
Méndez, Christian, Thelma Sierra, Vera Tiesler, and Andrea Cucina. 2009. Linear enamel hypoplasia at Xcambó, Yucatán, during the Maya Classic period: An evaluation of coastal marshland impact on ancient human populations. HOMO 60:343–358.
Montes, Javier. 2000. La práctica de la deformación cefálica intencional entre los pobladores prehispánicos de Palenque, Chiapas. B.A. diss., Escuela Nacional de Antropología e Historia, Instituto Nacional de Antropología e Historia.
Pallán, Carlos. 2009. Secuencia dinástica, glifos-emblema y topónimos en las inscripciones jeroglíficas de Edzná, Campeche (600–900 d.C.): Implicaciones históricas. Master diss., Universidad Nacional Autónoma de México.
Peña, Agustín, and Thelma Sierra Sosa. 2004. Cronología y contexto en las figurillas de terracota de la costa del Golfo de México en Xcambó, Yucatán. Los Investigadores de la Cultura Maya 12(1):210–225.
Price, T. Douglas, James H. Burton, Robert J. Sharer, Jane E. Buikstra, Lori E. Wright, Loa P. Traxler, and Katherine A. Miller. 2009. Kings and commoners at Copan: Isotopic evidence for origins and movement in the Classic Maya period. Journal of Anthropological Archaeology 29:15–32.
Quintal, Ella. 2000. Vírgenes e Ídolos: La religión en las manos del pueblo. Mesoamérica 39: 287–304.
Romano, Arturo. 1973. Los cráneos deformados de El Zapotal I, Veracruz. Antropología Física 1:57–64.
Romano, Arturo. 1977. Los cráneos deformados de El Zapotal. In Las esculturas en terracota de El Zapotal, Veracruz, ed. Nelly Gutiérrez and Susan K. Hamilton, 31. Mexico: Instituto de Investigaciones Estéticas, Universidad Nacional Autónoma de México.
Romano, Arturo. 1987. Iconografía cefálica maya. In Memorias del Primer Coloquio Internacional de Mayistas, 1413–1474. Mexico: Universidad Nacional Autónoma de México.
Scheuer, Louise, and Sue Black. 2000. Developmental juvenile osteology. San Diego: Academic Press.
Sharer, Robert J., and Loa P. Traxler. 2006. The ancient Maya. Stanford: Stanford University.
Sierra, Thelma. 1999a. Xcambó. Codiciado enclave económico del Clásico maya. Arqueología Mexicana 7(37): 40–47.
Sierra, Thelma. 1999b. Xcambo. Codiciado puerto del Clásico maya. El'Inaj Semilla de Maíz 10: 19–27. Mérida: Instituto Nacional de Antropología e Historia.

Sierra, Thelma. 2001. Los entierros de Xcambo. I'Inaj Semilla de Maíz 12:6–12. Mérida: Instituto Nacional de Antropología e Historia.

Sierra, Thelma. 2004. La arqueología de Xcambó, Yucatán, centro administrativo salinero y puerto comercial de importancia regional durante el Clásico. Ph.D. diss., Universidad Nacional Autónoma de México.

Sierra, Thelma, Andrea Cucina, T. Douglas Price, James H. Burton, and Vera Tiesler. 2013. Maya coastal production, exchange, life style and population mobility: A view from the port of Xcambó, Yucatan, Mexico. Unpublished paper, submitted to Ancient Mesoamerica.

Sotelo, Laura, and Carmen Valverde. 1992. Los Señores de Yaxchilán: Un ejemplo de felinización de los gobernantes mayas. Estudios de Cultura Maya 19:187–214.

Storey, Rebecca. 2005. Health and lifestyle (before and after death) among the Copan elite. In Copán: The history of an ancient Maya kingdom, ed. E. Wyllys Andrews and William L. Fash, 315–343. Santa Fe: School of American Research Press.

Suzuki, Shintaro, Andrea Cucina, Vera Tiesler, and Thelma Sierra Sosa. 2009. Morir en Xcambó: Evaluación de la mortalidad en un sitio costero maya del Clásico a partir del análisis histomorfológico. In XXII Simposio de Investigaciones Arqueolo ógicas en Guatemala 2008, ed. Juan Pedro Laporte, Bárbara Arroyo, and Héctor E. Mejía, 791–802. Guatemala: Ministerio de Cultura y Deportes, Instituto de Antropología e Historia, Asociación Tikal.

Taube, Karl A. 1992. The major gods of ancient Yucatan. Washington: Dumbarton Oaks.

Tiesler, Vera. 1997. Informe del estudio realizado en el sitio arqueológico Copán, Honduras. Unpublished report on file at the Centro de Investigaciones, Copán Ruinas.

Tiesler, Vera. 1998. La costumbre de la deformación cefálica entre los antiguos mayas: Aspectos morfológicos y culturales. Mexico: Instituto Nacional de Antropología e Historia.

Tiesler, Vera. 1999. Rasgos bioculturales entre los antiguos mayas: Aspectos arqueológicos y sociales. Ph.D. diss., Universidad Nacional Autónoma de México.

Tiesler, Vera. 2004. Vida y muerte de Janaab' Pakal de Palenque: Hallazgos bioarqueológicos recientes. In Janaab' Pakal de Palenque: Vida y muerte de un gobernante maya, ed. Vera Tiesler and Andrea Cucina, 37–67. Mexico: Universidad Nacional Autónoma de México, Universidad Autónoma de Yucatán.

Tiesler, Vera. 2005. Prácticas bioculturales y organización social en los sitios de Copán, Honduras, y Xcambó, Yucatán. Estudios de Antropología Biológica 12(1):103–117.

Tiesler, Vera. 2010. "Olmec" head shapes among the Preclassic period Maya and cultural meanings. Latin American Antiquity 21:290–311.

Tiesler, Vera. 2011. Becoming Maya: Infancy and upbringing through the lens of pre-Hispanic head shaping. Childhood in the Past 4:117–32.

Tiesler, Vera. 2012. Transformarse en maya. El modelado cefálico entre los mayas prehispánicos y coloniales. Mexico: Universidad Autónoma de Yucatán, Universidad Nacional Autónoma de México.

Tiesler, Vera. 2013. Informe del estudio realizado en el sitio arqueológico Yaxuná, Yucatán. Unpublished report on file at Universidad de las Américas, Cholula, Puebla.

Tiesler, Vera, and Ruth Benedict. 2001. Head shaping and dental decoration: Two biocultural attributes of cultural integration and social distinction among the ancient Maya. Paper presented at the 70th Annual Meeting of the American Associations of Physical Anthropology. Kansas.

Tiesler, Vera, and Andrea Cucina. 2008. Afinidades biológicas y prácticas bioculturales del sureste de Petén. In XXI Simposio de Arqueología de Guatemala 2007, ed. Juan Pedro Laporte, Bárbara Arroyo, and Hector Mejía, 589–600. Guatemala: Museo Nacional de Arqueología y Etnología, Ministerio de Cultura y Deportes, Instituto Nacional de Antropológia e Historia, Asociación Tikal.

Tiesler, Vera, and Andrea Cucina. 2010. La deformación craneana como emblema de identidad, etnicidad y reproducción cultural entre los mayas del Clásico. In Identidades y cultura material en la región maya, ed. Héctor Hernández and Marcos Pool Cab, 111–134. Mérida: Universidad Autónoma de Yucatán.

Tiesler, Vera, and Andrea Cucina. 2011. Filiación, relaciones interpoblacionales y enlaces culturales en las Tierras Bajas mayas durante el periodo Clásico. Estudios de la Cultura Maya 40:90–122.

Tiesler, Vera, and Andrea Cucina. 2012. Etnicidad, relaciones interpoblacionales y patrones migratorios a lo largo de la costa de Campeche durante los períodos Clásico y Posclásico. In Arqueología de la costa de Campeche: la época prehispánica, ed. Rafael Cobos, 67–95. Mérida: Universidad Autónoma de Yucatán.

Tiesler, Vera, Thelma Sierra, and Samuel Tejeda. 2002. Nutrición y condiciones de vida en la costa norte de la Península durante el Clásico: Una visión desde Xcambó, Yucatán. In Memorias del XV Simposio de Investigaciones Arqueológicas en Guatemala, ed. Juan Pedro Laporte, Héctor Escobedo, and Bárbara Arroyo, 752–762. Guatemala: Museo Nacional de Arqueología y Etnología, Ministerio de Cultura y Deporte, Instituto de Antropología e Historia, Asociación Tikal.

Tiesler, Vera, Andrea Cucina, Thelma Sierra, and Richard Meindl. 2004. Ancient Maya population structure and social dynamics. A palaeodemographic approach to the Classic coastal site of Xcambó, Yucatán. American Journal of Physical Anthropology: Annual Meeting Issue 2004, Supplement 37:194–195.

Tiesler, Vera, Andrea Cucina, Thelma Sierra, Marlene Falla, and Richard Meindl. 2005. Comercio, dinámicas biosociales y estructura poblacional del asentamiento costero de Xcambó, Yucatán. Los Investigadores de la Cultura Maya 13(2):365–372.

Tiesler, Vera, Arturo Romano, and Carlos Pallán. 2010. Las formas cefálicas en las vísperas del periodo Clásico: Implicaciones para el cambio social en el área maya. Los Investigadores de la Cultura Maya 18:83–96.

Torres-Rouff, Christina. 2002. Cranial vault modification and ethnicity in Middle Horizon San Pedro de Atacama, North Chile. Current Anthropology 43:163–171.

Tozzer, Alfred M. 1941. Landa's Relación de las Cosas de Yucatán. Cambridge: Peabody, Harvard.

Vargas, Ernesto. 1997. Tulum: Organización político-territorial de la costa oriental de Quintana Roo. Mexico: Instituto de Investigaciones Antropológicas, Universidad Nacional Autónoma de México.

Vargas, Ernesto. 2001. Itzamkanak y Acalan. Tiempos de crisis anticipando el futuro. Mexico: Instituto de Investigaciones Antropológicas, Universidad Nacional Autónoma de México.

Wanner, Isabel S., Thelma Sierra Sosa, Kurt W. Alt and Vera Tiesler. 2007. Lifestyle, occupation, and whole bone morphology of the pre-Hispanic Maya coastal population from Xcambó, Yucatan, Mexico. International Journal of Osteoarchaeology 17:253–268.

Webster, David, William T. Sanders, and Peter Van Rossum. 1992. A simulation of Copan population history and its Implications. Ancient Mesoamerica 3:185–197.

Webster, David, Ann Corinne Freter, and Nancy Gonlin. 2000. Copan: The rise and fall of an ancient Maya kingdom. Fort Worth: Harcourt College Publishers.

Whittington, Stephen L. 1989. Characteristics of demography and disease in low status Maya from Classic period Copan, Honduras. Ph.D. diss., Pennsylvania State University.

Wichmann, Søren. 1999. A conservative look at diffusion involving Mixe-Zoquean Languages. In Archaeology and language II: Archaeological data and linguistic hypotheses, ed. Blench, Roger & Matthew Spriggs, 297–323. London: Routledge.

Wyllie, Cherra E. 2002. Signs, symbols, and hieroglyphs of ancient Veracruz: Classic to Postclassic transition. Ph.D. diss., Yale University, New Haven.

Yépez, Rosaura. 2006. La práctica cultural de modelar la cabeza en dos culturas andinas del antiguo Perú: Paracas y Chancay. Un estudio de los procesos de significación de la cabeza modelada intencionalmente. Ph.D. diss., Universidad Nacional Autónoma de México.

Yépez, Rosaura. 2009. El simbolismo de la modificación cultural de la cabeza en la cultura andina de Paracas del antiguo Perú. Estudios de Antropología Biológica 14:526–543.

Chapter 10
Head Shaping During the Second Millennium. Postclassic and Postcontact Mesoamerica

10.1 Mesoamerican Cranial Modifications During the Postclassic Period

The onset of the second millennium was marked by the decline of many important Classic period nations, such as the Oaxaca Zapotec and the Lowland Maya. Oligarchic councils now operated in much of Central Mexico and in Yucatán and dual rulership appears to have replaced the more theocratic governments of prior Classic times. Evermore powerful, wide-ranging networks of *pochteca* merchants dominated the economic life towards and during the Postclassic era, while military orders gained prominence and, towards the close of the Postclassic period, culminated in the Aztec empire. Many parts of Mesoamerica experienced irrevocable social and cultural changes towards the second millennium A.D., a gradual replacement of political and economic organization, which was upheld by pan-Mesoamerican ideological shifts and a nahua-dominated streamlining of ritual life. If the unifying pressure of dominant ideology acted to assimilate the diverse worldviews, then it might have impoverished the real diversity, complexity, and even opposition of existing worldviews during the Postclassic period, as López-Austin (1989, pp. 470–471) states, for inner incongruences within ideological systems in general. This conjecture could also acquire relevance for cephalic modeling, should the increasing uniformity of techniques and cranial forms have responded to any sort of social pressure. In any case, at the end of the Postclassic, a uniform choice of short and broad head styles were produced by cradleboard devices, the type of apparatus still described by the Hispanic chroniclers during the sixteenth century. This tendency is most evident in those areas that staged diversity in headwear before the Postclassic period, such as Western Mexico and the Lowland Mayas.

10.1.1 Maya and Gulf Coast Cultures

Toward the close of the Classic period, the inhabitants of the Central and Southern Maya Lowlands suffered a disintegration process that would last for decades, even

V. Tiesler, *The Bioarchaeology of Artificial Cranial Modifications*, Interdisciplinary Contributions to Archaeology 7, DOI 10.1007/978-1-4614-8760-9_10,

centuries, and which took a dramatic form in most former regional centers. Here, the so-called "collapse" at the end of the Classic period is evidenced most patently in the material record. The construction and erection of stelae halts, centers are abandoned, vandalized, and finally destroyed. At the end of this process immense territories in the interior of the Petén had been abandoned altogether.

It is noteworthy for the subject of this volume, however, that the vanishing inland populations did not appear to have changed their preferences in head models during the difficult years that surrounded and followed the demise of their Classic inland hegemonies. Also, the very last settlers appear to have continued their ancestral head splinting traditions just as their forebearers centuries earlier (Tiesler 2012a, b). Such is the case, for example, in the segment of the Río de la Pasión population of Altar de Sacrificios that was dated to the last occupational phase towards the Early Postclassic period. Its late settlers seems to manifest a continuity of preferences with twice the number of tabular oblique to erect forms, just as the majority of Altareños had performed in their offspring all along the centuries that encompass the Classic period (Tiesler and Cucina 2012). Even considered on a regional scale, our results on the scrutinized Petén series lay out a panorama not of cultural substitution, but rather continuity and permanence. Here, the head models for newborns were sustained up to the very end, until the last settlers died or left for good (Tiesler 2012a; Tiesler and Cucina 2012).

Less affected by the collective catastrophe deep in the Petén basin were the northern lands of the Peninsula, where populations seemed to boom along the coast and connected inland (Sharer and Traxler 2006). Colonial chronicles allege that the Putun expansion reached its most important push during the domination of Chichén Itzá during the first half of the Postclassic era (Demarest et al. 2004; Sharer and Traxler 2006). Here, a new political–economic order establishes itself, linked to the cult of Quetzalcoatl and powerful pan-Mesoamerican merchant leagues (Ringle et al. 1998). This "new international order" was strongly mediated by the cultural territories that fringed the west coast of the Gulf of Mexico (Wyllie 2002). A growing importance in sea trade is expressed in the archaeological record by increased quantities of foreign materials and stylistic influences along the shores and inland Yucatán. As we have argued in the previous chapter, some of their carriers distinguished themselves by their distinctly top-flattened heads. They began to blend in with the coastal settlers of Yucatán and are also seen prominently in Chichén Itzá itself and its trader outpost, Isla Cerritos (Tiesler 2012). Also, other Postclassic period settlements around the Golf Coast, such as Isla de Sacrificios (Veracruz) or San Gervasio (Quintana Roo), show top-flattened head looks (Fig. 10.1).

Different from their inland Petén neighbors further south, Peninsular mothers seemed to have abandoned now, at the turn of the millennium, head splints in favor of cradleboarding, leading to the short and broad head form to dominate Postclassic looks almost everywhere in Mesoamerica. Mayapán, a city that after 1221 replaced Chichén Itzá as the ruling center and which in turn was abandoned during the fifteenth century, displays purely broad and short heads in its cranial record (Serafin 2010; Tiesler 2012a). This preference is already evident in Chichén Itzá itself, as well as in all examined east coast sites of our systematic survey. Such is also the case

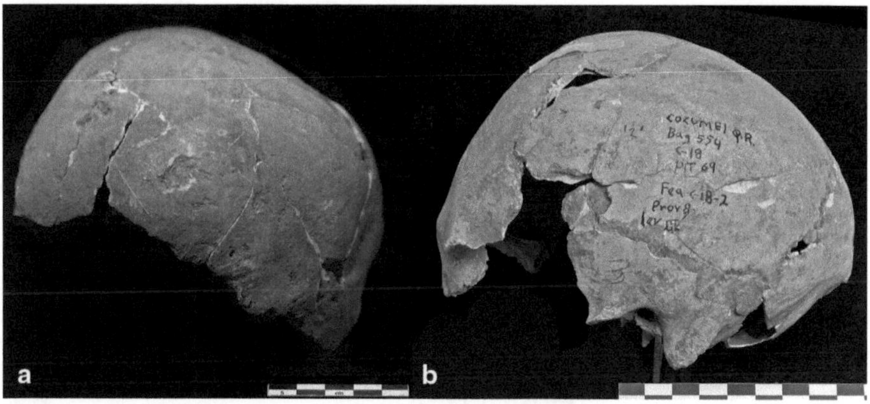

Fig. 10.1 Top-flattened crania from postclassic period Isla de Sacrificios, Veracruz (**a**), and San Gervasio, Quintana Roo (**b**). Profile view. (DAF/INAH; photo by V. Tiesler)

among inland dwellers fringing the eastern shore, such as Kohunlich, in southern Quintana Roo. Here, some 68 % of the inhabitants still bore tabular oblique forms ($N = 16$) during the Classic period. Afterwards, during the subsequent last phase of occupation, it is the erect form (69 % of $N = 13$) that is favored at the settlement (although the oblique modification still was known). This change in head looks is statistically significant ($p = 0.038$).

In the long run, although the Maya "collapse" and its population shifts did not lead to the abandonment of head modeling itself, it did result in a homogenization of techniques and forms (Fig. 10.2). This shift is apparent when confronting Late Classic patterns with those of the Early and Late Postclassic period from our regional data base (Tiesler 2012). As seen in Table 10.1, the use of head splints of all kinds was in decline in the Maya area and was abandoned altogether toward the second half of the Postclassic period, together with the use of the constriction wraps. This cultural replacement is statistically highly significant if we compare the proportion of erect and oblique types between the Early Postclassic and the Late Postclassic periods ($p = 0.001$). Here, we might muse how the spread of the new pan-Mesoamerican "international order," under the religious umbrella of the Kukulkan or Quetzalcoatl cult, might have streamlined the former tapestry of diverse preferences and their associated emblematic meanings, some of which were regionally specific (Ringle et al. 1998).

The different local expressions and trends in head modeling noted in this study are also echoed by other authors who describe and compare different cranial modifications of the Postclassic Maya series. T. Dale Stewart (1953, pp. 296–297) documents only erect shapes for the funeral population of Zaculeu for the Qankyak and Xinabahul Phases during the Postclassic era. Tabular erect models in the Postclassic mountain site of Mixco Viejo, close to present-day Guatemala City, likewise appear to be the only visible skull forms in all late skeletal population (Gervais 1989). Some of these profiles are shown with marked superior flattening, which is similar to the ones described in this survey for the coastal series from the Yucatán Peninsula.

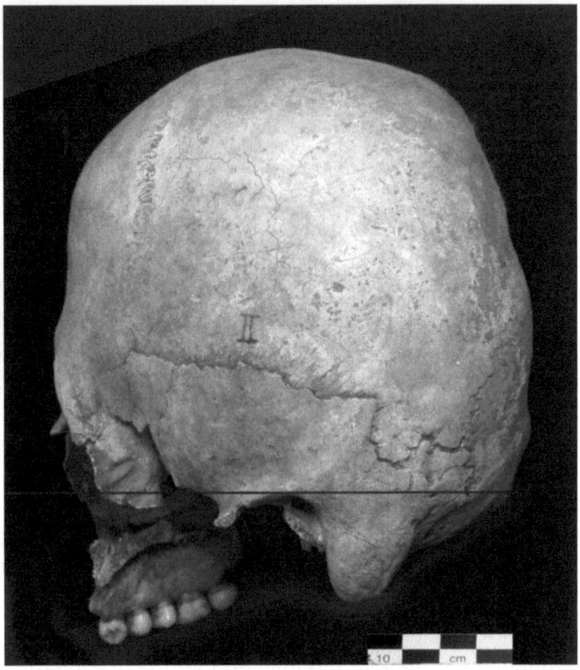

Fig. 10.2 Extreme frontal–occipital shortening in cranium from Argelia, Chiapas, Mexico. Profile view. (DAF/INAH; photo by V. Tiesler)

It is also noteworthy that a full third of the Postclassic crania included in our systematic survey from the Maya area (Tiesler 2012) show a sagittal sulcus, which is almost always associated with tabular erect modifications. Similar to the preceding millennium, the proportion of individuals with sagittal grooving appears slightly higher in the coastal communities and in the Maya Highlands than among inland Lowlanders. This sagittal groove commonly describes a slight depression that visibly sets apart the parietal bones of both sides, perhaps an inadvertent byproduct of the

Table 10.1 Patterns of head modification during the second millennium A.D.

	Early Postclassic, % (*N*)	Late Postclassic, % (*N*)	Colonial, % (*N*)
Frequency (%)	91.42 (70)	92.91 (127)	31.12 (77)
Frequency of bipolar head constriction	21.21 (33)	0 (70)	0 (11)
Frequency of sagittal sulcus	34.38 (32)	34.33 (67)	20 (10)
Tabular oblique/tabular erect	21.43 (51)	0 (115)	0 (21)
Conical variety of tabular erect	0 (47)	2.47 (81)	0 (11)
Average degree of modification (0–4)*	2.1 (64)	1.9 (118)	N.A. (4)

*Only adults scored

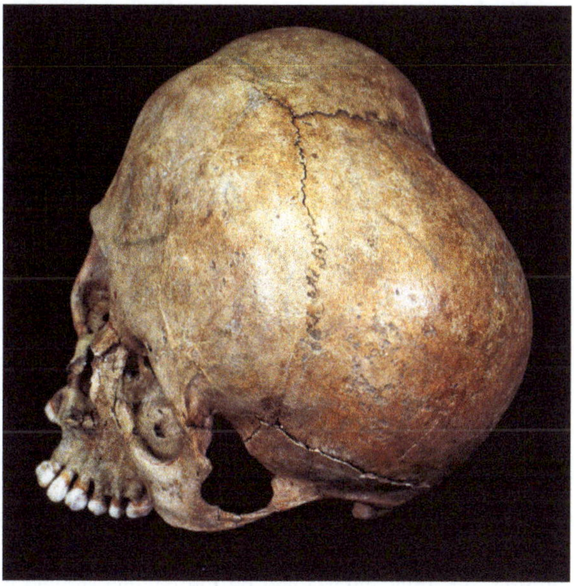

Fig. 10.3 Strong sagittal sulcus dividing parietals at the Argelia site (A-60) in the Valle de Angostura, Chiapas, Mexico. (DAF/INAH; photo by V. Tiesler)

compression crib restraints. However, in six of the 86 Postclassic grooves, this depression dramatically separates the calotte into two bipolar lobes and therefore appears to have held some form of emblematic role for practitioners and bearers (Fig. 10.3).

A trend towards homogenization is also felt among Gulf Coast cultures of Veracruz during the Postclassic period. However, here, this trend is not as clear as among Maya populations and, in fact, appears to follow different paths among Totonac and Huaxtec inhabitants, and still others, further south. Note that more than Totonac sculpture, monumental Huastec sculpture shows a marked penchant for reclined foreheads and straight occiput, as displayed by the sculpture known as the "adolescent of Tamuín (Montiel 2013)." Arturo Romano (1980) once identified this form with the conical cap of Quetzalcoatl that he believes was imitated by the tabular erect cephalic model. Our present survey of crania includes Totonacan individuals from Isla de Sacrificios and are dated to the Early Postclassic period (Romano 1965; Tiesler 2009). All but one of the crania shows tabular erect modeling both with and without superior flattening, following the trend described for coeval coastal Maya. West of this ceremonial island and towards the central Highlands lies the area of Maltrata (Mendoza and Lira 2005, pp. 258–259), which fell under Mexica power during the Late Postclassic. Here, only unconspicuous tabular erect skulls are documented, most of them slight lambdic flattenings.

North of Isla de Sacrificio, the coastline introduces to the cultural territories of the Huastec. Among its ancient occupations are the sites of Las Flores, Tamtok, Vista Hermosa, Pánuco, Isla del Ídolo, and Tamuín, all dated to the Postclassic period (Montiel 2013; Romano 1965; Tiesler 2009). These settlements delineated the vast northern fringes of Mesoamerica. Again, most, but not all, of their excavated crania display the effects of infant cradleboarding. The noticeable exception to this head look is posed by a number of skulls from Isla del Ídolo (Romano 1965, p. 11) and Vista Hermosa, which include visibly elongated specimens in intermediate and mimetic varieties (Montiel 2013). These forms are not attributable to any whole-body compressors but instead relate to ancestral head splinting traditions, by that time long forgotten in other parts of Mesoamerica. The fact that both series have been securely dated to the Late Posclassic era, make these exceptions to the general Mesoamerican panorama seem still more extraordinary, an aspect to be revisited in Sect. 10.1.3.

What were the preferences, what the shifts in head looks on the borderlands of Postclassic Mesoamerica in general? These are important, region-breeching questions but again, there are at present no regional coverages of shaping practices to provide meaningful "food for thought" on the interethnic socio-cultural complexities typically involved. For now, I will only touch briefly upon the much debated discussion of Mesoamerican influence and possibly direct occupation on the Pacific coastlines beyond El Salvador; namely, Nicaragua and the northwestern tip of Costa Rica, identified in the literature with the Gran Nicoya (see, for example, Braswell et al. 2002; Carmack and Salgado 2006; McCafferty and McCafferty 2011; Solís and Herrera 2011). Among other questions, scholarship has pondered here on the cultural dynamics entertained by the Nicoya settlers during the Sapoá period and their potential Mesoamerican origin (800–1350 A.D.).

Useful points of departure to further this debate from the perspective of artificially conferred head forms come from the Culebra Bay in northwest Costa Rica (Aguilar 2012; Solís and Herrera 2011). Here, recently exposed skeletal populations from the site of Jícaro display conspicuous body modifications in the form of dental filings and artificially modeled crania (Solís and Herrera 2011, pp. 13–14). These forms of physical embodiment starkly differ from other human remains of the Culebra Bay, which do not display any signs of such dental or cranial adaptation (see also Aguilar 2012). The authors note specifically the contrast in head form between ancestral contexts and sacrificial deposits (some with marks of defleshing), the latter presumed to stem from local folk with naturally rounded heads. Conversely, the skulls of the burial population presumed to accrue from Jícaro's migrant visitors express visible head reclination and elongation, some of them in a severe (pseudo-circular tabular oblique) form. Note that the literature identifies this and surrounding coeval settlements with chorotega occupations, both archaeologically and ethnohistorically. Historical accounts hold that the chorotega folk of the Gran Nicoya originated from southern Mexico, possibly the Soconusco area. After resettling, these were forced to pay tribute to the Aztecs centuries (Carmack and Salgado 2006; Solís and Herrera 2011; see also Braswell et al. 2002; McCafferty and McCafferty 2011).

So, who were these apparent newcomers with their ostentatious, yet anachronously elongated head shapes? Which cultural affinity, which geographic frame describes their original homelands best? At least the biocultural landscape of head modeling, which we just surveyed for the eastern Mesoamerican territories, would make Lowland Maya or Isthmic populations feasible canditades for further inquiries into this matter. It is clear that also the Pacific side of present-day Mexico, Guatemala, Honduras, and El Salvador hosted Maya settlements during the Classic era (Cobos and Sheets 1997). Such is the case of San Andrés, in the west of El

Salvador and in vicinity of Joya de Cerén, a settlement haunted by repeated volcanic eruptions during the first millennium A.D. (Cobos and Sheets 1997, pp. 38–41). Between 600 and 900 A.D., San Andrés functioned as the capital of a Maya polity with supremacy over the other establishments of Valle de Zapotitán, before losing importance during the tenth century A.D. From its central quarters, a male adult burial with strong fronto-occipital flattening and backward inclination of the forehead, was recovered, dated to the Late Classic (Cobos and Sheets 1997, p. 26). This shape is similar to the Lowland practices on the other side of the mountain ridges and to the shapes documented further east, at Nicoya's Jícaro site. Conversely, the Postclassic Period Mesoamerican Highland areas, not to speak of any Nahua folks, appear to me less likely venues for the origins of the resettled chorrotega folk.

10.1.2 Mexican Highlands

As in most of the areas already discussed earlier, Mesoamerican Highlanders either retained their cradling devices in their cultural repertoire or, among those few groups used to head splinting, gradually shifted to cradleboarding. This gradual change in body practices naturally led to a loss in distinctiveness in this tradition, prompting uniformity in short and broad head looks, one that is communicated not only by the cranial record but also by Postclassic period portraiture and, in fact, by the Hispanic chroniclers who still witnessed head modeling during the sixteenth century (Bautista 2004; Tiesler 2012a; Tiesler and Zabala 2011). Note that a number of scholars also refer to a drop in the frequency and degree of morphological alteration among the Late Postclassic Mexica (Dávalos 1965; Pereira 1999). I also wonder if the homologation of Postclassic era artificial morphology is perhaps the reason behind the lack of curiosity and therefore the dearth of scholarship on head flattening for this period.

A large percentage of artificially modified skulls recovered from Postclassic archaeological sites displays tabular erect models. The hundreds of skulls curated at the National Museum of Anthropology (Romano 1974, p. 206), show almost exclusively this form during this period. More site specific, but vague, is the information conveyed by Gómez et al. (1989) on Early Postclassic Tula, all of whose artificial modifications are characterized as tabular erect ($N = 11$). Lagunas (1989) reports analogous trends for Postclassic Cholula and Dávalos (1951) on Aztec Tlatelolco. Of the latter series, some 52 of 127 skulls were noted to possess cultural flattening. An extended study of Aztec period head-shaping practices ($N = 98$) by Dávalos and Romano (1965, pp. 76–79) documents 94 erect shapes, but also four bearing head reclination. The latter are from the sites of Zumpango, Santa Lucía Azcapozalco, and sites of unknown provenance from the state of Hidalgo. A single specimen from Late Postclassic Monte Albán was scored as tabular erect by Winter (1996) and others.

Similarly, short and broad are the skulls documented in the Western part of Mexico, which uniformly exhibit tabular erect shapes. Dávalos (1951; 1965, pp. 15–19) refers to the skull series of Tarascan Michoacán with this form. Macías (1989) identifies it, too together with sagittal constriction, in her description of a Tarascan population from a Postclassic ceremonial center at Huandacareo, Michocán. Also Pereira (1999) confirms a solid preference for tabular erect shapes for the Postclassic series under survey at El Palacio and Las Milpillas, Michoacán (see also Gervais 1989, p. 165). Pereira also adds important information on the practice of obelionic flattening in Western Mexico, which he argues makes its appearance in defined Late Classic contexts and persists in the cranial record also during the centuries of the Postclassic period.

10.1.3 Mesoamerican Head Practices at the Time of European Contact

Although uniform, native infant head practices were still widespread at the time of European contact during the late fifteenth and early sixteenth century. As said, five centuries into the second millennium A.D., artificial cranial modification had evolved into a rather uniform tradition (Romano 1974; Tiesler and Zabala 2011). The written testimonies of the people who still witnessed the performance of indigenous cradleboarding during the early sixteenth century add valuable information to the cranial record for this era (see also Chap. 5). In the novohispanic territories, in particular, colonists took an interest in the native cultural repertoires during early colonial times (Tiesler and Zabala 2011; see Chap. 5). These describe cranial modification consistently as broad and short (tabular erect) and refer exclusively to cradleboards and cribs as shaping devices (Dávalos 1951; Romano 1974; Tiesler 2011, 2012a). The sacrificial deposits of the island of Tenochtitlán, imposing urban capital of the late Aztec empire, are filled with tabular erect skull vaults (Berrelleza 1990). Other Late Postclassic sacrificial deposits, like Teopanzolco in the present state of Morelos, display broad, tabular erect crania with visible lambdic planes, together with round, unshaped specimens, which should come from the fringes of Mesoamerica according to the authors (González-Sobrino et al. 2001, pp. 526–527).

Possibly an exception to this panorama is the tradition of head compression among Postclassic era Gulf Coast cultures, as discussed by Yépez (2001, p. 59). She transcribes the native terminology among natives from Veracruz, who used either attributions such as "broad and flattened" (*patlachtic*) or alternatively, "column-like, roll-like thing" to designate the head form in Totonac languages. Note that the second term implies head elongation rather than shortening (Yépez 2001, p. 59). This last distinction is crucial for our review, as the Late Postclassic Huastec appear to have been one of the last strongholds of head splinting if we believe the chronological assignment (see also Sect. 10.1.1 in this Chapter).

Regarding the visible effects of compression, the reduction of the back of the head is stressed by a number of colonial writers who ascribe to it native attributions of bravery, courage, and high military rank (Anonymous 1977/1977, p. 145; Casas 1967 (sixteenth century), p. 177; Castillo 2001 (late sixteenth century), p. 99). Note that the compression effect on the forehead is described recurrently in terms not of flattening but of reduction (Hernández 2001/2001, p. 111; López de Gómara

1987/1987, p. 451). Uniformity of head form during the contact period is most conspicuous in those areas that utilized diversity in head wear before the Postclassical period, such as the Western Highlands or the Lowland Maya region (Pereira 1999; Tiesler 2012). As we have already pointed out earlier, a solid 90 % of late Postclassic Maya skulls show a tabular erect form. Now, diversity is evident only in the varieties of cradleboards and the daily performance of head shortening, and find their expressions in the different erect varieties in the cranial record. Now, the so-called tabular erect head varieties range from slight lamboid flattening to conical shapes, some few top-flattened silhouettes to extreme forms of antero-posterior reduction (Fig. 10.4; Table 10.1). An additional distinction is implied also simply by the absence of modeling, which is displayed, for instance, by some 10 % of the cranial vaults from the Maya area with none or only subtle traces of extrinsic modification.

Given the homogeneity in artificial head shaping in most of Mesoamerica's territories, we may assume that the cultural significance of Mesoamerican head compression was not as tightly related to its visible component during the second millennium A.D. as it had been in the first millennium in Mesoamerica, or as it still was in Inca Perú at that time (Lozada and Buikstra 2002; Tiesler and Zabala 2011). At least among Postclassic era Mesoamericans, it seems that head compression devices, more than a shaping instrument per se, functioned as protective and preventive measures implemented during the initial liminal stages of child upbringing (Duncan and Hofling 2011; Tiesler 2011). As we have argued in Chap. 6, Mesoamericans felt the concrete need to reduce the expansion of the little one's protruding back of the head, as head elongation was thought to interfere negatively with the harmonious functioning of a person's living essences and, therefore, was deemed harmful to health and even to life; this notion is also conveyed by the chroniclers who repeatedly associate occipital reduction to bravery and gallantry.

Apart from these fluid phenomenological and behavioral associations of cradling procedures during the early sixteenth century, flattened heads surely were imbued with more categorical notions of culturally attributed beauty and agreeability and, therefore, of Mesoamerican identity, as already argued in Chap. 6. In the centuries before Iberian contact, the broad headed "unilook" must have vaguely denoted a pan-Mesoamerican feeling of identity and personhood. Grégory Pereira (1999, pp. 167–168) brings home this point for the northwestern borderlands toward the Great Chichimeca by dissecting the Relación de Michoacán, written during the sixteenth century (Anonymous 1977/1977, p. 145). In it, the indigenus protagonists communicate the negative preconceptions that (Mesoamerican) Purepecha folk still harbored on the rounded (unshaped) head form of their incoming neighbors from the north, which they considered did not provide their human carriers with the credentials of courage and manhood.

Fig. 10.4 Colonial rendering of Tlazolteotl priest. The partially shaved head is shown with a reclined forehead and a flattened occiput. (Adapted from Trejo 2007, p. 21; drawing by B. Ceballos)

10.2 Head Shaping Practices During the Colonies

Beyond doubt, the European contact and conquest led to the most dramatic shifts that the long-standing Mesoamerican traditions had ever experienced. Now also, head flattening was started to be abandoned among the natives of the sixteenth century or was gradually substituted by alternative head practices (Tiesler and Oliva 2010; Tiesler and Zabala 2011). At the same time, it is intriguing to learn of the twists that shaping practices apparently took in postcontact Mesoamerica, a welcome point of departure to speculate on the underlying dynamics of native assimilation vs. resistance and cultural resilience in the postcontact era. What were the mechanisms of oppression that effectively triggered the abandonment of head practices? What kind of transformation did cradleboard uses undergo during the "Hispanization" process in the newly founded towns of the region? Did they follow the same transformation in the rural communities? For this study, I combine historical and craniological sets of data (see also Chap. 5, by Pilar Zabala; Tiesler and Oliva 2010; Tiesler and Zabala 2011).

10.2.1 European Assimilation Efforts

In the years that followed the discovery and conquest of the New Continent, it is clear that the Spanish Crown remained acutely alert to native beliefs and ways of life in Hispanic America. This awareness was not quite genuine or neutral but was instrumental instead to govern and control the dominated sectors of Hispanic society. In their efforts, the Iberian conquerors also recurred to religious arguments and specifically Catholic attitudes towards the human body (Tiesler and Zabala

2011, 2013). Most of the early colonial testimonies are clearly permeated by derogatory ethnocentric rhetoric when addressing autochthonous cultural heritage (see also Chap. 5). It is unsurprising in this vein that the Spanish Empire soon openly opposed and actively suppressed infant head compression in New Spain as well as in Perú, just like most other identity forging indigenous customs. The Lima and Quito Synodes, which date to the turn of the seventeenth century, forbade native Peruvian head modifications under punishment by law (A.G.I., Patronato 189, R. 40; Toledo 1929). The eradication efforts by the Spanish Crown were a convenient argument to justify the colonization measures that were part of a strategy of forced assimilation of all non-European sectors, a despotic claim to streamline culture conveniently and transform the social fabric to the needs of the Spanish crown.

Different from Peruvian chronicles, those of New Spain focus on the Mesoamerican cradleboarding traditions not so much by their visible result in the head—which is described bluntly as short and broad, almost as a given physical attribute (Sahagún 1989)—but target instead the methods used to produce the effect and the performance itself, which many of the Iberian writers, either expressly or implicitly, regard as immoral and corrupt (Tiesler and Zabala 2013). The "arbitrary" modification of the natural head form, which "has been created by God to mirror his own image" (Cieza de León 1984, p. 227), appears in the Spaniards' eyes as an aberration, to faith. Beyond sacrilege, the Iberians were quick to name also other reasons for suppressing cradleboarding in New Spain. They highlight superstition, health hazards, and suffering. This is the sense, Friar Diego de Landa (in Tozzer 1941, p. 125) conveys to the tradition among the Maya, of Yocatan, where. "the poor children's inconvenience and hazard was so great, that some were at risk [of losing their lives] (. . .) and when they had finished with the torture of flattening forehead and [back of the head], they took them to the priest to know the future and the craft [of their child] and to give it the name that it would have for the time of its childhood."

10.2.2 Resilience and Transformation

Despite the assimilation pressures exerted by the Spanish Crown, natives probably did not leave behind their native cultural heritage easily, including embedded traditions and old ways in general. These aspects relate to questions concerning the ways of hiding cradleboards, of accepting transformations in the body treatment of babies and their reconciliation with other elements of indigenous ideology. What persuaded natives to abandon or continue the practice of head flattening? What kind of circumstances and what types of suppression did effectively trigger the abandonment of this ancestral body tradition? Did customs undergo similar transformations in the urban settlements of Spaniards—and as swiftly—as in the native backwaters of colonization?

Instead of replying to these and other inquiries ad hoc and in a categorical manner, it is probably wise to approach the subject of native cultural adjustments by recalling some key features of native head practices at this point. Keep in mind that female caretakers, and predominantly the mothers, were in charge of the daily handling of their little ones' heads (Tiesler 2011; Tiesler and Zabala 2013). From this perspective, it is clear that the routine of daily head compression was subscribed to the female domain. A secluded domestic environment that separated societies by gender, such as the Andean or Mesoamerican nations, could easily conceal the practice from the eyes of the male-dominated spheres of European colonizers. For colonial Perú, Latcham proclaims for instance that "the Indians were very secretive in all these things [ritual ceremonies] and hid them from the Spaniards; not admitting anyone [to assist them] except for the family members of their blood line" (Latcham 1929, p. 544; Tiesler and Zabala 2013).

Paradoxically, there is also a strong public element involved in head compression, which relates to the visible head formation that the carrier would exhibit for the rest of his or her life, given its permanent nature of cultural head modifications. The colonial onlooker would surely grasp such features by perceiving the abundance of people with noticeably different shaped heads in a given location or region. In case there was, they would also recognize the diversity of the natives' head forms, as we may argue was the case in Inca Perú. This aspect institutes the second facet that is the key in evaluating the colonial impact on native head compression. This resides in the question of whether the visible morphological results of head flattening were important or not for the native practitioners. Put in context with Mesoamerican infant head practices, the lack of diversity and visibility, in general, may explain why the historic record soon falls silent on the custom and fails mention it in the colonial records after 1570.

A dozen postcontact skeletal collections from the Maya and Isthmus territories supplement the missing written information (Tiesler 2012a; Tiesler and Zabala 2011; Table 10.2). Although scarce and isolated, the cranial data at least does appear to echo the obliteration in the Iberian chronicles. The proportion of modeled skulls drops from over 90 % during the late Postclassic period (93 % of $N = 127$) to less than a third (31 % of $N = 77$) among native skeletons dated to colonial times. This decline in frequency is highly relevant statistically when both eras are compared in a pair-wise chi-square analysis (with $p = 0.000$). When present at all, colonial-period flattening appears less visible than before the contact and less diverse with only moderate expressions with or without sagittal grooves. Translated into the enactment of this body tradition, the combined evidence hints at the continued but declining use of cradleboards both in rural and urban settings during the sixteenth century.

The expression of head modeling among individuals buried at an early colonial churchyard in central Campeche, on the Yucatán Peninsula, is especially telling. This Hispanic cemetery was excavated in the year 2000 after revealing the foundations of Campeche's primitive church below its historical central plaza (Tiesler et al. 2010). The make-up of the multiethnic skeletal population includes natives, mestizos, and European and African newcomers (Price at al. 2012). In those individuals who were identified by dental morphology as natives or mestizos, only 3 out of 11 (27.3 %) still exhibited visible front-and-back flattening in their skull vault. If we take into consideration the direct interaction between natives and other sectors within the dense living quarters of this Hispanic sea town, we may assume the sense of rejection which the visible native head insignias must have caused in public. Here, the relevance and cultural identity that was once expressed by cranial flattening would have been destined to undergo a radical mental transformation in the oppressed carriers' eyes to denote exclusion and "otherness" in the Hispanic social fabric.

Table 10.2 Frequencies and types of cultural cranial modifications in colonial series in and around the Maya area

Archeological site, location, and time frame	Type of settlement	Number of artificially modified crania (*N* = number of examined individuals)	Type of modification
Central Plaza of Campeche, Campeche, Mexico (sixteenth and seventeenth century)	Urban	7 (45)	Tabular erect
Atrium of the San Francisco Monastery, Campeche, Campeche, Mexico (sixteenth and seventeenth century)	Urban	4 (6)	Tabular erect
Atrium of the Cathedral of Mérida, Yucatán, Mexico (sixteenth to eighteenth century)	Urban	0 (6)	No modification
Skull of caste war leader Bernadino Cen, Xuxub, Quintana Roo, Mexico (nineteenth century)	Rural	0 (1)	No modification
Church of Maxcanú, Yucatán, Mexico, (nineteenth century)	Rural	0 (1)	No modification
Sihó, Yucatán, Mexico (sixteenth to eighteenth century)	Rural	1 (1)	Tabular erect
Osumacinta, Chiapas, Mexico (sixteenth to eighteenth century)	Rural	0 (3)	No modification
Rock-shelter sanctuaries, Mensabak, Chiapas (probably sixteenth to eighteenth century?)	Rural	21 (21)	Tabular erect

This connotation introduces a third element involved in native head practices, one that certainly played a strong role in the process of acculturation: the compulsory notion of exodus and cultural failure that surely dominated the minds of many Native Americans during the first two centuries of colonization. Forced migration, aculturation, and a progression of deathly diseases doomed these people to catastrophic demographic decline and cultural breakdown. This reduced self-esteem surely worked in favor of the Spanish assimilation efforts, at least in the urban strongholds of the newly established Hispanic society (see Chuchiak 2006). The reduced proportion of modeled skulls during colonization demonstrates that the popularity of this head flattening was also in decline in Yucatán (Tiesler 2012a). A cemetery sample that is similar to that of Campeche, but dated more recently, was documented in the colonial atrium of the Cathedral of Mérida, excavated by an Instituto Nacional de Antropología e Historia (INAH) team (Tiesler et al. 2003). The approximately 20 burials recovered from the atrium did not show any signs of either of the two practices in spite of the indigenous ancestry that the majority of the skeletons, as represented probably represent by way of dental morphology.

Despite our inability to generalize on the results owing to the lack of more precise chronological information and sufficient sample size the urban populations stand out against the ratios from the native backwaters of colonization. There, studies of colonial Maya populations denote the continued presence of cranial modification (Havill et al. 1997; Saul 1980; Tiesler and Zabala 2011). It will be certainly necessary to expand the framework of these studies in order to assess more specific cultural and social trends, in particular between rural and urbanized Maya populations and between Mesoamerican and Andean populations in general. Naturally, this shift from public to private enactment is not only noticeable in head practices but also includes all visibly performed and expressed corporeal practices, such as religiously motivated human sacrifice before and after Iberian contact (Duncan 2005; Tiesler and Cucina 2010).

10.2.3 Last Strongholds of Native Head Practices in and Around Mesoamerica

Up to this point, we have discussed the transformations that native head practices underwent during early colonization and have confronted the indigenous roles of this tradition with the assimilation interests of the Iberian colonizers. But what do we know about the practitioners in the vast jungle stretches that remained isolated and out of reach of Hispanic opression and control well into and beyond the colonies? As we have learned from the cranial record of the rural backwaters of New Spain, here the abandonment of the practice should have occurred later than in the Iberian controlled urban areas (Tiesler 2012). Although in colonial townships, such as Mérida or Campeche, head shaping should have been eliminated within two or three generations, at the most, and then vanished gradually in their surrounding native communities, the impenetrable jungle of Southeastern Mexico continued to provide strongholds for the practice. The Lake Petén Itzá persisted, for example, as an independent native state until it was conquered in 1697 (Jones 1998). Here lies the Itzá site of Zacpetén, which was excavated as part of the *Proyecto Maya Colonial* during the 1990s. Skeletal screening documented the practices of tabular erect shaping in a Late Postclassic bone assemblage from this site (Op.1000; see Duncan 2005, p. 144; Duncan 2009, p. 350), and we presume that head flattening should have continued here well past the Iberian contact in the sixteenth century, especially given the independence of the Itzá sphere.

Also the Lacandon maintained a nearly independent lifestyle for centuries past the European contact. Back then, Lacandon folk still dwelled in isolated nomadic hamlets deep in the impenetrable forested areas of the Usumacinta Valley and further east within the Petén. Here, the continued practice of cradleboarding is confirmed until the turn of the twentieth century by Alfred C. Maudslay, who calls the attention to the receding foreheads among elderly locals. Maudslay specifies that "the extremely sloping forehead [of elder Lacandon males] was not quite so noticeable in the younger men, and it may be that the custom of binding back the forehead in infancy, which undoubtedly [was] obtained amongst the ancients, is being now abandoned" (Maudslay and Maudslay 1899, p. 236; see also Palka 2005). Still further west, also native Mixe from Coatlán in the Mexican state of Oaxaca are portrayed around that time with a similarly inclined forehead and a flattened occiput (Shattuck 1933, p. 29; Trias de Bes et al. 1928, p. 84).

Our recent ethnoarchaeological study of sacred rock-shelter shrines around Mensabak in the northern Lacandon area in Chiapas adds to this panorama. These ancient rock shelter sanctuaries were mapped between 2010 and 2013 as part of the ongoing Mensabak Archaeological Project (Palka 2005). The shrines accrued also human skeletal remains among which we documented some 20 crania, all of them with signs of cradleboard use. Given the relatively recent nature of the bones from the Lacandon cave sanctuaries, at least part of which is dated to after European contact, it is telling that all skulls still appear artificially shaped from cradle board use.

In closing, the colonial and postcolonial transformations in head flattening, either in the form of continuing and hiding, substituting, transforming, or abandoning it altogether, did not progress uniformly within the complex cultural tapestry of New Spain. As only one dominant head look persisted here at the time of contact, the Spaniards disapproved of not so much the distinct effect (as they did further south in Hispanic Perú), but targeted cradleboard use (see also Chap. 5 in this volume). Unsurprisingly, given the relative lack of emblematic meanings of Postclassic indigenous head modeling, the transformations that infant head practices suffered here after the Iberian conquest do not relate to their visible morphological outcome, which they probably did not hold in the centuries before either. Instead, we hold that the postcontact transformations in head compression should have targeted to supplant the protective role that cradleboard practices had always held. Now, swaddling rituals, the fire of the hearth, hot baths, cinder, wraps, covers, and massages of the newborn's body were deemed appropriate substitutes to retain the little one's heat, to stabilize its spiritual health, and guarantee its transformation into the person to be. Needless to say, many of these measures are still in place and part of today's rich Mesoamerican indigenous heritage (Duncan 2009, pp. 187–188; Guiteras 1986, p. 102; León-Pasquel 2005, pp. 128–136; Tiesler 2011; see also Chap. 5).

References

AGI Patronato 189, R. 40

Aguilar Ana Cristina. 2012. Género en los contextos funerarios de una aldea de los períodos Sapoá y Ometepe (800-1550 D.C.) en la Bahía de Culebra. Masters diss., Universidad de Costa Rica.

Anonymous. 1977 [1541]. Relación de Michoacán (trans: J. Tudela). Morelia: Balsas Editores.

Bautista, Josefina. 2004. Evidencias de deformación cefálica intencional en figurillas prehispánicas de México. Ph.D. diss. Universidad Nacional Autónoma de México.

Berrelleza, Juan Alberto. 1990. El sacrificio de niños en el Templo Mayor. Mexico: Instituto Nacional de Antropología e Historia.

Braswell, Geoffrey E., Silvia Salgado, Laraine A. Fletcher, and Michael D. Glascock. 2002. La antigua Nicaragua, la periferia sudeste de Mesoamérica y la región maya: interacción interregional (1-1522 d.C.). Mayab 15:19–39.

Casas, Bartolomé de las. 1967. Apologética historia sumaria, ed. Edmundo O'Gorman, 2 vols. México: Universidad Autónoma de México.

Castillo, Cristóbal del. 2001. Historia de la venida de los mexicanos y de otros pueblos e historia de la Conquista. Mexico: Consejo Nacional para la Cultura y las Artes.
Chuchiak IV, John F. 2006. Yaab uih yetal maya cimil: Colonial plagues, famines, catastrophes and their impact on changing Yucatec maya concepts of death & dying, 1570–1794. In Jaws of the underworld: Life, death, and rebirth among the ancient Maya, ed. Pierre Colas, Geneviéve LeFort and Bodil Liljefors Tersson. Acta Mesoamericana 16:3–18. Markt Schwaben: Anton Saurwein.
Cieza de León, Pedro. 1984. La crónica del Perú, ed. Manuel Ballesteros. Madrid: Historia 16.
Carmack, Robert M., and y Silvia Salgado. 2006. A World-Systems Perspective on the Archaeology and Ethnohistory of the Mesoamerican/Lower Central American Border. Ancient Mesoamerica 17:219–229.
Cobos, Rafael, and Payson Sheets. 1997. San Andrés y Joya de Cerén; Patrimonio de la Humanidad. San Salvador: Bancasa.
Dávalos, Eusebio. 1951. La deformación craneana entre los tlatelolca. B.A. and Master diss., Escuela Nacional de Antropología e Historia, Instituto Nacional de Antropología e Historia.
Dávalos, Eusebio 1965. Temas de antropología física. Mexico: Instituto Nacional de Antropología e Historia.
Dávalos, Eusebio, and Arturo Romano. 1965. Las deformaciones corporales entre los mexicas. In Temas de antropología física ed. Eusebio Dávalos Hurtado, 75–102. México: Instituto Nacional de Antropología e Historia.
Demarest, Arthur, Prudence M. Rice, and Don S. Rice (ed.). 2004. The Terminal Classic in the Maya Lowlands. Collapse, transition, and transformation. Boulder: University of Colorado.
Duncan, William N. 2005. The bioarchaeology of ritual violence in Postclassic El Petén, Guatemala (AD 950-1524). Ph.D. diss., Carbondale: Southern Illinois University.
Duncan, William N. 2009. Cranial modification among the Maya: Absence of evidence or evidence of absence?. In Bioarchaeology and identity in the Americas, ed. Kelly Knudson and Christopher M. Stojanowski, 177–193. Gainesville: University of Florida.
Duncan, William N., and Charles A. Hofling. 2011. Why the head? Cranial modification as protection and ensoulment among the Maya. Ancient Mesoamerica 22:199–210.
Gervais, Veronique. 1989. Déformations artificielles des crânes préhispaniques au Guatemala et au Mexique. Ph.D. diss., Caen University.
Gómez, Rosa María.1989. Los huaxtecos y la mutilación dentaria. In Estudios de Antropología Biológica, 4:599–607.
González-Sobrino, Blanca Z., Carlos Serrano, Zaid Lagunas, and Alejandro Terrazas. 2001. Rito y sacrificio humano en Teopanzolco, Morelos. Evidencias osteológicas y fuentes escritas. Estudios de Antropología Biológica 10:519–532.
Guiteras, Calixta. 1986. Los peligros del alma: Visión del mundo de un tzotzil. Mexico: Fondo de Cultura Económica.
Havill, Lorena M., Diane M. Warren, Keith P. Jacobi, Karen D. Gettelman, Della Collins Cook, and K. Anne Pyburn. 1997. Late Postclassic tooth filing at Chau Hiix and Tipu, Belize. In Bones of the Maya: Studies of ancient skeletons, ed. Stephen L. Whittington and David M. Reed, 89–104. Washington: Smithsonian Institution.
Hernández, Francisco. 2001 [1576]. Antiguedades de la Nueva España. Madrid: Dastin.
Jones, Grant D. 1998. The conquest of the last Maya kingdom. Stanford: Stanford University.
Lagunas, Zaid. 1989. Los antiguos habitantes de Cholula: Prácticas osteoculturales. Notas Americanas 11:8–50.
Latcham, Ricardo. 1929. Creencias religiosas de los antiguos peruanos. Santiago de Chile: Bacells.
León-Pasquel, María de Lourdes de. 2005. La llegada del alma: Lenguaje, infancia y socialización entre los mayas de Zinacantán. Mexico: Instituto Nacional de Antropología e Historia.
López-Austin, Alfredo. 1989. Cuerpo humano e ideología: Las concepciones de los antiguos nahuas. Mexico: Instituto de Investigaciones Antropológicas, Universidad Nacional Autónoma de México.
López de Gómara, Francisco. 1987 [1552]. La conquista de México. Madrid: Historia 16.

Lozada, María C., and Jane E. Buikstra. 2002. El Señorío de Chiribaya en la costa sur del Perú. Lima: Instituto de Estudios Peruanos.
Macías, Angelina. 1989. Los entierros de un centro ceremonial tarasco. Estudios de Antropología Biológica 4:531–559.
Maudslay, Anne C., and Alfred Percival Maudslay. 1899. A glimpse at Guatemala, and some notes on the ancient monuments of Central America. London: John Murray.
McCafferty, Geoffrey y, Sharisse McCafferty. 2011. Bling Things: Ornamentation and identity in Pacific Nicaragua. In Identity crisis: Archaeological perspectives on social identity, ed. Lindsey Amond Sen-Meyer, Nicole Angel, and Sean Pickotag, 243–252. Calgary: Chacmool Archaeological Association, University of Calgary.
Mendoza, Eira, and Yamile Lira. 2005. La población posclásica del valle de Maltrata, Veracruz. Evidencias osteológicas. Estudios de Antropología Biológica 12:751–766.
Montiel, Mireya. 2013 Modificaciones corporales en la huaxteca prehispánica. Master diss., Universidad Nacional Autónoma de México.
Palka, Joel W. 2005. Unconquered Lacandon Maya. Ethnohistory and archaeology of indigenous culture change. Gainesville: University Press of Florida.
Pereira, Grégory. 1999. Potrero de Guadalupe: anthropologie funéraire d'une communauté pré-tarasque du nord du Michoacán, Mexique. Oxford: British Archaeological Research International Series 816.
Price, T. Douglas, James H. Burton, Andrea Cucina, Pilar Zabala, Robert Frei, Robert H. Tykot, and Vera Tiesler. 2012. Isotopic studies of human skeletal remains from a sixteenth to seventeenth century AD churchyard in Campeche, Mexico: Diet, place of origin, and age. Current Anthropology 53:396–433.
Ringle, William, Tomás Gallareta Negron, and George J. Bey III. 1998. The return of Quetzalcoatl: Evidence of the spread of a world religion during the Epiclassic period. Ancient Mesoamerica 9:183–232.
Romano, Arturo. 1965. Estudio morfológico de la deformación craneana en Tamuín, S.L.P., y en la Isla del Ídolo, Veracruz. Serie de Investigaciones 10. México: Instituto Nacional de Antropología e Historia.
Romano, Arturo. 1974. Deformación cefálica intencional. In Antropología física, época prehispánica ed. Juan Comas, 197–227. México: Instituto Nacional de Antropología e Historia.
Romano, Arturo. 1980. Appendix 3, The skull from El Pajón, Chiapas. In Pampa El Pajón: An Early estuarine site, Chiapas, Mexico, ed. María de la Cruz Paillés, New World Archaeological Foundation, 95–114. Provo: Brigham Young University.
Saul, Frank. 1980. The human skeletal remains from Tankah, México. In On the edge of the sea, Appendix II, Washington D.C: Universidad de Harvard.
Sahagún, Bernadino de. 1989. Historia general de las cosas de Nueva España. Mexico: Porrúa.
Serafin, Stanley. 2010. Bioarchaeological Investigation of Violence at Mayapan. Ph.D. Diss., Tulane University.
Sharer, Robert J., and Loa P. Traxler. 2006. The ancient Maya. Stanford: Stanford University.
Shattuck, G.C. 1933. The Peninsula of Yucatan, Medical, biological, metereological and sociological studies. Washington, D.C.: Carnegie Institution of Washington, no. 431.
Solís, Felipe, and Anayensy Herrera. 2011. Mesoamericanos en la Bahía de Culebra, noroeste de Costa Rica. Cuadernos de Antropología 21:1–31.
Stewart, Thomas D. 1953. Skeletal remains from Zaculeu, Guatemala. In The ruins of Zaculeu, Guatemala, Vol. 1, ed. Richard B. Woodbury and Aubrey S. Trik. Richmond, 295–311. Richmond: William Byrd.
Tiesler Vera. 2009. Informe del estudio esquelético, realizado en las colecciones del Estado de Veracruz, México. Unpublished report on file at the DAF/INAH, Mexico.
Tiesler, Vera. 2011. Becoming Maya: Infancy and upbringing through the lens of pre-Hispanic head shaping. Childhood in the Past 4:117–32.

Tiesler, Vera. 2012a. Transformarse en maya. El modelado cefálico entre los mayas prehispánicos y coloniales. México: Universidad Nacional Autónoma de México, Universidad Autónoma de Yucatán.

Tiesler, Vera. 2012b. Studying cranial vault modifications in ancient Mesoamerica. Journal of Anthropological Sciences 90:1–26.

Tiesler, Vera, and Andrea Cucina. 2010. Sacrificio, tratamiento y ofrenda del cuerpo humano entre los mayas peninsulares. In El sacrificio humano en la tradición religiosa mesoamericana ed. Leonardo López-Luján and Guilhem Oliver, 195–226. Mexico: Instituto Nacional de Antropología e Historia, Universidad Nacional Autónoma de México.

Tiesler, Vera, y Andrea Cucina. 2012. Etnicidad, relaciones interpoblacionales y patrones migratorios a lo largo de la costa de Campeche durante los períodos Clásico y Posclásico. In Arqueología de la costa de Campeche: la época prehispánica, ed. Rafael Cobos, pp. 67–95. Mérida: Universidad Autónoma de Yucatán.

Tiesler, Vera, and Iván Oliva 2010. Identity, alienation, and integration: Body modifications in the early colonial population from Campeche. In Natives, Europeans and Africans in colonial Campeche: History and archaeology, ed. Vera Tiesler, Pilar Zabala, and Andrea Cucina, pp. 130–151. Gainesville: University of Florida.

Tiesler, Vera, and Pilar Zabala. 2011. El modelado artificial de la cabeza durante la Colonia. Una tradición maya en el espejo de las fuentes históricas. Estudios de Cultura Maya 38:75–96.

Tiesler, Vera, and Pilar Zabala. 2013. Survival and abandonment of indigenous head shaping practices in Iberian America after the European contact. Submitted as a chapter contribution to the volume: Bioarchaeology of contact, colonial encounters and colonialism, ed. Melissa S. Murphy and Haagen D. Klaus. Gainesville: University Press of Florida.

Tiesler, Vera, Pilar Zabala, and Agustín Peña. 2003. Vida y muerte en Mérida durante los siglos XVI y XVII. Rescate de las osamentas del atrio de la Catedral. In Mérida. Miradas múltiples. Investigaciones de antropología social, arqueología e historia, ed. Francisco Fernández and José Fuentes, pp. 41–55. Mérida: Cámara de Diputados.

Tiesler, Vera, Pilar Zabala, and Andrea Cucina (eds.). 2010. Natives, Europeans, and Africans in colonial Campeche, Mexico. History and archaeology. Gainesville: University Press of Florida.

Toledo, Francisco de. 1929. Ordenanzas de Don Francisco de Toledo, Virrey del Perú 1569–1581 ed. R. Levillier, Madrid: Imprenta de Juan Pueyo.

Tozzer, Alfred M. 1941. Landa's Relación de las Cosas de Yucatán. Cambridge: Peabody Museum, Harvard.

Trejo, Silvia. 2007. Xochiquétzal y Tlazoltéotl. Diosas mexicas del amor y la sexualidad. Arqueología Mexicana 15(87):18–25.

Trias de Bes, Luis, Pedro Bosch Gimpera, Jose de C. Serra-Ráfols, D. José Batista, and D. Alberto del. 1928. Las razas humanas. Su vida, sus costumbres, su arte, Vol. 2. Barcelona: Instituto Gallach.

Winter, Marcus. 1996. Entierros humanos de Monte Albán, dos estudios, Oaxaca: Centro INAH, Oaxaca.

Wyllie, Cherra E. 2002. Signs, symbols, and hieroglyphs of ancient Veracruz: Classic to Postclassic transition. Ph.D. diss., Yale University.

Yépez, Rosaura. 2001. El modelado cefálico intencional en los pobladores prehispánicos del barrio teotihuacano de La Ventilla. Exploración 1992–1994. Master diss., Universidad Nacional Autónoma de México.

Chapter 11
Conclusions: New Perspectives For Studying Head-shaping Practices in Mesoamerica

This book concludes with an interpretational synthesis on Mesoamerican head-shaping traditions, their ideological undercurrents, and their meanings. Also some methodological issues are to be addressed; namely, when differentiating artificial shapes from those that are not. The issues raised in the following paragraphs, are intended to set the stage for future innovative technical, conceptual, and empirical approaches in the study of this body tradition in this and other cultural areas.

11.1 Synthesis

Here, I wish to round up this work by disclosing a brief summary of the most noticeable results and arguments of this work, revolves around facets that deserve broader reflections and traces arguments and perspectives in Mesoamerican research that still await scrutiny by the academic community. As posited in the beginning of this book, any deeper understanding of the roles of cephalic modeling traditions among native Mesoamericans requires a culturally sensitive, *emic* consideration of ideological systems and the role they assign to the human body (Chaps. 2 and 6). A bridging analytical unit in this work was that of the "archaeological individual" and finally the "social individual" (see Chaps. 2 and 6), who interacts with other members of society at different levels and in different material, sensual, and intellectual spheres.This is expressed by the modalities of the practice's performance, as witnessed in the skeletal record and complementary information. In this framework, head modeling is subscribed to the ideological sphere, its shared enactment of customs, worldviews, and beliefs, which reflect on—and are influenced by—more basic aspects of social life, either by promoting social permanence or change (López-Austin 1989).

I began by exploring the role of head shaping in the course of everyday infant life and as a means of integration in Mesoamerican society. The semantisation of head modeling according to body functions, as performance and in its emblematic role, delineate three ideological dimensions that follow indigenous concepts. First, there is the "organoplastic" role which assigns a protective function to the practice. Occipital flattening was believed to avert potential harm that could materialize in the occipital prominence, a delicate part of the human anatomy, which could endanger

V. Tiesler, *The Bioarchaeology of Artificial Cranial Modifications*, 257
Interdisciplinary Contributions to Archaeology 7, DOI 10.1007/978-1-4614-8760-9_11,

the physical integrity of newborns. These corporeal correlates still gain more strength as we consider that the head, with its outer insignia, was held by many Mesoamerican groups as a metaphor for designating the "self" and personhood and, still beyond, as a model for the indigenous universe. Analogously, the outer "looks" of the head and its anatomical constituents epitomized deeply embedded worldviews and longstanding traditions. This holds true also for the second angle of conferring meaning to head-shaping in this work, i.e., its daily performance. Its quotidian enactment held a critical role in protecting the baby and prepared it for later integration into society.

Less resilient than the first two components is the third, emblematic facet of Mesoamerican head flattening, which confers significance on the morphology that resulted from the compression, the culturally crafted head looks, so-to-speak. Here I used the concept of visible assimilation of desirable cultural forms, some with the added value of emulated sacred powers. As witnessed from the cranial and figurative record across Mesoamerica, these attributions appear to be conservative albeit less enduring than the phrenological and behavioral connotations. Their myriad meanings in the systems of Mesoamerican ideology appear to shift across the tapestry of its landscapes and also change over time, although nothing hints at their exclusive or distinctive use for status reasons. in the form they were displayed among the ynea.

For the purposes of a systematic survey and an integrated interpretation of the Mesoamerican cranial record, I explore basic morphological dynamics, implied in artificial cranial-vault compression (Chap. 3), and detailed, systematic scoring criteria (Chap. 4) to meet the needs of this regionally anchored study. It was essential to untangle the devices and combination of compression kits involved in most modifications, as witnessed in the material record. A systematic survey of some hundred post-contact written testimonies from Hispanic America, compiled and commented upon by my historian colleague, Pilar Zabala, complements the other approaches. In Chapter 5, she fiddles out the nuanced messages conveyed by Mesoamerican sources to those that document Andean, Caribbean, and Amazonian head modification traditions during colonization and beyond.

Part II puts the analytical and conceptual framework developed in Part I to work. The examination of the regional record with general conceptual and analytical parameters traces trajectories for cephalic modeling that transcend the areas and eras of Mesoamerican cultural evolvement. This survey (with comments) of conservative, yet dynamic, native Mesoamerican head shaping, aspiring to go beyond the comprehension of its formal, and technical expressions, conveying a deeper understanding of the cultural contexts within the social and political situations that substantiated head shaping and head shapes in different groups, communities, and times. It is clear that culturally crafted head forms epitomize deeply embedded cultural dynamics that characterize each epoch; shifts in head practices and their visible results are also prone to materialize broader cultural change that appears to accentuate during times of social change.

In our panorama of Preclassic period practices, such is the case for the emblematic pear-shaped skull forms that are so prominently represented in Olmec and Olmecoid portraitures. These appear to reify sacred powers that were venerated by populations on the Gulf Coast. However, the regional distribution of artificially heightened and narrowed skulls and their mortuary patterning confirms that, even then, the cephalic modeling was not associated with patterns of social differentiation or infant gender. Toward the first millennium A.D., the head forms in vogue in the era B.C. are slowly replaced in the iconographic—and probably also in the cranial—record of Mesoamerica. Together with their visible ideological identification, the new cranial forms that women were imprinting on the infants' heads express broad notions of active, gender-related cultural reproduction and ethnogenesis.

By the Classic period, Mesoamerica had turned into a diverse kaleidoscope of cranial models and regional preferences for certain techniques and head shapes are apparent in than areas (Chap. 8). We have taken a closer look at the shifts these preferences experienced on a local and regional scale, referring to a database of 1,200 Classic Maya skulls (Chap. 9). The different head looks among families and neighborhoods, as recognized among the rural and urban folks of the Copán Valley in Honduras, or those of the small merchant port of Xcambó on the northern coast of Yucatán, convey a uniquely human side of body traditions and their human carriers in terms of residence, mobility, and social change. In both settlements, although distinct, the internal distribution of the cultural forms underlines the role of artificially conferred cranial forms as signifiers of family tradition, gender, communal identity and ethnicity.

Beyond mere population and kin affinity, it is argued for the Classic period Maya that the head forms vaguely communicated different identifications with the sacred, perhaps also more explicitly certain selected protector gods. These identifications could have propelled cohesion and well-being for families, communities, or specific professional avocations, as we have posited for the merchant folk displaying superior flattening, mirroring the particular head shape of their patron god. In general, these ties to sacred forces would explain the local internal diversity in cephalic shapes that are found repeated over and over in the settlement populations we analyzed. However, local diversities also shift across the Maya cultural landscape; preferred head looks differ between the Petén corridor and the Usumacinta basin, between Mixe-Zoque territories and the highlands to the south and east. These transitions express the cultural plurality and trace the vibrant linguistic and most probably also ethnic mosaic during the first millennium of our era.

The close of the Classic period is expressed in a more substantial shift in head forms in those Mesoamerican areas that staged diversity before the close of the first millennium. On a pan-regional scale, a homologation process toward erect shapes occurs, some of them expressed by superior flattening. After the decline of Teotihuacán, this transition goes along with the installment of an increasingly powerful pan-regional military and trade leagues toward the close of the first millennium, which embody the new "international" social order (Chap. 10). These changes are anticipated and expressed by increasing divergence in head form along the eastern, and then western, shores of Yucatán many centuries before the decline of the Classic hegemonies, potentially associated with the gradual migration of ever more powerful merchant groups and, with them, a propagation of new political, economic, and ideological elements.

During the Middle Postclassic period, the vectors of these modifying dynamics faded in favor of a still more uniform cultural repertoire of the practice. At that time, the custom continued to be practiced, however with none or less of the Classic period emblematic connotations expressed. Also the Totonac people of Veracruz and Western Mexican cultures experience changes in the performance and technicity of ancestral head modeling, aspects that have been surveyed by my colleagues. Like in the Maya Lowlands, tabular erect forms gain prominence and then uniformity. A noted exception to this pan-Mesoamerican trend is evident on the northeastern border areas of Mesoamerica. Archaeologically retrieved crania from the Huastec region communicate that anachronous head splinting and head elongation continued to be practiced here up to the final stages of pre-Hispanic life.

The Iberian conquest of Mesoamerica leads to dramatic transitions in the corporeal traditions enacted on the heads of newborns. Once again, the visible expression of the tradition gains prominence, at least in the eyes of the Spanish oppressors and of those natives who were exposed directly to the forces of assimilation (Chap. 10). If flattened head profiles signaled integration and belonging in the millennia before the contact, in the years to follow the Conquest, artificially flattened heads should have transmitted connotations of alterity and exclusion now, especially in the shared urban environments of Spanish villas, where the native tradition appears to disappear swiftly from the public view. Unsurprisingly, the last strongholds of native cradleboarding are relegated to the impenetrable southeastern hinterlands of Mesoamerica, far from the spheres of Iberian oppression. But also less isolated indigenous communities, those that were to abandon head flattening, showed and still show their strong Mesoamerican heritage by keeping with the original protective functions of cradleboard practices, assured by swaddling routines, baths, wraps, covers, and massages of the newborn's body. Beyond doubt, these denote *longe durée*, core elements of the native cultural heritage, related to the head in the native phrenology of vital and harmful body centers and the performance of protective measures on babies (Tiesler and Zabala 2013).

11.2 Flattened and Unflattened Heads

At this point, one of the issues that has not received a clear answer is the categorical absence of head flattening. If the majority of Mesoamericans displayed a culturally modified cranial vault, why did a certain proportion of the indigenous folks not show any change at all? What were the motives that caused some mothers to not mold the heads of their infants? This idea is compelling, given that a visible proportion of most of Mesoamerica's populations did not exhibit it and that this ratio would remain stable for more than a millennium, at least in the systematically surveyed Maya area. Some authors have offered a solution to this apparent dilemma by distinguishing between "intentional modifications" vs. "non-intentional modifications" (Saul 1972; but see also Duncan 2009, p. 182). On the down side, this distinction lumps slight lambdic flattening (and other cradleboard imprints for that matter) together with those skulls that do not show any modification, and thereby separates them from those deemed modified on purpose. I do not adhere to this logic (as argued already in Chaps. 1, 2, and 3), given the multifold connotations and meanings of head flattening and head treatments in general, at least in the Mesoamerican sphere.

However, the dilemma persists and apparently cannot be resolved ad hoc. What distinguishes the mothers who did not subject their children to cephalic modeling at a young age? Several scenarios come to mind to account for the absence. One could be a simple oversight at the time of practice; another could be any severe complication that did not allow continuation of the procedure. Still one other possible conjecture could explain absence of head modeling, at least among Classic period Maya populations: natural head form could have identified the individuals or families who showed affinity to those sacred powers with no artificial head conformation (such as the Death God A of the Classic Period Maya pantheon, for example). Another explanation relates to the duration and force of compression, which in some children might have been applied too swiftly to generate a permanent imprint on the head, which is also due to the "rebound" effect in the child's growth after the removal of the presses (Chap. 3).

After all, it might well be that the reason for non-modified heads will never be determined. However, it seems improbable that this lack of treatment was sought for emblematic purposes, as the proportion of skulls with no cultural modification remains constant over the centuries A.D., at least in the surveyed record of the Maya area. Neither does the registry contain elements that would allow these individuals to be related to a determined social sector, as argued in Chap. 9. Unmodified individuals show an equivalent scope of funerary attires as shaped persons of the same time and cultural area. The conjecture that these infants were placed in these devices and yet showed no subsequent effects still awaits scrutiny, but does not find support in the registry, which shows a variety of degrees of modifications in all periods.

The last aspect, i.e. different degrees of head flattening, gains importance when confronting all crania with slight to very slight compression planes. The classification used in our systematic survey of Maya crania distinguishes all degrees marked as '0' and '0.25' as absent ("not artificially modified"). The latter score denotes those slight changes in cranial curvature that might stem from active flattening but just as well could be the product of posture. Given the complexity involved in native head practices and in reconstructing their permanence in living babies from a skeletal substate of mostly adult skulls, it is clear that all classification efforts cannot be more than approximations toward measuring and patterning of behaviors sustained by the ancient head-molding practitioners. No categorical distinction, however well-founded, will probably be sufficient to account for the real complexity of the processes involved in head shaping and their expression in growing bony substrate. The information transmitted by the crania from the Maya area must, therefore, as with all bioarchaeological data, be considered as inherently incomplete.

This in mind, the methodological constraints that are implied in distinguishing between people with and without artificial head shaping make patent that this categorical division may in fact be arbitrary and even misleading in the context of most Mesoamerican cultures, as we have argued already in Chap. 6. Here, instead of dichotomic expression, the visible manifestations of head modeling express a fluid continuity in forms and degrees. These may range from unmodified to extreme artificial configurations even within single-site series. Beyond methodological limitations and the range of individual routines in its performance, I think that this notion expresses the underlying conditions of enactment and their meanings, some of which did not adhere to the visible outcome of the procedure but to the protection and formation of the infant. William Duncan (2009, pp. 187–188) comes to a similar conclusion when speaking of two different cephalic practices, one being protective and "incorporating" (deemed omnipresent in pre-Hispanic society), the other one identifying head modeling per se. This idea, in turn, envisions conveys the living dimensions of the cephalic modification as a vibrant tradition that could well integrate more than one single procedural routine, performed on the head and body, a custom well in the bosom of society. In a nutshell, this view suggests then that the practice(s), beyond the associated diverse yet coherent meanings and motivations, was enacted on an everyday basis and therefore prone to be influenced by a host of fortuitous circumstances and ranges of particular individual experiences, skills and family ways.

11.3 Perspectives

As with all bioarchaeological investigations that seek to resolve unknowns and to delve deeper into causal questions of indigenous cultural conduct from the (biocultural or bioarchaeological) intersection of disciplines, this work has benefitted by bridging the academic fields. In this endeavor, I have attempted to put the regional and local results in perspective with broader biological and methodological, ideological, social, and historic contexts that might have accompanied or conditioned the trends documented in an effort to breach the scales of recognition and aspects of social enactment. These are documented for the dynamics of population mobility among the settlers of Copán and Xcambó and those that touch upon the stark rupture implied by the Maya collapse, which sent Classic high culture into oblivion for good. The specific historical glimpses our results confer on historical processes, as expressed by the shifting demographic patterns, and choices of different head looks, are revealing indeed, considering the hazardous, ideological quality assigned to infant head models in most of the literature. I follow from this experience that cephalic modification, as a long-lasting means of social reproduction, is capable of expressing conditions well beyond the specific culturally motivated behavior for which it is commonly taken. These correlates touch upon the dynamic socioeconomic and political bases of Mesoamerican society itself, during the different stages of development.

The analytical units of context-informed burials, harboring human remains, were particularly informative in the regional survey of the Maya area, together with chronological phases and periods, regional and site demarcations, neighborhoods (as recognized by patio compounds) and patio units for complex sites such as urban Maya Copán, in Honduras. These categories, such as the relationship we have established between the forms and their potential expressions of sacred forces, should have been important also in other cultural spheres of Mesoamerica where artificial head shapes were formally diverse and visibly recognized, such as in central Veracruz, Teotihuacán, Western Mexico, or in Oaxaca.

In this vein, I also expect that the so-called "emblematic" associations between the practitioners, the bearers, and the sacred might offer fresh points of departure to examine potential religious affiliations in Mesoamerica and also in the Andean sphere. The rich Mesoamerican patrimony of anthropomorphic imagery of humans and mythical beings would surely accord with such an approach in a number of its territories. It would also be worthwhile—given the cohesion and continuity that characterize the native worldview and rituals—to examine more closely if some of the cultural mechanisms operating in Mesoamerican head practices, such as occipital reduction of emulation of patron deities, might prove relevant also in other cultural spheres, like the Andes. Here, a slowly growing corpus of research has started to establish, beyond ethnicity and distinction, ties to cultural emulation, family traditions, and totem veneration (Blom et al. 1998; Lozada and Buikstra 2002; Nado et al. 2012; Purizaga 1991; Yépez 2006, 2009).

Also a systematic cartography of Mesoamerican head forms remains to be established for different moments of its millenary past. I hope that future works may fill in the gaps in the information presented in this work. A greater systematic spatial and temporal coverage would, likewise, benefit efforts towards an understanding of broader Mesoamerican cultural expression and social evolvement. Still today, Mesoamericanist scholarship relies almost exclusively on the material produce of humans, but not on the cultural information that is communicated by humans through their own bodies. This idea by itself, including other permanent body works, ornamentation, and cosmetics, delineates a venue of research that has not been fully explored so far in Mesoamerican studies. Apart from a lack of awareness and poor conservation, this situation is probably also due to the scarcity of systematic mortuary information. This identifies a more general problem, related to the breach and pertinence of information on the ancient repositories of the dead. These data are described with much detail in certain site reports and publications, but are scarce or absent in others, most conspicuously in the case of undocumented private collections or isolated looted specimens. In the practice, these inconsistencies sadly come to constrain the possibilities of nuanced comparisons and interpretations of most of the early skeletal series.

One line of investigation on cephalic modification that I think is particularly apt for tearing down the barriers of disciplinary divides is the theoretically informed and contextualized evaluation of geographic proveniences of individuals with different head forms. Chemical studies based on oxygen and strontium isotopes in dental and bone tissues reveal tandem information and have already demonstrated their great potential in studies on Andean head-shaping practices (Blom 2005; Price et al. 2008, 2009; Torres-Rouff 2002). Also in Mesoamerica, isotopic research on the geographic origins of human carriers of different cephalic forms holds much promise, such as the provenance study of individuals with modified skulls in a multiethnic colonial series of Campeche (Price et al. 2012) or those of a host of other, pre-Hispanic settlement communities (Buikstra et al. 2004; Price et al. 2008; Sierra et al. 2013).

Similarly encouraging is the provenience profile of Xambó's settlers who carry different culturally conferred head forms, as synthesized in this work and described in detail elsewhere (Sierra et al. 2013). The underlying importance of this and other old and new scientific tools lies in the fact that they allow us to enrich archaeological reconstruction with additional data-sets. This has been achieved also in similarly concocted multi- and inter-disciplinary approaches on Mesoamerica and Mesoamericans, which examine the functioning of ideology and family relations, on domestic organization, mobility, and migration. Work of this sort, such as the archaeological dissection and charting of neighborhoods and artisanal workshops in Teotihuacán led by Linda Manzanilla (2006, 2009; see also Arnauld et al. 2012), or that of Houston et al. (2006) on ancient Maya bodies and body notions—and naturally others along this track—are enormously encouraging, as they emphasize the potential of balanced inter- and trans-disciplinary research beyond the constraints of individual approaches or disciplinary confines.

It is clear that the theoretically informed combination and confrontation especially of bioarchaeological data-sets offer a rich sphere in which to establish a more human understanding of the past, a balance between information and synthesis, a dialogue between science and the humanities. The present work has joined for this purpose morphological and contextual archaeological data, artistic and historical (written) information on heads and skulls. Their interpretation within cogent regional ideological frames, transcends toward the complex cultural undercurrents and provides a general understanding of what kept this body tradition alive over the centuries and millenia.

References

Arnauld, M. Charlotte, Linda R. Manzanilla, and Michael E. Smith (ed.). 2012. The neighborhood as a social and spacial unit in Mesoamerican cities. Tucson: University of Arizona Press.

Blom, Deborah E. 2005. Embodying borders: Human body modification and diversity in Tiwanaku society. Journal of Anthropological Archaeology 24:1–24.

Blom, Deborah E., Benedikt Hallgrimsson, Linda Keng, Maria C. Lozada, and Jane E. Buikstra. 1998. Tiwanaku "colonization": Bioarchaeological implications for migration in the Moquena Valley, Perú. World Archaeology 30:238–261.

Buikstra, Jane E., T. Douglas Price, Lori E. Wright, and James A. Burton. 2004. Tombs from the Copan acropolis: a life history approach. In Understanding Early Classic Copan, ed. Ellen E. Bell, Marcello, A. Canuto, and Robert J. Sharer, 191–212. Philadelphia: University of Pennsylvania Museum of Archaeology and Anthropology.

Duncan, William N. 2009. Cranial modification among the Maya: Absence of evidence or evidence of absence?. In Bioarchaeology and identity in the Americas, ed. Kelly Knudson and Christopher M. Stojanowski, 177–193. Gainesville: University of Florida.

Houston, Stephen, David Stuart, and Karl A. Taube. 2006. The memory of bones: Body, being and experience among the Classic Maya. Austin: University of Texas.

López-Austin, Alfredo. 1989. Cuerpo humano e ideología: Las concepciones de los antiguos nahuas. México: Instituto de Investigaciones Antropológicas, Universidad Nacional Autónoma de México.

Lozada, María C., and Jane E. Buikstra 2002. El Señorío de Chiribaya en la costa sur del Perú. Lima: Instituto de Estudios Peruanos.

Manzanilla, Linda. 2006. La producción artesanal en Mesoamérica. Arqueología Mexicana 80:28–35.

Manzanilla, Linda. 2009. Corporate life in apartment and barrio compounds at Teotihuacan, Central Mexico: Craft specialization, hierarchy, and ethnicity. In Domestic life in prehispanic capitals, a study of specialization, hierarchy, and ethnicity, ed. Linda Manzanilla and Claude Chapdelaine, 21–42. Ann Arbor: University of Michigan.

Nado, Kristin, Sara J. Marsteller, Laura M. King, Blair M. Daverman, Christina Torres-Rouff, and Kelly J. Knudson. 2012. Examining local social identities through patterns of biological and cultural variation in the Solcor Ayllu, San Pedro de Atacama, Chile. Chungara 44(12):341–357.

Price, T. Douglas, James H. Burton, Paul D. Fullager, Lori E. Wright, Jane E. Buikstra, and Vera Tiesler. 2008. Strontium isotopes and the study of human mobility in ancient Mesoamerica. Latin American Antiquity 19:167–180.

Price, T. Douglas, James H. Burton, Robert J. Sharer, Jane E. Buikstra, Lori E. Wright, Loa P. Traxler, and Katherine A. Miller. 2009. Kings and commoners at Copan: Isotopic evidence for origins and movement in the Classic Maya period. Journal of Anthropological Archaeology 29(1):15–32.

Price, T. Douglas, James H. Burton, Andrea Cucina, Pilar Zabala, Robert Frei, Robert H. Tykot, and Vera Tiesler. 2012. Isotopic studies of human skeletal remains from a sixteenth to seventeenth century AD churchyard in Campeche, Mexico: Diet, place of origin, and age. Current Anthropology 53:396-433.

Purizaga, Medardo. 1991. El rito del nacimiento entre los Inca. Lima: Universidad de San Martín de Porres.

Saul, Frank P. 1972. The human skeletal remains of Altar de Sacrificios. Papers of the Peabody Museum of Archaeology and Ethnology 63(2). Cambridge: Harvard University.

Sierra, Thelma, Andrea Cucina, T. Douglas Price, James H. Burton, and Vera Tiesler. 2013. Maya coastal production, exchange, life style and population mobility: A view from the port of Xcambó, Yucatan, Mexico. Unpublished paper, submitted to Ancient Mesoamerica.

Torres-Rouff, Christina. 2002. Cranial vault modification and ethnicity in Middle Horizon San Pedro de Atacama, North Chile. Current Anthropology 43(1):163–171.

Yépez, Rosaura. 2006. La práctica cultural de modelar la cabeza en dos culturas andinas del Antiguo Perú: Paracas y Chancay. Un estudio de los procesos de significación de la cabeza modelada intencionalmente. Ph.D. diss., Universidad Nacional Autónoma de México.

Yépez, Rosaura. 2009. El simbolismo de la modificación cultural de la cabeza en la cultura andina de Paracas del antiguo Perú. Estudios de Antropología Biológica 14:526–543.

Index

V. Tiesler, *The Bioarchaeology of Artificial Cranial Modifications*,
Interdisciplinary Contributions to Archaeology 7, DOI 10.1007/978-1-4614-8760-9,

MIX
Papier aus verantwortungsvollen Quellen
Paper from responsible sources
FSC® C105338

If you have any concerns about our products,
you can contact us on
ProductSafety@springernature.com

In case Publisher is established outside the EU,
the EU authorized representative is:
Springer Nature Customer Service Center GmbH
Europaplatz 3, 69115 Heidelberg, Germany

Printed by Libri Plureos GmbH
in Hamburg, Germany